新生物学丛书

新生物学年鉴 2014

《新生物学年鉴 2014》编委会 主编

科学出版社
北京

内 容 简 介

《新生物学年鉴 2014》为一本由《新生物学丛书》6 位编委组稿并把关、各地大学及研究所骨干研究人员撰写，包括植物功能性状研究、植物蛋白激酶在生态学领域的应用、免疫生物学研究、神经科学研究，以及哺乳动物全脑神经测量等内容，共 6 篇文章。这些文章的内容均为撰写者的最新研究成果，因此年鉴可以在一定程度上体现出我国生物学领域的发展现状。

《新生物学年鉴 2014》秉承了之前两本的制作理念，并扩充了涉及的生物学子学科，有更多的中国生物学研究领域的中坚力量参与撰写，务求使相关领域的研究人员获得第一手权威的综述性文章。本书适合各相关领域的高年级本科生、研究生、专业研究人员学习参考。对于希望了解生物学发展现状的科研爱好者，本书也可作为很好的阅读材料。

图书在版编目（CIP）数据

新生物学年鉴. 2014 /《新生物学年鉴 2014》编委会主编. —北京：科学出版社，2015.2
（新生物学丛书）
ISBN 978-7-03-043260-5

Ⅰ. ①新⋯ Ⅱ. ①新⋯ Ⅲ. ①生物学–文集 Ⅳ. ①Q-53

中国版本图书馆 CIP 数据核字（2015）第 017996 号

责任编辑：岳漫宇 / 责任校对：韩 杨
责任印制：肖 兴 / 封面设计：北京美光制版有限公司

科学出版社 出版
北京东黄城根北街 16 号
邮政编码：100717
http://www.sciencep.com

北京通州皇家印刷厂 印刷
科学出版社发行 各地新华书店经销

*

2015 年 2 月第 一 版 开本：787×1092 1/16
2015 年 2 月第一次印刷 印张：13 1/2 插页：8
字数：300 000

定价：98.00 元
（如有印装质量问题，我社负责调换）

《新生物学丛书》专家委员会

主　　任：蒲慕明

副 主 任：吴家睿

专家委员会成员（按姓氏汉语拼音排序）

昌增益	陈洛南	陈晔光	邓兴旺	高　福
韩忠朝	贺福初	黄大昉	蒋华良	金　力
康　乐	李家洋	林其谁	马克平	孟安明
裴　钢	饶　毅	饶子和	施一公	舒红兵
王　琛	王梅祥	王小宁	吴仲义	徐安龙
许智宏	薛红卫	詹启敏	张先恩	赵国屏
赵立平	钟　扬	周　琪	周忠和	朱　祯

《新生物学年鉴 2014》编委会

主　　编：蒲慕明
副 主 编：吴家睿
编　　委（按姓氏汉语拼音排序）
　　　　　　杜久林　中国科学院上海生命科学研究院神经科学研究所
　　　　　　刘小龙　中国科学院上海生命科学研究院生物化学与细胞生物学
　　　　　　　　　　研究所
　　　　　　骆清铭　华中科技大学生物医学工程系
　　　　　　马克平　中国科学院植物研究所
　　　　　　汤其群　复旦大学上海医学院生物化学与分子生物学系
　　　　　　薛红卫　中国科学院上海生命科学研究院植物生理生态研究所

丛 书 序

当前，一场新的生物学革命正在展开。为此，美国国家科学院研究理事会于 2009 年发布了一份战略研究报告，提出一个"新生物学"（New Biology）时代即将来临。这个"新生物学"，一方面是生物学内部各种分支学科的重组与融合，另一方面是化学、物理、信息科学、材料科学等众多非生命学科与生物学的紧密交叉与整合。

在这样一个全球生命科学发展变革的时代，我国的生命科学研究也正在高速发展，并进入了一个充满机遇和挑战的黄金期。在这个时期，将会产生许多具有影响力、推动力的科研成果。因此，有必要通过系统性集成和出版相关主题的国内外优秀图书，为后人留下一笔宝贵的"新生物学"时代精神财富。

科学出版社联合国内一批有志于推进生命科学发展的专家与学者，联合打造了一个 21 世纪中国生命科学的传播平台——《新生物学丛书》。希望通过这套丛书的出版，记录生命科学的进步，传递对生物技术发展的梦想。

《新生物学丛书》下设三个子系列：科学风向标，着重收集科学发展战略和态势分析报告，为科学管理者和科研人员展示科学的最新动向；科学百家园，重点收录国内外专家与学者的科研专著，为专业工作者提供新思想和新方法；科学新视窗，主要发表高级科普著作，为不同领域的研究人员和科学爱好者普及生命科学的前沿知识。

如果说科学出版社是一个"支点"，这套丛书就像一根"杠杆"，那么读者就能够借助这根"杠杆"成为撬动"地球"的人。编委会相信，不同类型的读者都能够从这套丛书中得到新的知识信息，获得思考与启迪。

《新生物学丛书》专家委员会
主 任：蒲慕明
副主任：吴家睿
2012 年 3 月

前　言

当前，生命科学正处于革命性变化的前夜，"新生物学时代"已经到来。创立《新生物学丛书》的一个主要目的，就是要及时总结当下最新的科研成果，展望未来的发展方向。为此，《新生物学丛书》专家委员会决定打造中国的生命科学年度报告——《新生物学年鉴》（以下简称《年鉴》）。《年鉴》以综述性文集的形式出版，系统总结国内外生命科学的研究成果，重点追踪和评述近年来生命科学各个领域的研究热点和前沿进展。《年鉴》的编委每年从《新生物学丛书》专家委员会中产生，负责撰写或组织相关领域的专家学者进行评述。

《年鉴》每年出版一本，涉及的研究领域都是当今生命科学最热点的前沿领域，参与文章撰写的所有人员均是活跃在科研第一线的权威专家学者，每篇文章都可以很好地体现出相关领域的前沿进展。我们希望能够将《年鉴》打造成与美国 *Annual Review* 系列一样高水平的中文版综述平台，及时地报道生命科学的成果和进步。

第一本《新生物学年鉴 2012》于 2013 年 1 月顺利出版，涉及了生物学的 7 个领域，共 12 篇文章。《新生物学年鉴 2013》于 2013 年 12 月出版，扩充了涉及的生物学子学科，涉及 9 个领域，共 10 篇文章。

在以上两本年鉴的基础上，2014 年的《新生物学年鉴 2014》也邀请了相关领域的 6 位专家组织并撰写。本书涉及生物学与生态学的 6 大领域，包括植物功能性状研究、植物蛋白激酶在生态学领域的应用、免疫生物学研究、神经科学研究，以及哺乳动物全脑神经测量等内容，共 6 篇文章。通过这些精彩的综述性文章，可以了解到该学科的前沿发展、国际动向，开阔了视野，更为中国科学研究人员在该领域中的发展树立了信心。

本《年鉴》从动物到植物、从微观到宏观，全面地介绍了新生物学的多个热点学科的最新动向。适合各相关领域的高年级本科生、研究生，以及专业研究人员学习参考。对于希望了解生物学发展现状和动态的科研爱好者，本书也是一本很好的阅读材料。

《新生物学年鉴 2014》编委会
2014 年 1 月

目 录

丛书序

前言

植物功能性状研究进展 ··· 1
　　　　　　　　　　　　　　　　　　刘晓娟、马克平

植物中的蛋白磷酸化 ··· 24
　　　　　　　　　　　　　　　　　　薛红卫、谭树堂、晏萌

抗原受体 V(D)J 基因重排 ··· 77
　　　　　　　　　　　　　　　　　　邓子木、许可、刘小龙

脂肪细胞的发育分化与代谢 ·· 104
　　　　　　　　　　　　　　　　　　黄海艳、汤其群

以透明脑，观澄明心：斑马鱼神经功能研究进展 ······································· 141
　　　　　　　　　　　　　　　　　　尚春峰、穆宇、杜久林

基于直接测量的哺乳动物全脑神经——血管网络精细结构及三维可视化 ·············· 171
　　龚辉、李向宁、袁菁、吕晓华、李安安、陈尚宾、杨孝全、曾绍群、骆清铭

图版

* 策划编辑：马克平　中国科学院植物研究所

植物功能性状研究进展

作　　者：刘晓娟　马克平

中国科学院植物研究所　北京

- 1. 植物功能性状全球分布格局及性状间的关联和平衡 / 3
- 2. 功能性状沿环境梯度的分布格局以及变化规律 / 6
- 3. 植物功能性状与功能多样性 / 9
- 4. 功能性状与群落物种共存机制及群落动态变化 / 10
- 5. 功能性状与谱系亲缘关系 / 11
- 6. 功能性状与生态系统功能 / 13
- 7. 功能性状与干扰 / 14
- 8. 展望 / 15

摘要

植物功能性状是指植物体具有的与其定植、存活、生长和死亡紧密相关的一系列核心植物属性，且这些属性能够显著影响生态系统功能，并能够反映植被对环境变化的响应。越来越多的研究表明，相比大多数基于植物分类和数量的研究，植物功能性状在种群、群落和生态系统尺度上，都已成为解决重要生态学问题的可靠途径。本文回顾了植物功能性状研究的发展历程，总结了近10年来基于植物功能性状研究的前沿科学问题，包括功能性状的全球分布格局和内在关联，沿环境梯度的变化规律，功能多样性的定义及应用，与群落物种共存机制和群落动态变化的关系，与系统发育的关系，对生态系统功能的影响，以及对各类干扰的影响和响应。尽管功能性状的研究已经延伸到生态学领域的各个方面，有利地推动了对各个前沿科学问题研究的发展，仍然有很多值得关注和着重研究的方向。本文也对未来基于植物功能性状的研究，从性状测量和选取、研究方法及研究方向上提出了展望。并指出在当前全球气候变化背景下，功能性状也可应用于指导生物多样性保护和生态系统管理政策的制定。我国幅员辽阔，天然的数据库给未来基于植物功能性状的研究提供了广阔的平台。

关键词

功能性状、群落构建、群落动态、环境、功能多样性、谱系关系、生态系统功能、干扰

"功能性状"(functional trait)第一次正式出现在植物学期刊中的时间大约是在20世纪末，被定义为能够强烈影响生态系统功能，以及能够反映植被对环境变化响应的核心植物属性(core plant trait)，这些属性是大多数植物具有的共有或常见性状(Diaz et al., 1998; Diaz et al., 1999; Weiher et al., 1999)。继而，生态学家们认识到过去很多仅从植物分类学角度无法理解的生态学关键问题，却能够从基于功能性状的研究中得到很好的解释。由此给出了更为详细准确的定义，即"植物功能性状"是指对植物体定植、存活、生长和死亡存在潜在显著影响的一系列植物属性，且这些属性能够单独或联合指示生态系统对环境变化的响应，并且能够对生态系统过程产生强烈影响(Cornelissen et al., 2003; Reich et al., 2003)。

最常见的植物功能性状分类包括：形态性状(morphological trait)和生理性状(physiological trait)、营养性状(vegetative trait)和繁殖性状(regenerative trait)、地上性状(aboveground trait)和地下性状(belowground trait)、影响性状(effect trait)和响应性状(response trait)(Lavorel and Garnier, 2002)，以及后来被广泛接受的软性状(soft trait)和硬性状(hard trait)(Cornelissen et al., 2003)。软性状通常指相对容易快速测量的植物性状，如繁殖体大小、形状、叶片面积，树高等；而硬性状是指相比软性状而言，更能准确地代表植物对外界环境变化的响应，却很难直接大量测量的一类植物性状，如叶片光合速率、植物耐寒性、耐阴性等。软性状和硬性状之间有着很好的关联，硬性状对于环境的变化往往和软性状紧密相连，如利用种子大小和形状(软性状)可以很好地推测种子在土

壤种子库中的持久力(硬性状)。不同的植物功能性状和植物体新陈代谢的特定阶段,以及生态系统的特定功能密切相关。大量研究表明木质密度和叶片干物质与生态系统的生物化学循环及植物的生长速率显著相关,种子大小影响着植物体的扩散和定植能力,树高、比叶面积和冠层结构与植物邻体之间的竞争能力有关,木质密度、树体结构和种子大小则与植物的耐受能力紧密相关(Lavorel and Garnier, 2002; Westoby et al., 2002; Cornelissen et al., 2003; Diaz et al., 2004; Westoby and Wright, 2006; Wright et al., 2007; Wright et al., 2010)。

过去 10 年间,基于功能性状的研究已经横跨从个体水平到生态系统水平,延伸到了生态学研究的各个领域,特别受到关注的前沿科学问题包括:①功能性状之间的内在联系及权衡关系(trade-off);②功能性状在个体之间、物种之间,以及地域之间是否存在差异及造成差异的原因;③功能性状沿环境梯度的分布格局及变化规律(气候、地形、土壤养分等);④基于功能性状的功能多样性研究;⑤通过功能性状来研究局域群落物种共存机制(结构及格局)及群落动态变化规律(存活率、更新率、生长率、死亡率等);⑥功能性状与系统发育的关系;⑦以功能性状为手段,研究植被对生物多样性及生态系统功能的影响;⑧功能性状对全球气候变化的响应和指示作用;⑨功能性状与植物入侵;⑩功能性状对干扰的响应(包括土地利用变化、火烧、放牧等);⑪功能性状对于森林经营和生态系统可持续发展的作用。本文将就当前生态学家关注的前沿问题总结回顾过去 10 年间的发展历程,并展望植物功能性状未来最有潜力的研究方向,以供国内相关领域的同行参考。

1. 植物功能性状全球分布格局及性状间的关联和平衡

1.1 重要功能性状的全球分布格局研究

随着科学家对全球环境变化研究的深入,过去的 10 年间,对于能够响应和影响环境变化的重要植物功能性状在全球尺度上的研究受到了更大的关注。其中最著名的叶片经济型谱(leaf economics spectrum),即通过对全球从北极到热带 175 个地点收集的 219 科 2548 种植物的 6 种能够很好代表叶片形态、结构及生理特性的性状进行分析,发现这 6 种叶片性状间存在着普遍性的规律,即在养分循环速率间保持平衡。但是气候对叶片性状及性状间关系的影响在全球尺度上并不明显(Wright et al., 2004)。同年,Reich 等通过对全球 452 个地区 1280 种植物的叶片氮、磷含量与温度和纬度梯度之间关系的研究,提出了叶片氮磷含量的全球分布格局。即越靠近热带,随着温度的升高和生长季的延长,植物叶片氮磷含量下降,氮磷比升高。而形成此格局的原因和植物生理化学计量及土壤基质的地理格局密切相关(Reich and Oleksyn, 2004)。除对叶片性状的全球格局研究外,Moles 等(2007)收集了来自全球 11 481 个物种和地点的种子质量数据,通过与纬度梯度的回归分析发现从赤道到纬度 60°的地区之间,种子质量有 320 倍的差异,但与纬度梯度不成线性相关。并且在热带边缘处，种子质量呈 7 倍的骤变，这种差异及骤变主要由于植物生活型和植被类型的不同而造成的。随后，Moles 等(2009)又收集了全球 7084 种植物及其分布地点最大树高的数据。分析发现最大树高在南北半球沿纬度梯度都有十分明显的变化，赤道附近的平均最大树高是北寒带平均最大树高的 29 倍，南温带的 31 倍，且在南北半球的变化趋势无异。和种子全球分布类似，在热带边缘处，也同样发现了 2.4 倍的平均最大树高骤变，再次证明物种在热带和温带间生活策略的转变。此外，最大树高变异跨度比较大的植物通常都生长在干冷和低生产力地区。Chave 等(2009)通过对全球尺度上包括木质密度(wood density)，机械强度(mechanical strength)，解剖特征(anatomical feature)以及分枝特征(clade-specific feature)等树干性状之间，以及这些性状与生长率、死亡率及地理梯度的关联分析，建立了全球树干经济型谱(wood economics spectrum)。

1.2 功能性状间的内在联系和平衡关系

物种的功能性状之间存在着各种各样的联系，其中最普遍的就是功能性状间的平衡关系(trade-off)。这种平衡关系是经过自然筛选后形成的性状组合，也称之为"生态策略"(ecological strategy)，即物种沿着一定的生态策略轴排列于最适应或最具竞争力的位置(Westoby et al., 2002; Diaz et al., 2004; Wright et al., 2007)。研究植物性状间的平衡关系不仅能够使生态学家了解植物生态策略在不同环境内和环境间的差异，更可以深入探索生态位分化和物种共存的内在机制，而且有助于理解生态系统的净初级生产力和营养循环的发展和变化。这些平衡关系主要包括叶片性状间、叶片与枝条及树干性状间、繁殖性状与数量间、繁殖性状与幼苗叶片间、叶片与根性状间的平衡。

对 7 个新热带森林中 2134 种木本植物的 7 种功能性状间的关系分析发现，种子大小、果实大小，以及树高间存在着正相关，叶片大小和果实大小正相关，和木质密度负相关(Wright et al., 2007)。Westoby 等(2002)也定义了几个重要的平衡维度，如叶片单位面积质量与叶寿命、种子大小和数量、叶面积与小枝大小，以及树高与很多性状间的平衡，而这些平衡维度在不同气候带、同一景观类型的不同区域间及同一区域内的共存物种间都有变化，由此推测出同一地点内的物种共存其实是多种生活策略的稳定组合。其实，早在 1998 年，Westoby 就已经定义过叶片-树高-种子(leaf-height-seed, LHS)这个重要的性状平衡维度。他首先提出了比叶面积、树高和种子大小是最能体现物种生存策略的重要性状，即植物的生存策略可由它们处于三者组成的三维空间内的位置来表达。随后这个概念无论在草原还是森林生态系统中都得到了证明(Golodets et al., 2009; Laughlin et al., 2010)。

种子作为植物生活史中唯一可以移动的阶段，种子大小、性状和数量，以及后来形成的幼苗性状间都紧密相关(Moles et al., 2004; Moles and Westoby, 2004, 2006)。Moles 等对于种子性状的研究做出了卓越的贡献，如通过对 128 种植物种子重量和数量及其他不同生活史阶段的性状进行相关分析发现，植物并不是简单地通过产生一系列庞大数量的小种子或者小数量的大种子来维持生存策略，种子质量应该与其他不同阶段的植物性状相结合而形成一定的生活策略，如植株高度、幼苗存活率、林冠面积、生产和成熟的时间长度等(Moles and Westoby, 2006)。对种子大小和幼苗性状分阶段的研究发现，大种子植物的幼苗有较高的存活率，这种关系在幼苗建立的最早阶段比较显著，随着时间的推移，对幼苗的存活率起重要作用的还有光照强度和单位面积产种量等其他因素(Moles and Westoby, 2004)。

枝干在植物生活史中起着很重要的作用，包括支撑地上组织、保存水分养分，以及传导树液，因而和树干相关的性状往往和植物的生理和结构平衡密切相关，如木质密度、导水率等(van Gelder et al., 2006; Chave et al., 2009; Poorter et al., 2010)。通过对玻利维亚热带森林中 30 种物种包括树冠透光率、木质密度、抗弯强度、抗压强度、韧性及最大树高等多种性状的研究发现，共存的植物物种间木质密度有显著差异，且木质密度和枝干的强度和硬度显著正相关。树高和木质密度的比值是衡量枝干安全度的重要指标。耐阴性物种具有比较高的木质密度以增加在林下层的存活率；先锋树种木质密度低以适应在林窗内的快速生长(van Gelder et al., 2006)。而对其中更多物种的研究发现，树种的枝干性状主要有两种平衡关系，一种是导管体积和数量间的平衡，另外一种是组织类型间的平衡；前一种平衡可以反映水分传导能力和支撑安全度，后一种平衡则反映了植物对储存和枝干强度的分配比例。而且植物生长率总是和导管直径和导水率成正相关关系，而和木质密度成负相关关系(Poorter et al., 2010)。也有研究对测量木质密度的不同方法进行比较，发现枝条密度和树干密度显著正相关，因而可以利用非破坏性的方法获得枝条密度来代替通常使用的带有破坏性的测量木质密度的方法(Swenson and Enquist, 2008)。

植物生长率和死亡率之间的平衡被认为是植物生活史中最重要的平衡轴。而生长率死亡率间的平衡往往由植物的内在属性、光获取及干扰来共同决定(Poorter et al., 2008; Kraft et al., 2010; Wright et al., 2010)。因而研究与植物扩散能力、资源获取能力、耐阴性、抗压性等相关的功能性状与生长率、死亡率之间的关系都有助于了解植物的生活史，以及物种共存的真谛。通过对来自 5 个热带森林中 240 个物种 4 种植物性状和树木生长率、死亡率的研究发现，随着木质密度和种子体积的增大，树木的生长率和死亡率都降低；而随着树高的增加，生长率增加，死亡率降低。其中木质密度是生长率和死亡率的最好指示性状(Poorter et al., 2008)。在此基础上，Wright 等(2010)进一步将这 5 个森林中的物种划分为不同的生长阶段，更细致地研究了 5 种功能性状和植物生长率和死亡率之间的关系。结果发现幼树性状与生长率、死亡率之间的关系密切，而到成年树后趋势逐渐减弱。也同样证明了木质密度是指示植物生长和死亡率的最好性状。

植物需要控制生长、开花、传粉和结果的时间以合理分配资源。时间配置的不同不仅会影响到植物母株的生长，也会影响到子代的存活率(Bolmgren and Cowan, 2008; Kushwaha et al., 2011)。对瑞典北温带中 572 种植物开花时间、种子大小、扩散方式、植株高度及生活型的调查研究发现，对于多年生草本植物，大种子及比较低矮的植物开花时间早，而一年生的植物却发现了相反的趋势(Bolmgren and Cowan, 2008)。通过对印度干旱热带森林中 24 种植物连续两年的开花、结果、叶片萌发和凋落的时间进行观察，以及其与木质密度、叶片面积和资源利用率的关系研究发现，植物通过改变各个阶段的起始时间来适应不同年份的恶劣天气以保证后代的数量(Kushwaha et al., 2011)。

2. 功能性状沿环境梯度的分布格局以及变化规律

植物功能性状与环境关系的研究由来已久，植物功能性状决定着植物的生长、存活和繁殖，因此在物种沿环境梯度的分布格局中起着很重要的作用。研究两者之间的关系，不仅有助于生态学家研究生态系统功能及群落物种共存的原理，而且为预测全球气候变化对植物分布的影响提供了有效的研究手段。物种间形态和生理性状的迥异使植物形成了一系列生活策略。这些策略最终在不同的生态系统和生物群间表现为沿基础资源轴排列，资源轴的一端聚集了资源快速获取的生活策略，另一端则聚集资源高度保存的生活策略(Diaz et al., 2004)。资源轴即代表了各种各样的环境因子。而在不同的尺度上，对植物功能性状分布起决定性影响的环境因子是不同的。功能性状在特定地点的分布往往是从大尺度到小尺度层层过滤，多重因子共同作用后的结果。大部分的研究都证实，在全球尺度或大尺度上，气候因子对植物功能性状的分布起决定性作用；在中等尺度上，土地利用和干扰起主要作用；而在小尺度或局地范围内，地形因子和土壤因子决定着性状的分布。

2.1 与气候之间的关系

大尺度或是区域尺度上气候和性状关系的研究，多是关于气候梯度下，植被类型间的性状差异或是性状对不同气候类型的响应(Ackerly, 2004; Thuiller et al., 2004)。气候因子一般包含温度(如年均温、生长积温、最高月均温等)，降水(降水量、潜在蒸散、空气湿度)，光照(太阳辐射)等。

温度往往和降水、光照等多种因子共同决定着植物性状的分布。就全球气候分布看，往往在高纬度气温相对低的地方降水量也比较小。气温的变化，特别是生长季的温度及其持续时间的长度，是影响植物生长所需的重要因子。低温限制了光合作用所需要的碳吸收及重要元素在植物体内的流动，决定了叶片很多生理性状的地理分布(Reich and Oleksyn, 2004)。而温度的高低也影响了水黏滞度、细胞渗透率，以及新陈代谢速率，使得有机质的分解和矿化速率发生变化，从而改变植物叶片和根系可吸收的营养物质的成分和含量，植物的形态生理结构及化学物质含量也会发生适应性变化(Aerts and Chapin, 2000)。此外，气温和地球各个地质时期的干扰也影响了土壤肥力的分布，最终影响到位于土壤上层植物的功能性状分布。对全球 2500 多种维管束植物的研究显示，气温越高、越干旱、太阳辐射越强的地方，植物每单位面积的叶片质量和叶氮含量越高，叶寿命越短，光合能力越弱(Wright et al., 2005)。对中国北部 404 个地点 800 多种植物的性状研究表明，植物生活型、叶形和光合途径都沿生长季温度和降水梯度显著变化(Meng et al., 2009)。而对全球 452 个地点 1280 多种植物的观察比较发现，越靠近赤道，随着温度的增高和生长季的延长，叶片氮磷含量减小，氮磷比增加。且这种分布格局对针叶林、草地植物、禾本科植物及木本植物都是相似的(Reich and Oleksyn, 2004)。对全球范围内 182 个地点收集的 558 种阔叶树种和 39 种针叶树种的性状和气候间的关系比较发现，年最大月均温高且太阳辐射强的地方，叶片较厚，每单位面积的叶干物质高(Niinemets, 2001)。对澳大利亚豆科大豆属(*Glycine*)34 种植物的研究表明，种子重量和温度及太阳辐射紧密相关，而和降水量无显著关系。在纬度较低的地带，随着温度升高和植物新陈代谢加快，种子重量变大；同样太阳辐射增大，光合产物增多，种子重量也变大(Murray et al., 2004)。

降水也是决定植物物种分布和群落组成的一个重要因子，特别是在水分散失很快的地区，降水是制约植物多样性最重要的因子。因此，沿降水梯度，植物选择快速吸收水分还是更多地保存水分也成为了一种策略。水分的保存可以通过有效利用有限的水分资源或者通过避免干旱的策略来实现(Markesteijn and Poorter, 2009)。所以一般情况下，耐干旱的物种叶片小而窄，比叶面积较低而木质密度偏高(Niinemets, 2001; Poorter and Markesteijn, 2008)；不耐旱的物种则更多地选择落叶来减少干旱季节的叶片蒸散，以及低的木质密度在茎干中储存水分(Markesteijn and Poorter, 2009)。对澳大利亚东南部 46 个地点 386 种多年生植物叶片宽度、比叶面积及林冠层高度沿降水和土壤全磷梯度的研究发现，这 3 个性状都和降水量呈正相关关系(Fonseca et al., 2000)。通过对非洲西部热带雨林沿降雨梯度的分析发现，该地区超过 80%的物种性状分布格局和年降雨量显著相关。而有性状组成的生活策略中，起决定性作用的是耐荫性和抗旱性。木质密度和叶片落叶程度是物种沿降水梯度分布的最好指示性状(Maharjan et al., 2011)。对澳大利亚东部 4 个地点 70 多种多年生植物的一些生理性状和降水量及土壤肥力的关系研究发现，光合能力、叶片暗呼吸速率、气孔导水率及叶氮磷含量和降雨量都表现出显著相关性。降水量小的干旱地区，叶氮磷含量和叶片暗呼吸速率高，而光合作用效率和气孔导水率低(Wright et al., 2001)。巴拿马 4 个低地森林的研究结果则表明，随着降雨量的增大，群落的常绿树种增多，林冠层叶片单位重量的氮含量和光合速率增大；而在降雨量小的地方，植物具有更多寿命较短的叶片，且叶片磷含量高(Santiago et al., 2004)。

光照强度或光量的获取是制约植物生长的重要因子，特别是在荫蔽环境较大的热带森林中尤其突出。喜阳和喜阴植物无论在林冠形状、叶片结构、生理特征和枝干密度上都差异很大(Niinemets, 2006; Rozendaal et al., 2006)。太阳辐射量的不同使得植物的高度和树冠结构形成差异。而树冠结构包括枝型、叶序、节间长度，以及枝条角度，这些都决定了叶片的数量、分布和几何结构。进而可以影响到光合作用效率和与此相关的生理形态性状(Sterck and Bongers, 2001; Poorter et al., 2003)。在光线不足的环境中，植物树冠宽大且叶层结构薄，以最大限度地截取光量和减少自我荫蔽；而生活在光线充足环境中的植物，生长迅速、树冠垂直结构深、水平宽度和树干直径小，以竞争更多的能量。先锋植物的生长方式也与此类似，需要在光照充足的生境中迅速生长至其邻体无法竞争的高度，首先占据光照资源。对利比亚 53 种热带雨林树种的树高、树冠厚度和直径与光照关系的分析发现，树高和光照显著正相关。在光照充足的地方，树木形态细高(Poorter et al., 2003)。对全球 20 个地点 208 种本本植物的研究显示，在光线充足的生境中，植物叶片暗呼吸速率、比叶面积、光合能力以及叶寿命都显著高于光线较弱生境中的植物(Wright et al., 2006)。

2.2 与地理环境的关系

地理环境一般指植物所处的地理位置(经纬度)和地形因子(海拔、坡度、坡向)的环境特征。植物所处的地理位置对其功能性状的影响途径主要是通过不同经纬度地区气候(温度、降水及光照条件)和土壤的差异来实现，结果会形成不同类型的功能群，即对地带性环境因子具有相同或相似响应机制的植物组合(冯秋红等，2008；孟婷婷等，2007)。沿纬度梯度，从热带到温带到寒带，植物叶片、种子、树高等功能性状都表现出很大的差异(Wright et al., 2004; Moles et al., 2007; Moles et al., 2009)。并且全球大部分阔叶林中，叶片形态性状和氮磷含量在各个植被类型间，以及相同植被类型内都有明显差异。对中美洲和南美洲2456个树种木质密度和地理环境因子的研究发现，随着海拔的增高，木质密度显著降低；而在低海拔的不同区域内，湿热森林中的木质密度要显著低于干旱森林。其他位于不同纬度地带的森林间木质密度也有很大差异(Chave et al., 2006)。

地形因子对功能性状的影响则主要表现在小尺度上，随着海拔、坡度、坡向的不同，水热光照条件都会随之发生变化，因而使得植物形成不同功能性状组合的生活策略。随着海拔的升高，平均气温会降低，降水会减少，大气压变低，太阳辐射增强，植物的生长季变短。这样的环境条件会过滤掉一些抗压能力低的物种(Cornwell et al., 2006; Cornwell and Ackerly, 2009)。通过对瑞士森林和草原中120种植物13种功能性状沿5个海拔样带的研究发现，在草原生态系统中，随着海拔的升高，风媒植物减少，而虫媒植物增多，瘦果减少，蒴果增多；特别是在低海拔地区，种子传播的方式多为动物传播和风力传播。在森林生态系统中也发现了类似的结果(Pellissier et al., 2010)。对在夏威夷群岛上分布范围很广的植物 *Metrosideros polymorpha* 在4种不同海拔高度的树干导管直径的比较发现，在高海拔地区的植株树干导管直径显著小于中等海拔中的直径(Fisher et al., 2007)。对新西兰30个地点24种草地植物叶和根性状沿海拔梯度变化的研究表明，高海拔比低海拔地区叶和根中氮浓度显著偏低，叶片偏厚，根系偏粗(Craine and Lee, 2003)。随着坡度和坡向的变化，太阳辐射量、垂直光强，以及土壤水分、养分含量都会发生很大的变化。对日本南部屋久岛上常绿阔叶林群落中物种性状的研究发现，山脊上的植物叶氮含量和比叶面积显著高于中坡和坡底植物(Hanba et al., 2000)。对美国加利福尼亚海岸群落中311个地点灌木丛木本植物的性状研究发现，在物种水平上，植物叶面积在南坡的变化显著大于北坡；而在群落水平上，叶面积大小和光照强度呈显著正相关(Ackerly et al., 2002)。

2.3 与土壤营养之间的关系

相比气候和地理因子，土壤和植物性状之间的研究更多集中在局域尺度的同种植被类型内。土壤肥力是决定群落物种组成的主要因子之一。而植物因个体迥异又能反作用于土壤营养循环。一般来说，尽管同一地点内的物种生活策略千差万别(Fonseca et al., 2000; Wright et al., 2004)，但生长在土壤肥沃环境中的植物会产生大量的枯落物，这些枯落物产生的大量营养元素返回到土壤中，从而土壤可以维持高的肥力水平；与之相反，生活在贫瘠环境中的物种，会把更多的营养保存在寿命长且抗性高的植物组织中，产生的凋落物少，土壤肥力水平也无法提高(Aerts and Chapin, 2000)。植物和土壤之间的这种作用和反馈关系是生长与营养留存之间的一种平衡，而这种平衡是通过将植物的功能性状排列在土壤肥力轴上来实现的(Aerts and Chapin, 2000; Wright et al., 2004)。能够快速吸收养分、生长迅速但是寿命较短的植物通常出现在营养富足的土壤中，与此相关的性状特征为高的叶面积、比叶面积、光合速率、呼吸速率等；而在营养供给不足的土壤中，能够保留养分的性状特征往往更重要，如小的叶面积、比叶面积，高的木质密度、比根长、根系深度等。通过对全球99个地点474种植物比叶面积、叶氮磷浓度和氮磷比沿土壤梯度的分布研究发现，土壤肥力与植物比叶面积和叶氮磷浓度成正比；土壤肥力更多地决定了植物叶片性状，而气候条件则更多地决定了植物的生活型(Ordonez et al., 2009)。对非洲南部近海地带银叶树属(*Leucadendron*)88种植物的研究表明，贫瘠土壤中的植物叶面积偏小，且种子扩散能力低，生长缓慢(Thuiller et al., 2004)。而利用控制实验对拟南芥(*Arabidopsis thaliana*)在养分高、养分低、中等酸碱度、碱性土壤中的性状变化的研究发现，不同基因型间的植物叶片性状和叶寿命表现出显著差异。高的土壤养分会使植物的比叶面积偏高，种子偏大。在中等酸碱度且高养分土壤中的植株产种量最大(Bonser et al., 2010)。

3. 植物功能性状与功能多样性

功能多样性(functional diversity)是指影响生态系统功能的物种或有机体所具有的功能性状的大小、范围及分布(Diaz and Cabido, 2001; Petchey and Gaston, 2002)，其表示方法随着对"功能"定义的不同而不同，最有代表性的功能多样性指数包括：①群落中的功能群数量；②功能性状形成的多维空间中，两两物种的距离之和；③功能性状树中所有枝长的总和(Petchey and Gaston, 2002)。也有人进一步将基于连续功能性状数据的功能多样性指数分为：功能丰富度(functional richness)、功能均匀度(functional evenness)、功能相异度(functional divergence)及功能发散度(functional dispersion)(Villeger et al., 2008; Laliberte and Legendre, 2010)。

相比只包括物种有或无的物种多样性，功能多样性更能直接体现植物体在生态系统中所起的作用。因而一经提出，马上受到关注，用以预测和解释若干重要的生态学问题，如：①功能多样性沿纬度梯度的变化规律；②基于功能多样性的种间竞争机制及群落结构形成机制；③功能多样性对生态系统过程和功能的影响。长白山阔叶红松林的功能丰富度随着演替阶段而下降，而群落间功能多样性的差异则是遗传和环境共同作用形成的(么旭阳等, 2014)。而对亚热带森林的类似研究，却未能在不同的演替阶段间发现功能多样性的显著差异，功能均匀度反而更加显著(Bohnke et al., 2014)。对南北美洲5亿多木本植物的研究发现，赤道附近的功能多样性要高于温带地区，并且在局地范围内，功能多样性在热带地区比随机模拟的要高，而在温带地区比随机模拟的要低，也说明了生境过滤限制了温带森林群落的功能性状分布(Swenson et al., 2012)。也有学者将个体水平的功能性状纳入功能多样性指数的计算中，发现其比通常基于物种水平的功能多样性指数对群落格局的预测更加敏感准确(Albert et al., 2012)。来自阿根廷亚热带森林功能多样性对森林碳储量的预测研究表明，基于树高和木质密度计算的功能多样性具有最佳预测能力(Conti and Diaz, 2013)。来自29个草原生态系统的研究结果显示，几种功能性状相结合的功能多样性指数和谱系多样性指数一起构成了解释生态系统-生物多样性关系的强力途径(Flynn et al., 2011)。而综合全球众多关于生物多样性丧失对土地利用的响应的研究表明，功能多样性的变化规律和物种多样性的变化规律并无明显差异(Flynn et al., 2009)。

功能多样性是组成生物多样性众多元素中关键且重要的一员，但是它也是复杂和多面的。目前，还没有广为接受的普适性的功能多样性测度方法。因此，在预测和评估生态系统的不同功能时应选择最合适的功能多样性指数。

4. 功能性状与群落物种共存机制及群落动态变化

4.1 与群落物种共存机制的关系

功能性状沿纬度梯度或在大尺度上的分布规律大都源于小尺度上局地群落内的功能性状差异。有研究将木本植物的功能性状差异分离为群落间和群落内两部分，发现功能性状不仅沿土壤水分梯度分布，而且在同一土壤梯度内的共存物种间也存在着差异(Ackerly and Cornwell, 2007)。研究和比较不同的局域群落内功能性状的分布格局、共存机制及动态变化有助于理解大尺度物种分布格局的成因(Wright et al., 2004)。

在局地群落内，功能性状通过影响物种沿环境梯度的排列、种间竞争，以及群落内的资源分配而影响群落的物种共存格局。具有相似功能性状的物种间具有相似的资源需求，竞争更加激烈，因而在群落中的多度不高；而另一方面，具有最适合某类生境功能性状的物种或种群得以在该生境下大量存活繁殖，因而在群落中会具有很高的多度(Mouillot et al., 2007)。进一步研究表明，决定物种在群落中有无和多少的功能性状是不一定相同的，因此在用功能性状验证不同的群落形成机制时，应同时将物种的存在与多度数据纳入影响因素中(Cingolani et al., 2007)。

近年来大量涌现出通过分析功能性状在空间和时间上的离散格局来验证和量化生境过滤、生态位分化，以及中性过程在群落物种共存中的相对重要性的研究。支持生境过滤理论的研究，往往集中于验证基于物种水平的功能性状沿环境梯度的分布，以及功能性状的差异造成的竞争作用；支持生态位分化理论的研究，往往集中于验证功能性状在不同的群落间存在显著差异，以及群落内共存物种间的性状距离比随机模型模拟的性状格局要更加发散；而支持中性理论的研究，则往往集中于验证群落的实际功能性状格局是否和模拟的由扩散限制引起的空间聚集格局相吻合(Kraft et al., 2008; Swenson and

Enquist, 2009; Kraft and Ackerly, 2010; Lebrija-Trejos et al., 2010; Paine et al., 2011; Adler et al., 2013; Liu et al., 2013)。这些大量研究的结果也表明,三种理论在解释群落物种共存机制时都起着比较重要的作用,未来还需更有力的证据来验证和量化其相对重要性。

4.2　与群落动态变化的关系

群落的动态变化由群落内植物个体的繁殖、存活、生长、死亡等一系列动态过程组成,这些动态过程取决于植物体在群落中的适合度(fitness)。已有研究表明植物体具有的形态性状和生理性状及这些性状的组合都显著影响植物体在群落中的表现,即适合度。因此,通过植物功能性状来探讨群落动态变化规律已经成为有力且可靠的方式(Enquist et al., 2007; Poorter et al., 2008; Wright et al., 2010)。相比大量关于功能性状与群落格局的研究,功能性状与群落动态变化的研究远远不够,特别是生活史周期比较长的森林群落,需要很长时间来获取动态数据。

植物通过调节自身的资源配置来保持在群落中的最大适合度。具体来说,生活在资源相对丰富环境中的植物体具有较高的扩散和定植能力,以快速抢夺生存所必须的资源,因而植物体会将更多的能量分配给繁殖和生长,从而形成一系列经济型的器官结构。如大量快速扩散的小种子、光合速率较高且比叶面积较大的叶片、生长迅速而木质密度较低的植株茎干及较短的生命周期等。与之相反,生活在资源限制环境中的植物体会将更多的能量分配给维持植物在不利环境中生存和抵抗所需的物质储存,如少量大体积的种子、肥厚的叶片、坚硬的茎干等,这些植物体通常存活率高、生长缓慢、耐受能力强且死亡率低(Reich et al., 2003; Westoby and Wright, 2006; Poorter et al., 2008; Chave et al., 2009; Wright et al., 2010; Ruger et al., 2012)。这些理论推测得到了实践证明,无论是在地理位置不同的热带森林群落(Poorter et al., 2008; Ruger et al., 2012)、植物体不同的生活史时期(Wright et al., 2010)、径级大小不同的个体间(Sterck et al., 2012),还是分别基于个体和物种水平的研究(Martinez-Vilalta et al., 2010; Herault et al., 2011; Sterck et al., 2012)都有案例发表。

5. 功能性状与谱系亲缘关系

近年来,越来越多的科学家将植物进化史(evolutionary history)和谱系亲缘关系(phylogenetic relationship)带入到生态学研究中,这使得生态学家能将短时期内局地尺度的生态学过程和全球尺度进化史上发生的生态学过程联系在一起,用于解决群落物种共存和生物多样性的形成机制。实际上,生境过滤、中性过程,以及谱系亲缘关系已经逐渐成为解释群落物种共存的三大核心理论。特别是将植物间的亲缘关系和功能性状整合在一起,不仅可以增强对多尺度上生物区系分布和功能的理解,而且可以帮助我们预测物种间相互作用及其对生态系统和进化过程的影响(Cavender-Bares et al., 2009)。

植物功能性状往往用来检验植物生长繁殖和资源获取等生态策略是否相同。如果两个物种生态位重叠，它们应该具有相似的生理、生殖和防御性状。然而只有一小部分性状能够用来验证植物沿环境梯度的分布，证明生境过滤作用对形成多样性群落的贡献（McGill et al., 2006; Westoby and Wright, 2006）。如何准确地测量这些有限的性状，特别是种间性状的差异，限制了基于功能性状研究的发展。而过去的 10 年间，随着利用植物 DNA 序列重建谱系关系逐渐变得容易，基于谱系方法来验证群落物种共存理论的研究得到了越来越多的应用，物种间的谱系关系同样被用来验证它们的生态位重叠（Webb, 2000; Cavender-Bares et al., 2004; Losos, 2008）。这的确弥补了由于性状数据获取难而带来的不足，特别是对于那些有显著谱系信号（phylogenetic signal）的性状。而基于谱系的方法也同样证明了生境过滤，以及密度制约在促进物种共存中的重要作用（Kraft and Ackerly, 2010; Uriarte et al., 2010; Hardy et al., 2012）。基于谱系的研究都预先假定谱系距离能很好地指示植物种间的生态位重叠程度；然而，如果群落内谱系聚集或高度分化，谱系距离便不能很好地指示物种间的生境相似性（Cadotte et al., 2009; Mayfield and Levine, 2010）。由此可见，单独基于两者的研究都有一定的缺陷。在验证群落物种共存原理时，应该将基于测量多物种多个体植物性状的方法和基于分子技术建立谱系关系的方法相结合。

将功能性状与谱系结合的研究主要集中在：①利用系统发育比较方法（phylogenetic comparative method）去除种间水平比较时物种性状或环境变量样本非独立的影响，以得到排除谱系关系后物种性状间或物种与环境变量间的真实关系（Felsenstein, 1985; Bolmgren and Cowan, 2008）。如利用系统发育独立比较分析（phylogenetically independent contrast）和广义估计方程（generalized estimating equation）对瑞典东南部北温带植物开花时间、种子重量和植株高度之间的关系进行了分析（Bolmgren and Cowan, 2008）。②比较基于性状方法和基于谱系方法得到结果的异同。如通过对哥斯达黎加干旱热带森林中多个尺度上，功能性状与谱系关系沿主要生活策略梯度分布格局异同的比较发现，当同一群落内共存物种的性状同时发生功能聚集和发散时，谱系关系对植物功能性状分布的指示作用会减小；并且当群落谱系发散时，谱系信号也不能用来推断植物性状的分布格局（Swenson and Enquist, 2009）。Prinzing 等（2008）将荷兰植物数据库中物种分布，16 种功能性状分布，以及物种谱系关系数据相结合进行研究，发现两种不同的趋势，即在区域尺度上，谱系保守使得来自同一谱系的物种间性状高度相似；而在小尺度或群落尺度上，谱系保守只发生在其中的少数谱系上，群落内物种间的功能性状相异性状很高。③将性状与谱系，或性状、谱系、环境相结合共同解释群落物种共存机制。如利用圭亚那热带雨林中 668 个物种 17 种功能性状，以及这些物种的谱系关系，对群落形成过程的探究发现，共存物种比随机产生的物种间具有更高的功能相似性和谱系相似性，很好地验证了生境过滤在形成群落物种多样性中的重要作用（Baraloto et al., 2012）。对亚马逊 $25hm^2$ 热带雨林中 1100 多种植物的性状分布和谱系关系分布格局的多个尺度分析发现，在小尺度到中等尺度上，生境过滤对群落构建（community assembly）起重要作用。基于两者的比较也发现，由于谱系数据没有性状数据精确，因而基于谱系关系的分析虽能验证生境过滤作用的存在，却很难验证由于密度制约效应造成的均匀分布效应。而基于性状的方法却可以同时发现两个过程的存在(Kraft and Ackerly, 2010)。通过对阿尔及利亚东南部 10 000hm² 沿海沼泽地内 97 个地点 10 种植物性状、谱系关系、土壤因子和空间相对位置的调查，并利用性状结合谱系，环境结合空间的方法进行分析，发现生境过滤对群落构建的作用，并能具体到究竟在植物进化史的哪个时间段，发生了由环境过滤作用形成的性状聚集(Pavoine et al., 2011)。

6. 功能性状与生态系统功能

大部分关于生物多样性与生态系统功能的研究都还是利用物种丰富度作为预测变量。近年来，科学家逐渐发现植物体在生态系统中所起的作用，可能是由于该植物体具有的某类功能性状使得它可以获取更多的有效资源；而不同植物体功能性状的相异性使得它们具有不同的获取和利用资源的策略，从而减少了生态位的重叠，增加了多样性及维持着生态系统功能的稳定性(Diaz et al., 2004; Petchey and Gaston, 2006; Roscher et al., 2012; Roscher et al., 2013)。因此，利用功能性状的多度、多样性和分布能更好地解释和预测生物多样性对生态系统功能的影响。

近几十年的研究表明多样性高的生态系统具有高的生产力，机制之一便是由于物种多样性高的生态系统通常也具有丰富的功能性状多样性，而功能性状和物种获取资源的能力息息相关，使得物种能够获得更多的资源(养分、水、传粉者、共生菌等)，并且允许更多的竞争者共存在丰富的资源内(Reich et al., 2004; Roscher et al., 2012; Roscher et al., 2013)。虽然在补偿效应中很难分辨生态位分化与物种间的交互作用，但是由于物种功能性状的种内和种间变异能够更有效地描述物种间的相互关系，越来越多的证据表明基于功能性状的研究方法有助于阐明生物多样性效应的潜在机制(卜文圣和马克平, 2014)。

来自德国不同草原多年的控制实验数据表明，群落功能性状组成相比物种丰富度能更好地解释群落生产力和凋落物分解速率，特别是基于群落水平的性状平均值能够解释不同多样性水平下群落生产力差异的 80%(Mouillot et al., 2011; Roscher et al., 2012)。对欧洲 3 个不同草原地带功能性状与地下微生物对群落生产力和养分储存和循环的研究显示，功能性状能显著解释地上生物量和地下土壤有机质含量的差异。具有高的比叶面积和叶片氮含量，以及低的叶片干物质含量的植物通常能产生更多的地上生物量，但这类植物所处的土壤环境通常碳氮含量低，反之则相反(Grigulis et al., 2013)。在美国明尼苏达州草原生态系统中却发现植物功能性状对群落生产力的解释力不如谱系关系的解释力高，也说明未来需要探索关键的植物功能性状并与植物的进化关系结合来共同解释群落生产力的差异(Cadotte et al., 2009)。而对巴拿马热带森林的研究显示，功能性状在预测森林碳储量时，在自然生态系统和控制实验系统间存在很大差异，因此不能简单根据自然森林的单一结果进行碳储量恢复的人工林种植(Ruiz-Jaen and Potvin, 2011)。也有研究表明，同一生态系统过程受到功能性状组合的影响，而某一重要功能性状则在多个生态系统过程中起着关键作用(de Bello et al., 2010)。

7. 功能性状与干扰

干扰对生态系统的影响机制一直是保护生态学关注的热点。传统的通过分析群落物种组成和变化来探究这些机制的方法已经不能满足需要，而功能性状研究的逐渐兴起使得生态学家可以量化和预测干扰对生物多样性格局的影响，从而有助于制定生物有效保护和资源可持续利用策略。尽管很多理论研究认为中度干扰可以促进生态系统的可持续发展，但这一理论并未在实际研究中得到大量证实(Mouillot et al., 2013)。因此，利用植物功能性状来解释和预测受干扰生态系统中的系列变化是必要的。

7.1 火

火被认为是影响陆地生态系统的全球性干扰，是影响生态系统中植被动态的重要因子之一。它主要将生态系统中的有机物质转化为无机物质，不仅可以改变群落物种组成，也强烈影响着群落的功能性状结构。火灾后，草原群落中物种功能性状的改变趋向于个体变小，C4植物和豆科植物增加(Diaz et al., 2001; Noy-Meir and Kaplan, 2002)，而功能性状数量的变化则取决于火干扰的强度和频度(Briggs et al., 2002; Heisler et al., 2003)。在北美高草草原连续22年的火干扰控制实验结果表明，火的干扰，加快了群落中植物功能性状范围的收敛，而收敛的程度取决于火烧的频度(Spasojevic et al., 2010)。而对于冠层火干扰后的群落恢复最有利的则是具有高的再次萌芽能力和能保留一定土壤种子库的植物物种(Pausas et al., 2004)。

7.2 放牧

除火以外，通过家畜或野生有蹄类动物的放牧被认为是第二大影响植被动态的干扰行为，特别是对于植被面积和生态系统生产力的影响。而放牧对植被的干扰程度不仅受到生态系统中植物功能性状差异的强烈影响，同时也影响着功能性状自身的变化。放牧和植物功能性状的这种相互作用的结果将反馈在生态系统的物种组成及养分循环上。对草原生态系统的研究表明，放牧程度低的群落中植株个体更加高大、产种量更高、开花时间更晚；叶片干物质含量、叶面积、株高及开花时间对放牧的影响最大(Louault et al., 2005)；而在全球尺度，综合了涉及197个典型群落的研究结果发现，放牧对植被的影响程度受到植被功能性状类型的显著影响，普遍趋于一年生大于多年生，个体矮小大于个体高大，植株倒伏大于直立，匍枝和环状结构大于束状结构(Diaz et al., 2007)。

7.3 植物入侵

全球气候变化及频繁的人类活动加速了入侵植物对生态系统的干扰和对生物多样性的影响。对于植物群落可入侵性产生的原因已有几十年的研究历史，多归因于本土植物有利于入侵种入主群落、自然或人为干扰、时空上的资源波动，以及群落或种群内存在可利用的空缺生境(Levine and D'Antonio, 1999; Davis et al., 2000; Mack et al., 2000; Bruno et al., 2003)。越来越多的研究发现，与植物开花、扩散、定植、萌芽、生长、抵抗啃食、调节生物量分配等能力相关的功能性状都和物种的入侵能力密不可分(Kolar and Lodge, 2001; Grotkopp et al., 2002; van Kleunen and Richardson, 2007)。科学家常常通过在较大尺度上寻找入侵植物所具有的共同功能性状，比较入侵物种和本土物种在局地尺度上的性状差异，以及环境异质空间下的入侵植物在功能性状上的区别来探究影响植物入侵的机制。来自自然观测和控制实验的大量结果都表明，高的植物结实率和种子萌芽速率是增加入侵干扰的重要性状(Daehler, 2003; Chrobock et al., 2011)；来自全球的综合数据也表明，入侵植物和本土植物相比，往往具有更大的株高、叶面积、枝条数及生长率(van Kleunen et al., 2010; Lamarque et al., 2011)；而入侵植物和非入侵植物相比，具有和入侵生态系统本土植物更相似的功能性状(Knapp and Kuehn, 2012)。这些研究都使得功能性状成为推测植物潜在入侵性的有效途径。

7.4 土地利用变化

土地利用的变化显著影响着生态系统功能和生物多样性，特别是在快速发展的地区，原来复杂的自然生态系统逐渐被简单的农田、道路、建筑等取代。越来越多的研究发现，功能性状可以代表有机体在生态系统中的作用，以及可以用来评估有机体丧失后的生态影响。热带和温带生态系统的研究表明，利用植物功能性状多样性来评估土地利用对生态系统与生物多样性关系的影响和利用植物体的丰富度来评估是同样重要的(Flynn et al., 2009)。

此外，叶片干物质含量和比叶面积被认为和生态系统抵抗干扰的韧性显著相关(Bernhardt-Roemermann et al., 2011)；幼树比叶面积和叶绿素较干扰前含量降低，叶片氮含量增加(闵鹏等, 2013)。

8. 展 望

基于功能性状的研究在近 20 年的发展使得生态学家从新的角度重新来审视复杂的生态学过程。从个体、种群、群落到生态系统，功能性状已被证明是有效可靠的手段来探索各类生态学前沿问题。特别是近几年来，随着测量技术的发展及大尺度数据电子化的进程，功能性状更是成为预测全球气候变化，以及生物多样性丧失对陆地生态系统影响的重要工具。尽管这些研究成果有力地推动了生态科学的发展，但细观每一方面都仍然有很多值得未来大力发展和探索的空白区域。

首先，在功能性状的选取和测量方面，绝大部分的研究都是偏向于选择易于大量测量的软性状，尽管软性状和硬性状紧密相关，但并不是可以完全代替。这也导致了很多研究结果的相悖性及大量无法解释的部分。未来需要更多与植物新陈代谢相关的硬性状（如生理性状）的测量来弥补目前的不足。此外，对地上部分功能性状的研究也远远大于地下性状，造成了研究的不平衡。然而，地下部分性状对于研究植物的生长发育、生态系统的养分循环、地上地下生态系统的相互作用至关重要。全球气候变化背景下，物候性状也是未来值得关注的重要功能性状，其季节性波动对群落动态有着强烈影响。而随着激光雷达技术在生态学中的逐渐推广应用，快速且大范围测量各类形态性状成为可能。

其次，在功能性状的研究方法上，应结合谱系亲缘关系。大部分的功能性状被认为和植物的系统发育史息息相关，可以说是先天遗传因素和后天环境因素共同作用的结果。因此未来在利用功能性状探索各类生态学问题时，应同时将谱系分析整合进来，以增强解释力。此外，近年来全球性大型野外研究平台的建立，如全球森林监测网络（ForestGEO）和全球森林生物多样性的生态系统功能实验网（TreeDivNet），包括中国亚热带森林生物多样性与生态系统功能实验平台（BEF-China）等（马克平，2013），以及生物多样性信息学的发展，如全球生物多样性数据网络（GBIF）和植物性状数据库（TRY）等，使得利用功能性状研究国家尺度、洲际尺度、全球尺度等大尺度上生态学问题成为热点。也有越来越多的研究尝试用基于个体水平的功能性状重新检验基于物种水平功能性状发现的结果，往往会得到不同的答案，也是未来值得关注的方向。

总体而言，在基于功能性状的研究方向上，有几个科学问题值得特别重视：①全球气候变化背景下功能性状在时间尺度上的变化规律；②功能性状与群落动态的关系；③功能性状与种间竞争的关系；④地上地下功能性状的相互作用及其对群落结构和动态的影响；⑤功能多样性与谱系多样性对生态系统功能的影响；⑥功能性状分布的聚散性在不同植被类型间的差异和全球的变化规律。

最后，基于功能性状的研究也应该与生物多样性保护和生态系统管理紧密结合。如根据功能性状的指示作用来制定生物多样性保护策略、植被恢复方案、保护区设定规则，以及生态环境规划方案等。

我国幅员辽阔，纬度梯度明显，有着丰富的植被和生境资源，这些资源带来丰富的生物多样性，为利用功能性状从局地尺度到大尺度研究提供了得天独厚的条件。

致谢

感谢国家重大科学研究计划（2014CB954104）和国家自然科学基金项目（31300353）对本项工作的资助。

参 考 文 献

卜文圣, 马克平. 2014. 基于功能性状探讨生物多样性与生态系统功能关系. //马克平. 中国生物多样性保护与研究进展 X. 北京: 气象出版社, 175-182

冯秋红, 史作民, 董莉莉. 2008. 植物功能性状对环境的响应及其应用. 林业科学, 44(4): 125-131.

马克平. 2013. 生物多样性与生态系统功能的实验研究. 生物多样性, 21: 247-248

孟婷婷, 倪健, 王国宏. 2007. 植物功能性状与环境和生态系统功能. 植物生态学报, 31(1): 150-165.

么旭阳, 胡耀升, 刘艳红. 2014. 长白山阔叶红松林不同群落类型的植物功能性状与功能多样性. 西北农林科技大学学报, 42(3): 77-84.

闵鹏, 齐代华, 贺丽等. 2013. 四川山矾幼树叶功能性状在毛竹干扰下的响应. 林业科学, 49(3): 70-77.

Ackerly D D, Cornwell W K. 2007. A trait-based approach to community assembly: partitioning of species trait values into within- and among-community components. Ecology Letters, 10: 135-145.

Ackerly D D, Knight C A, Weiss S B, et al. 2002. Leaf size, specific leaf area and microhabitat distribution of chaparral woody plants: contrasting patterns in species level and community level analyses. Oecologia, 130: 449-457.

Adler P B, Fajardo A, Kleinhesselink A R, et al. 2013. Trait-based tests of coexistence mechanisms. Ecology Letters, 16: 1294-1306.

Aerts R, Chapin F S. 2000. The mineral nutrition of wild plants revisited: A re-evaluation of processes and patterns. Advances in Ecological Research, Vol30, 30: 1-67.

Albert C H, de Bello F, Boulangeat I, et al. 2012. On the importance of intraspecific variability for the quantification of functional diversity. Oikos, 121: 116-126.

Baraloto C, Hardy O J, Paine C E T, et al. 2012. Using functional traits and phylogenetic trees to examine the assembly of tropical tree communities. Journal of Ecology, 100: 690-701.

Bernhardt-Roemermann M, Gray A, Vanbergen A J, et al. 2011. Functional traits and local environment predict vegetation responses to disturbance: a pan-European multi-site experiment. Journal of Ecology, 99: 777-787.

Bohnke M, Krober W, Welk E, et al. 2014. Maintenance of constant functional diversity during secondary succession of a subtropical forest in China. Journal of Vegetation Science, 25: 897-911.

Bolmgren K, Cowan P D. 2008. Time-size tradeoffs: a phylogenetic comparative study of flowering time, plant height and seed mass in a north-temperate flora. Oikos, 117: 424-429.

Bonser S P, Ladd B, Monro K, et al. 2010. The adaptive value of functional and life-history traits across fertility treatments in an annual plant. Annals of Botany, 106: 979-988.

Briggs J M, Knapp A K, Brock B L. 2002. Expansion of woody plants in tallgrass prairie: A fifteen-year study of fire and fire-grazing interactions. American Midland Naturalist, 147: 287-294.

Bruno J F, Stachowicz J J, Bertness M D. 2003. Inclusion of facilitation into ecological theory. Trends in Ecology & Evolution, 18: 119-125.

Cadotte M W, Cavender-Bares J, Tilman D, et al. 2009. Using Phylogenetic, Functional and Trait Diversity to Understand Patterns of Plant Community Productivity. PLoS ONE, 4(5): e5695.

Cavender-Bares J, Ackerly D D, Baum D A, et al. 2004. Phylogenetic overdispersion in Floridian oak communities. American Naturalist, 163: 823-843.

Cavender-Bares J, Kozak K H, Fine P V A, et al. 2009. The merging of community ecology and phylogenetic biology. Ecology Letters, 12: 693-715.

Chave J, Coomes D, Jansen S, et al. 2009. Towards a worldwide wood economics spectrum. Ecology Letters, 12: 351-366.

Chave J, Muller-Landau H C, Baker T R, et al. 2006. Regional and phylogenetic variation of wood density across 2456 neotropical tree species. Ecological Applications, 16: 2356-2367.

Chrobock T, Kempel A, Fischer M, et al. 2011. Introduction bias: Cultivated alien plant species germinate faster and more abundantly than native species in Switzerland. Basic and Applied Ecology, 12: 244-250.

Cingolani A M, Cabido M, Gurvich D E, et al. 2007. Filtering processes in the assembly of plant communities: Are species presence and abundance driven by the same traits? Journal of Vegetation Science, 18: 911-920.

Conti G, Diaz S. 2013. Plant functional diversity and carbon storage - an empirical test in semi-arid forest ecosystems. Journal of Ecology, 101: 18-28.

Cornelissen J H C, Lavorel S, Garnier E, et al. 2003. A handbook of protocols for standardised and easy measurement of plant functional traits worldwide. Australian Journal of Botany, 51: 335-380.

Cornwell W K, Ackerly D D. 2009. Community assembly and shifts in plant trait distributions across an environmental gradient in coastal California. Ecological Monographs, 79: 109-126.

Cornwell W K, Schwilk D W, Ackerly D D. 2006. A trait-based test for habitat filtering: Convex hull volume. Ecology, 87: 1465-1471.

Craine J M, Lee W G. 2003. Covariation in leaf and root traits for native and non-native grasses along an altitudinal gradient in New Zealand. Oecologia, 134: 471-478.

Daehler C C. 2003. Performance comparisons of co-occurring native and alien invasive plants: Implications for conservation and restoration. Annual Review of Ecology, Evolution and Systematics, 34: 183-211.

Davis M A, Grime J P, Thompson K. 2000. Fluctuating resources in plant communities: a general theory of invasibility. Journal of Ecology, 88: 528-534.

de Bello F, Lavorel S, Diaz S, et al. 2010. Towards an assessment of multiple ecosystem processes and services via functional traits. Biodiversity and Conservation, 19: 2873-2893.

Diaz S, Cabido M, Casanoves F. 1998. Plant functional traits and environmental filters at a regional scale. Journal of Vegetation Science, 9: 113-122.

Diaz S, Cabido M, Zak M, et al. 1999. Plant functional traits, ecosystem structure and land-use history along a climatic gradient in central-western Argentina. Journal of Vegetation Science, 10: 651-660.

Diaz S, Cabido M. 2001. Vive la difference: plant functional diversity matters to ecosystem processes. Trends in Ecology & Evolution, 16: 646-655.

Diaz S, Hodgson J G, Thompson K, et al. 2004. The plant traits that drive ecosystems: Evidence from three continents. Journal of Vegetation Science, 15: 295-304.

Diaz S, Lavorel S, McIntyre S, et al. 2007. Plant trait responses to grazing - a global synthesis. Global Change Biology, 13: 313-341.

Diaz S, Noy-Meir I, Cabido M. 2001. Can grazing response of herbaceous plants be predicted from simple vegetative traits? Journal of Applied Ecology, 38: 497-508.

Enquist B J, Kerkhoff A J, Stark S C, et al. 2007. A general integrative model for scaling plant growth, carbon flux, and functional trait spectra. Nature, 449: 218-222.

Felsenstein J. 1985. Phylogenies and the Comparative Method. American Naturalist, 125: 1-15.

Fisher J B, Goldstein G, Jones T J, et al. 2007. Wood vessel diameter is related to elevation and genotype in the Hawaiian tree *Metrosideros polymorpha* (*Myrtaceae*). American Journal of Botany, 94: 709-715.

Flynn D F B, Gogol-Prokurat M, Nogeire T, et al. 2009. Loss of functional diversity under land use intensification across multiple taxa. Ecology Letters, 12: 22-33.

Flynn D F B, Mirotchnick N, Jain M, et al. 2011. Functional and phylogenetic diversity as predictors of biodiversity-ecosystem- function relationships. Ecology, 92: 1573-1581.

Fonseca C R, Overton J M, Collins B, et al. 2000. Shifts in trait-combinations along rainfall and phosphorus gradients. Journal of Ecology, 88: 964-977.

Golodets C, Sternberg M, Kigel J. 2009. A community-level test of the leaf-height-seed ecology strategy scheme in relation to grazing conditions. Journal of Vegetation Science, 20: 392-402.

Grigulis K, Lavorel S, Krainer U, et al. 2013. Relative contributions of plant traits and soil microbial properties to mountain grassland ecosystem services. Journal of Ecology, 101: 47-57.

Grotkopp E, Rejmanek M, Rost T L. 2002. Toward a causal explanation of plant invasiveness: Seedling growth and life-history strategies of 29 pine (*Pinus*) species. American Naturalist, 159: 396-419.

Hanba Y T, Noma N, Umeki K. 2000. Relationship between leaf characteristics, tree sizes and species distribution along a slope in a warm temperate forest. Ecological Research, 15: 393-403.

Hardy O J, Couteron P, Munoz F, et al. 2012. Phylogenetic turnover in tropical tree communities: impact of environmental filtering, biogeography and mesoclimatic niche conservatism. Global Ecology and Biogeography, 21: 1007-1016.

Heisler J L, Briggs J M, Knapp A K. 2003. Long-term patterns of shrub expansion in a C-4-dominated grassland: Fire frequency and the dynamics of shrub cover and abundance. American Journal of Botany, 90: 423-428.

Herault B, Bachelot B, Poorter L, et al. 2011. Functional traits shape ontogenetic growth trajectories of rain forest tree species. Journal of Ecology, 99: 1431-1440.

Knapp S, Kuehn I. 2012. Origin matters: widely distributed native and non-native species benefit from different functional traits. Ecology Letters, 15: 696-703.

Kolar C S, Lodge D M. 2001. Progress in invasion biology: predicting invaders. Trends in Ecology & Evolution, 16: 199-204.

Kraft N J B, Ackerly D D. 2010. Functional trait and phylogenetic tests of community assembly across spatial scales in an Amazonian forest. Ecological Monographs, 80: 401-422.

Kraft N J B, Metz M R, Condit R S, et al. 2010. The relationship between wood density and mortality in a global tropical forest data set. New Phytologist, 188: 1124-1136.

Kraft N J B, Valencia R, Ackerly D D. 2008. Functional traits and niche-based tree community assembly in an amazonian forest. Science, 322: 580-582.

Kushwaha C P, Tripathi S K, Singh K P. 2011. Tree specific traits affect flowering time in Indian dry tropical forest. Plant Ecology, 212: 985-998.

Laliberte E, Legendre P. 2010. A distance-based framework for measuring functional diversity from multiple traits. Ecology, 91: 299-305.

Lamarque L J, Delzon S, Lortie C J. 2011. Tree invasions: a comparative test of the dominant hypotheses and functional traits. Biological Invasions, 13: 1969-1989.

Laughlin D C, Leppert J J, Moore M M, et al. 2010. A multi-trait test of the leaf-height-seed plant strategy scheme with 133 species from a pine forest flora. Functional Ecology, 24: 493-501.

Lavorel S, Garnier E. 2002. Predicting changes in community composition and ecosystem functioning from plant traits: revisiting the Holy Grail. Functional Ecology, 16: 545-556.

Lebrija-Trejos E, Perez-Garcia E A, Meave J A, et al. 2010. Functional traits and environmental filtering drive community assembly in a species-rich tropical system. Ecology, 91: 386-398.

Levine J M, D'Antonio C M. 1999. Elton revisited: a review of evidence linking diversity and invasibility. Oikos, 87: 15-26.

Liu X J, Swenson N G, Zhang J L, et al. 2013. The environment and space, not phylogeny, determine trait dispersion in a subtropical forest. Functional Ecology, 27: 264-272.

Losos J B. 2008. Phylogenetic niche conservatism, phylogenetic signal and the relationship between phylogenetic relatedness and ecological similarity among species. Ecology Letters, 11: 995-1003.

Louault F, Pillar V D, Aufrere J, et al. 2005. Plant traits and functional types in response to reduced disturbance in a semi-natural grassland. Journal of Vegetation Science, 16: 151-160.

Mack R N, Simberloff D, Lonsdale W M, et al. 2000. Biotic invasions: Causes, epidemiology, global consequences, and control. Ecological Applications, 10: 689-710.

Maharjan S K, Poorter L, Holmgren M, et al. 2011. Plant Functional Traits and the Distribution of West African Rain Forest Trees along the Rainfall Gradient. Biotropica, 43: 552-561.

Markesteijn L, Poorter L. 2009. Seedling root morphology and biomass allocation of 62 tropical tree species in relation to drought- and shade-tolerance. Journal of Ecology, 97: 311-325.

Martinez-Vilalta J, Mencuccini M, Vayreda J, et al. 2010. Interspecific variation in functional traits, not climatic differences among species ranges, determines demographic rates across 44 temperate and Mediterranean tree species. Journal of Ecology, 98: 1462-1475.

Mayfield M M, Levine J M. 2010. Opposing effects of competitive exclusion on the phylogenetic structure of communities. Ecology Letters, 13: 1085-1093.

McGill B J, Enquist B J, Weiher E, et al. 2006. Rebuilding community ecology from functional traits. Trends in Ecology & Evolution, 21: 178-185.

Meng T T, Ni J, Harrison S P. 2009. Plant morphometric traits and climate gradients in northern China: a meta-analysis using quadrat and flora data. Annals of Botany, 104: 1217-1229.

Moles A T, Ackerly D D, Tweddle J C, et al. 2007. Global patterns in seed size. Global Ecology and Biogeography, 16: 109-116.

Moles A T, Falster D S, Leishman M R, et al. 2004. Small-seeded species produce more seeds per square metre of canopy per year, but not per individual per lifetime. Journal of Ecology, 92: 384-396.

Moles A T, Warton D I, Warman L, et al. 2009. Global patterns in plant height. Journal of Ecology, 97: 923-932.

Moles A T, Westoby M. 2004. Seedling survival and seed size: a synthesis of the literature. Journal of Ecology, 92: 372-383.

Moles A T, Westoby M. 2006. Seed size and plant strategy across the whole life cycle. Oikos, 113: 91-105.

Mouillot D, Graham N A J, Villeger S, et al. 2013. A functional approach reveals community responses to disturbances. Trends in Ecology & Evolution, 28: 167-177.

Mouillot D, Mason N W H, Wilson J B. 2007. Is the abundance of species determined by their functional traits? A new method with a test using plant communities. Oecologia, 152: 729-737.

Mouillot D, Villeger S, Scherer-Lorenzen M, et al. 2011. Functional structure of biological communities predicts ecosystem multifunctionality. PLoS ONE, 6(3): e17476.

Murray B R, Brown A H D, Dickman C R, et al. 2004. Geographical gradients in seed mass in relation to climate. Journal of Biogeography, 31: 379-388.

Niinemets U. 2001. Global-scale climatic controls of leaf dry mass per area, density, and thickness in trees and shrubs. Ecology, 82:453-469.

Niinemets, U. 2006. The controversy over traits conferring shade-tolerance in trees: ontogenetic changes revisited. Journal of Ecology, 94: 464-470.

Noy-Meir I, Kaplan D. 2002. Species richness of annual legumes in relation to grazing in Mediterranean vegetation in northern Israel. Israel Journal of Plant Sciences, 50: S95-S109.

Ordonez J C, van Bodegom P M, Witte J P M, et al. 2009. A global study of relationships between leaf traits, climate and soil measures of nutrient fertility. Global Ecology and Biogeography, 18: 137-149.

Paine C E T, Baraloto C, Chave J, et al. 2011. Functional traits of individual trees reveal ecological constraints on community assembly in tropical rain forests. Oikos, 120: 720-727.

Pausas J G, Bradstock R A, Keith D A, et al. 2004. Plant functional traits in relation to fire in crown-fire ecosystems. Ecology, 85: 1085-1100.

Pavoine S, Vela E, Gachet S, et al. 2011. Linking patterns in phylogeny, traits, abiotic variables and space: a novel approach to linking environmental filtering and plant community assembly. Journal of Ecology, 99: 165-175.

Pellissier L, Fournier B, Guisan A, et al. 2010. Plant traits co-vary with altitude in grasslands and forests in the European Alps. Plant Ecology, 211: 351-365.

Petchey O L, Gaston K J. 2002. Functional diversity (FD), species richness and community composition. Ecology Letters, 5: 402-411.

Petchey O L, Gaston K J. 2006. Functional diversity: back to basics and looking forward. Ecology Letters, 9: 741-758.

Poorter L, Bongers F, Sterck F J, et al. 2003. Architecture of 53 rain forest tree species differing in adult stature and shade tolerance. Ecology, 84: 602-608.

Poorter L, Markesteijn L. 2008. Seedling traits determine drought tolerance of tropical tree species. Biotropica, 40: 321-331.

Poorter L, McDonald I, Alarcon A, et al. 2010. The importance of wood traits and hydraulic conductance for the performance and life history strategies of 42 rainforest tree species. New Phytologist, 185: 481-492.

Poorter L, Wright S J, Paz H, et al. 2008. Are functional traits good predictors of demographic rates? Evidence from five Neotropical forests. Ecology, 89: 1908-1920.

Prinzing A, Reiffers R, Braakhekke W G, et al. 2008. Less lineages - more trait variation: phylogenetically clustered plant communities are functionally more diverse. Ecology Letters, 11: 809-819.

Reich P B, Oleksyn J. 2004. Global patterns of plant leaf N and P in relation to temperature and latitude. Proceedings of the National Academy of Sciences of the United States of America, 101: 11001-11006.

Reich P B, Tilman D, Naeem S, et al. 2004. Species and functional group diversity independently influence biomass accumulation and its response to CO_2 and N. Proceedings of the National Academy of Sciences of the United States of America, 101: 10101-10106.

Reich P B, Wright I J, Cavender-Bares J, et al. 2003. The evolution of plant functional variation: Traits, spectra, and strategies. International Journal of Plant Sciences, 164: S143-S164.

Roscher C, Schumacher J, Gubsch M, et al. 2012. Using plant functional traits to explain diversity-productivity relationships. PLoS ONE, 7(5): e36760.

Roscher C, Schumacher J, Lipowsky A, et al. 2013. A functional trait-based approach to understand community assembly and diversity-productivity relationships over 7 years in experimental grasslands. Perspectives in Plant Ecology, Evolution and Systematics, 15: 139-149.

Rozendaal D M A, Hurtado V H, Poorter L. 2006. Plasticity in leaf traits of 38 tropical tree species in response to light; relationships with light demand and adult stature. Functional Ecology, 20: 207-216.

Ruger N, Wirth C, Wright S J, et al. 2012. Functional traits explain light and size response of growth rates in tropical tree species. Ecology, 93: 2626-2636.

Ruiz-Jaen M C, Potvin C. 2011. Can we predict carbon stocks in tropical ecosystems from tree diversity? Comparing species and functional diversity in a plantation and a natural forest. New Phytologist, 189: 978-987.

Santiago L S, Kitajima K, Wright S J, et al. 2004. Coordinated changes in photosynthesis, water relations and leaf nutritional traits of canopy trees along a precipitation gradient in lowland tropical forest. Oecologia, 139: 495-502.

Spasojevic M J, Aicher R J, Koch G R, et al. 2010. Fire and grazing in a mesic tallgrass prairie: impacts on plant species and functional traits. Ecology, 91: 1651-1659.

Sterck F J, Bongers F. 2001. Crown development in tropical rain forest trees: patterns with tree height and light availability. Journal of Ecology, 89: 1-13.

Sterck F J, Martinez-Vilalta J, Mencuccini M, et al. 2012. Understanding trait interactions and their impacts on growth in Scots pine branches across Europe. Functional Ecology, 26: 541-549.

Swenson N G, Enquist B J, Pither J, et al. 2012. The biogeography and filtering of woody plant functional diversity in North and South America. Global Ecology and Biogeography, 21: 798-808.

Swenson N G, Enquist B J. 2008. The relationship between stem and branch wood specific gravity and the ability of each measure to predict leaf area. American Journal of Botany, 95: 516-519.

Swenson N G, Enquist B J. 2009. Opposing assembly mechanisms in a Neotropical dry forest: implications for phylogenetic and functional community ecology. Ecology, 90: 2161-2170.

Thuiller W, Lavorel S, Midgley G, et al. 2004. Relating plant traits and species distributions along bioclimatic gradients for 88 Leucadendron taxa. Ecology, 85: 1688-1699.

Uriarte M, Swenson N G, Chazdon R L, et al. 2010. Trait similarity, shared ancestry and the structure of neighbourhood interactions in a subtropical wet forest: implications for community assembly. Ecology Letters, 13: 1503-1514.

van Gelder H A, Poorter L, Sterck F J. 2006. Wood mechanics, allometry, and life-history variation in a tropical rain forest tree community. New Phytologist, 171: 367-378.

van Kleunen M, Richardson D M. 2007. Invasion biology and conservation biology: time to join forces to explore the links between species traits and extinction risk and invasiveness. Progress in Physical Geography, 31: 447-450.

van Kleunen M, Weber E, Fischer M. 2010. A meta-analysis of trait differences between invasive and non-invasive plant species. Ecology Letters, 13: 235-245.

Villeger S, Mason N W H, Mouillot D. 2008. New multidimensional functional diversity indices for a multifaceted framework in functional ecology. Ecology, 89: 2290-2301.

Webb C O. 2000. Exploring the phylogenetic structure of ecological communities: An example for rain forest trees. American Naturalist, 156: 145-155.

Weiher E, van der Werf A, Thompson K, et al. 1999. Challenging Theophrastus: A common core list of plant traits for functional ecology. Journal of Vegetation Science, 10: 609-620.

Westoby M, Falster D S, Moles A T, et al. 2002. Plant ecological strategies: Some leading dimensions of variation between species. Annual Review of Ecology and Systematics, 33: 125-159.

Westoby M, Wright I J. 2006. Land-plant ecology on the basis of functional traits. Trends in Ecology & Evolution, 21: 261-268.

Westoby M. 1998. A leaf-height-seed (LHS) plant ecology strategy scheme. Plant and Soil, 199: 213-227.

Wright I J, Ackerly D D, Bongers F, et al. 2007. Relationships among ecologically important dimensions of plant trait variation in seven Neotropical forests. Annals of Botany, 99: 1003-1015.

Wright I J, Reich P B, Cornelissen J H C, et al. 2005. Assessing the generality of global leaf trait relationships. New Phytologist, 166: 485-496.

Wright I J, Reich P B, Westoby M, et al. 2004. The worldwide leaf economics spectrum. Nature, 428: 821-827.

Wright I J, Reich P B, Westoby M. 2001. Strategy shifts in leaf physiology, structure and nutrient content between species of high- and low-rainfall and high- and low-nutrient habitats. Functional Ecology, 15: 423-434.

Wright J P, Naeem S, Hector A, et al. 2006. Conventional functional classification schemes underestimate the relationship with ecosystem functioning. Ecology Letters, 9: 111-120.

Wright S J, Kitajima K, Kraft N J B, et al. 2010. Functional traits and the growth-mortality trade-off in tropical trees. Ecology, 91: 3664-3674

* 策划编辑：薛红卫　中国科学院上海生命科学研究院植物生理生态研究所

植物中的蛋白磷酸化

作　者：薛红卫　谭树堂　晏　萌

中国科学院上海生命科学研究院植物生理生态研究所

▶ 1. 植物蛋白磷酸化概述 / 25
▶ 2. 蛋白磷酸化在植物生长发育中的功能 / 37
▶ 3. 蛋白磷酸化在植物激素信号转导中的作用 / 44
▶ 4. 蛋白磷酸化在植物-环境相互作用中的功能 / 51
▶ 5. 总结与展望 / 55

摘要

磷酸化(phosphorylation)是一种重要的蛋白质翻译后修饰方式，一般通过蛋白激酶(protein kinase)和蛋白磷酸酶(protein phosphatase)对底物(substrate)蛋白的特定氨基酸进行调控，在真核生物的生命活动中发挥关键作用。研究表明磷酸化修饰调控蛋白质的构象、活性、稳定性和亚细胞定位等，因而对底物蛋白发挥功能具有重要决定作用。20世纪80年代末至今，随着分子生物学的发展，模式植物拟南芥(*Arabidopsis thaliana*)正向遗传学的兴起，生物化学、蛋白质组学和结构生物学的发展，植物蛋白磷酸化的研究取得了重要发展，人们对蛋白磷酸化在植物生长发育过程中的功能有了相当的了解。特别是近10年以来，对蛋白磷酸化的研究表明其参与了植物细胞间信息交流、胞内信号转导、免疫反应和表观遗传调控等许多重要过程，在植物生长发育和环境响应等方面发挥了重要调节作用。人们也发现除蛋白激酶和磷酸酶对特定底物的调控外，一些磷酸结合蛋白也参与了相关信号转导过程。我们将对植物蛋白激酶和蛋白磷酸酶的分类、生物化学特性和重要功能作系统阐述，并总结蛋白磷酸化相关生理过程调控的分子机制，注重体现其中一些重要发现的研究策略与技术手段，以为今后的蛋白磷酸化研究工作提供借鉴。

关键词

翻译后修饰、磷酸化、蛋白激酶、蛋白磷酸酶、底物、信号转导、配体、受体、结构域、拟南芥

1. 植物蛋白磷酸化概述

1.1 蛋白磷酸化修饰

1953年，Watson与Crick发现了脱氧核糖核酸(DNA)的双螺旋结构，由此揭开了现代分子生物学的序幕。现在，"DNA-RNA-蛋白质"的"中心法则"几乎是分子生物学领域的常识，但是蛋白质要行使功能仍需要不同的加工和修饰。在翻译完成后，蛋白质会发生如磷酸化、糖基化、乙酰化、甲基化等翻译后修饰(post-translational modification)，其中，可逆磷酸化(reversible phosphorylation)是一种最常见的修饰，由蛋白激酶和蛋白磷酸酶负责调控(图版Ⅰ图1)。目前的研究表明蛋白磷酸化是一种十分普遍的调控机制，几乎参与了每个生命活动过程和整个生命周期。质谱分析数据显示人类蛋白质的70%可以被磷酸化，而且一般具有多个磷酸化位点。在人类、酵母与植物等真核生物中，蛋白磷酸酶与蛋白激酶的总量可以占到基因组编码蛋白质的2%~4%。拟南芥基因组中大约编码了150个左右的蛋白磷酸酶，与约1050个蛋白激酶(Kerk et al., 2008)。普度大学在其PlantsP数据库(http://plantsp.genomics.purdue.edu/index.html)中对拟南芥的蛋白激酶与蛋白磷酸酶作了系统的分类及芯片信息的整合，基本明确拟南芥基因组中有130个蛋白磷酸和989个蛋白激

酶。而在水稻中有 1508 个蛋白激酶(Rice Kinase Database, RKD; http://rkd.ucdavis.edu)。从近年来的磷酸化组分析结果来看，植物全蛋白质组的大量蛋白质都可以被磷酸化，考虑到植物基因组所编码的大量蛋白激酶与蛋白磷酸酶，不排除植物与人类等具有相当比例的蛋白质受磷酸化调控(参见综述 Uhrig et al., 2013)。

真核生物的蛋白磷酸化可以发生在线粒体、叶绿体、胞质、细胞核、高尔基体，甚至于细胞外。从目前的研究报道和磷酸化组结果来看，尽管植物基因组中并未编码典型的酪氨酸激酶(tyrosine kinases)，但植物蛋白质组中却有相当数量的蛋白质可以在酪氨酸位点上发生磷酸化修饰。植物蛋白质组的磷酸化位点中，丝氨酸(serine, Ser, S)约为 84%~86%，苏氨酸(threonine, Thr, T)约为 10%~12%，酪氨酸(tyrosine, Tyr, Y)约为 2%~4%，这样的比例与其他真核生物比较类似(参见综述 Uhrig et al., 2013)。

1.2 蛋白激酶

蛋白激酶负责催化底物蛋白的磷酸化反应，其利用 ATP 或 GTP 作为磷酸基团的供体，将 γ 位上的磷酸基团转移到底物蛋白氨基酸的羟基上，是一种共价修饰(图版 I 图 1)。基于磷酸化主要发生在丝氨酸、苏氨酸和酪氨酸的羟基上，可以将激酶分为 3 大类：丝氨酸/苏氨酸蛋白激酶(serine/threonine kinase, STK)、酪氨酸蛋白激酶(tyrosine kinase, TK)、双重底物蛋白激酶(dual specificity kinase, DSK)。此外，植物中还有组氨酸蛋白激酶(histidine kinase, HK)。在植物中最主要的是 STK，近年来的研究表明酪氨酸的磷酸化也具有十分重要的作用，光信号转导中的 PIF(phytochrome interacting factor)与光敏色素(phytochrome, PHY)(Nito et al., 2013)、油菜素内酯(brassinosteroid, BR)信号中的 BRI1(brassinosteroid insensitive 1)(Jaillais et al., 2011)和 BAK1(BRI1 associated receptor kinase 1)(Oh et al., 2010; Jaillais et al., 2011)、抗病反应中 EFR(EF-Tu receptor)(Macho et al., 2014)，以及 BIK(botrytis-induced kinase 1)(Lin et al., 2014)都可以发生酪氨酸位点上的磷酸化，表明酪氨酸磷酸化修饰参与到多个信号途径中。植物中的 HK 主要是乙烯受体 ETR(ethylene response 1)类与细胞分裂素受体 AHK(*arabidopsis* histidine kinase)类，也具有十分重要的功能。

根据氨基酸序列的同源性与对底物选择的生化特性，蛋白激酶可以分为 9 个大类：分别为酪氨酸蛋白激酶(TK)，类酪氨酸蛋白激酶(tyrosine kinase like, TKL)，STE 类蛋白激酶(与酵母中 ste 类激酶同源)，AGC 类蛋白激酶(命名来自 PKA、PKG、PKC)，Ca^{2+} 或钙调素依赖蛋白激酶(calmodulin dependent protein kinase, CaMK)，CMGC 类蛋白激酶[主要包括细胞周期依赖蛋白激酶(cyclin dependent protein kinase, CDK)、有丝分裂原蛋白激酶(mitogen activated protein kinase, MAPK)、糖原表面激酶(glucogen synthase kinase, GSK)和细胞周期样蛋白激酶(CDC like kinase, CLK)]，受体鸟苷酸环化酶(receptor guanylate cyclase, RGC，无激酶活性)类蛋白激酶，I 型酪蛋白激酶(casein kinase 1, CK I, CK1)和非典型蛋白激酶(如 HK 等)。植物的基因组中编码了 1286 个蛋白激酶(参考 PlantsP 数据)，基本上具有一般真核生物所具有的激酶类群，但由于类受体激酶(receptor like kinase, RLK)的大量基因倍增，其分类与动物中稍有不同。根据 PlantsP 数据库的分类，可以将植物的蛋白激酶分为 5 个大类：类受体激酶（跨膜蛋白激酶），类 ATN1/CTR1/EDR1/GmPK6 蛋白激酶，I 型酪蛋白激酶，非跨膜蛋白激酶（主要包括 CDK、MAPK、GSK、AGC、CK2 和 CDPK 等）和其他。植物中 CDK 类与酵母、动物中功能类似，在此不单独介绍。

1.2.1 RLK 与 RLCK

类受体激酶（RLK）与类受体胞质激酶（receptor-like cytoplasmic kinase, RLCK）是植物中重要的一个类群，拟南芥基因组大约编码了 420 个 RLK、200 个 RLCK（Nodine et al., 2011）。RLK 一般具有胞外配体结合结构域、跨膜结构域和胞内激酶结构域 3 个部分，介导了细胞外信号到胞内的转导，不仅参与了植物激素等体内信号的感知，也参与环境因子（如病原菌等体外信号）的感知。根据 RLK 胞外域和胞内域的序列差异，大致可以分为 46 个亚家族（Shiu and Bleecker, 2003；部分有功能报道的 RLK 可参考图版 I 图 2）。不同的亚家族之间成员数目差异很大，其中，在胞外结构域中含有富含亮氨酸的重复序列（leucine rich repeat, LRR）的 LRR-RLK 是最庞大的一个亚家族，其有 15 个小类，包括 233 个成员。与 RLK 相比，RLCK 亚家族成员不具有典型的胞外结构域，有一些成员甚至缺少跨膜结构域而定位在胞质中。

陆生植物营固着生活，面临着巨大的生存压力，有效地感受环境的变化从而调控生长发育就显得十分重要。RLK 作为植物一类重要的蛋白激酶，具有信号识别与感知功能，一般处于信号转导途径的上游，它们通过介导不同的信号途径参与了多种生长发育和换届响应过程（图版 I 图 2），包括配子体发育、花粉-柱头间的相互识别、茎和根顶端分生组织的分化、激素信号转导（Belkhdir et al., 2006）、细胞形态建成、表皮细胞分化、细胞生长（Hematy and Hofte, 2008）、顶端分生组织干细胞的维持（Aichinger et al., 2012）、植物-病原菌间的相互作用（Afzal et al., 2008）和伤害反应等（Shiu and Bleecker, 2001a）。RLK 作为信号的起始，功能多样，我们将在本章第 2、3 节中着重介绍。

1.2.2 CDPK、CaMK 与 CBL-CIPK

Ca^{2+} 是一种重要的第二信使，相应地，Ca^{2+} 相关的蛋白激酶作为信号的中间传输者，参与到众多生命活动中，如植物向性反应、气孔运动和免疫反应。植物中的 Ca^{2+} 调控蛋白激酶包括 CDPK（calcium dependent protein kinase）、可与钙调素（calmodulin, CaM）结合的 CaMK（calmodulin dependent protein kinase）和 CBL-CIPK（calcineurin B-like protein, CBL interacting protein kinase）。尽管植物中存在钙调素，但相关的 CaMK 的研究报道较少，一般认为植物与酵母或动物的机制有所不同，CaM 与激酶合二为一，进化出 CDPK，直接感受钙浓度的变化。本文着重对 CDPK 和 CBL-CIPK 作详细的介绍。

CDPK（拟南芥中一般称作 CPK）既具有一个可以结合 Ca^{2+} 的钙调素类结构域，又具有一个激酶结构域，可以感知特定信号引起的胞内 Ca^{2+} 浓度的变化，并进一步通过磷酸化下游底物，将相关信号传导下去。植物基因组编码了大约 30 个 CDPK，根据序列同源性可以分为 4 个亚家族，I~IV。拟南芥中的研究表明，CDPK 广泛地参与植物对外界环境的响应，如对干旱和病原菌的侵染等（Hamel et al., 2013）。CDPK（包括 CPK3、4、6、

10 和 11)可以磷酸化重要的阴离子通道蛋白 SLAC1/SLAH3，参与 ABA 途径，影响气孔运动及对干旱胁迫的响应(Mori et al., 2006; Zhu et al., 2007; Zou et al., 2010; Geiger et al., 2010, 2011; Brandt et al., 2012)。Jen Sheen 实验室对拟南芥 CDPK 家族的成员作了系统性研究，发现多个成员在先天免疫中发挥了重要作用(Boudsocq et al., 2010)。最近的研究表明，在植物免疫反应中，CDPK 可以直接磷酸化 NAPH 氧化酶，控制活性氧(ROS)的产生(Kobayashi et al., 2007; Gao et al., 2013)。CDPK 不仅参与植物对环境的响应，也直接参与了生长发育过程的调控，CPK17 与 CPK34 调控了花粉管的生长(Myers et al., 2009)；质膜定位的 CDPK 相关激酶 CRK5(CDPK related kinase 5)通过磷酸化生长素输出载体 PIN2 影响生长素的极性运输，参与向重性反应和侧根发生的调控(Rigó et al., 2013)。

植物 CBL 是重要的钙信号感应蛋白，通过与 CIPK 互作调控底物的磷酸化。calcineurin 是一个 Ca^{2+}/钙调蛋白结合的蛋白磷酸酶，从酵母到动物都十分保守。calcineurin 由 A、B 两个亚基所组成，calcineurin A 为催化亚基，B 为调控亚基，具有与钙调蛋白的同源结构域，含有 EF hand 基序，结合 Ca^{2+}。植物的 CBL 与动物或酵母中的同源性较低，包括 10 个成员，最早发现于植物的盐响应途径中(Liu et al., 1998; Kudla et al., 1999)。植物中 CBL 并不是作为磷酸酶行使功能，而是通过与新的一类蛋白激酶即 CIPK 互作调控下游信号(Shi et al., 1999)。拟南芥基因组编码了 25 个 CIPK 成员，其在氨基酸序列上与酵母的 SNF1(sucrose non-fermenting 1)和动物的 AMPK(AMP dependent kinase)同源，因此也有分类将其称为 SnRK3。拟南芥的 *SOS2*(salt over sensitive 2)编码了蛋白激酶 CIPK24(Guo et al., 2001)，Ca^{2+} 结合后促进形成 CBL-CIPK 复合体，进一步磷酸化下游因子，从而调控逆境响应。Weihua Wu 实验室对筛选获得的低钾敏感突变体 *lks1*(low-K^+-sensitive 1)的分析表明 *LKS1* 编码 CIPK23，CBL1/CBL9-CIPK23 通过磷酸化 K^+ 通道蛋白 AKT1(*arabidopsis* K^+ transporter 1)调控了 K^+ 吸收和气孔运动过程(Xu et al., 2006; Li et al., 2006; Cheong et al., 2007; Lee et al., 2007)。基于 PP2C(type-2C protein phosphatase)直接调控 AKT2，且 PP2C 可以与 CIPK 形成复合体，PP2C/CBL-CIPK 可能作为一个开关调控了 K^+ 通道的开放，在气孔开闭中起重要调控作用(具体将在 4.2 节，磷酸化参与植物-环境作用部分进行详细阐述)。

CBL2 与 CBL3 参与了液泡膜上 H^+-ATP 酶(H^+-ATPase)的活性调控，影响拟南芥的生长发育(Tang et al., 2012)。*cbl2 cbl3* 双突变体表现出叶片坏死，果荚变短，结实率低等缺陷，但是没有任何一个 CIPK 的单突变体出现类似表型，表明 CBL 可能通过与特定的 CIPK 相互作用参与具体生物学过程的调控，这方面还有待更系统的遗传学与生物化学研究。

1.2.3 AGC

AGC 激酶是真核生物中普遍存在的一类非常保守的蛋白激酶，既可以感知信号，也可以作为信号的中间传输者，在细胞分裂、信号转导等方面发挥重要的调控功能。拟南芥 AGC 家族包括 39 个成员，根据动物中的分类，可以分为磷脂依赖的蛋白激酶 PDK1(3'-phosphoinositide-dependent kinase 1)、AGCV I、AGCV II、AGCV III 和其他共 5 类(Bögre et al., 2003; Garcia et al., 2012)。PDK1 具有与磷脂分子结合的 PH(pleckstrin homology)结构域,可以通过磷酸化其他 AGC 成员并调控其活性,在 AGC 家族中居于核心地位。利用拟南芥进行的研究表明 PDK1 可以磷酸化 S6K1(S6 kinase 1)、PID(pinoid)、OXI1(oxidative signal-inducible 1)等成员,参与众多生长发育或环境响应的调控(Zegzouti et al., 2006)。AGCVIII在拟南芥中有 23 个成员,分为 AGC1、AGC2、AGC3、AGC4 4 个亚进化分枝(Galvan-Ampudia et al., 2007)。AGC1 类的 D6PK(D6 protein kinase)在生长素的极性运输中具有十分重要的作用,可以磷酸化生长素输出受体 PIN(pin-formed)蛋白而调控其活性,对于植物正常生长发育和向性必不可少(Zourelidou et al., 2009; Willige et al., 2013; Barbosa et al., 2014; Zourelidou et al., 2014)。AGC2 类的 OXI1 参与了根毛发生(Anthony et al., 2004)和逆境反应(Rentel et al., 2004; Forzani et al., 2011; Camehl et al., 2011)。AGC3 类的 PID 也可以磷酸化生长素输出载体 PIN 蛋白调控其极性定位,从而影响生长素的极性运输(向基性运输)并调控胚胎发育、向重力反应和向光性反应等(Zegzouti et al., 2006; Michniewicz et al., 2007; Li et al., 2011; Ding et al., 2011)。AGC4 类的向光素 PHOT1(phototropin 1)和 PHOT2 是一类蓝光受体,介导了植物向光性生长、发育、气孔开闭、叶绿体的趋光性运动等(Christie et al., 2007)。此外,S6K1/2 参与了细胞周期的调控(Henriques et al., 2010)。

1.2.4 MAPK

MAPK 是一个非常保守的信号模块(module),由 MAPKKK(MAPK kinase kinase,也称作 MAP3K)磷酸化并激活 MAPKK(MAPK kinase,也称作 MAP2K),MAPKK 进一步磷酸化并激活 MAPK,通过这样的磷酸化级联(cascade)放大信号,最终由 MAPK 磷酸化下游效应蛋白,完成信号传递。此外,MAPK 也受到 MKP(MAPK phosphatase)类磷酸酶的负调控作用(参见综述文章 Rodriguez et al., 2010)。

拟南芥基因组编码了 60 个 MAPKKK,10 个 MAPKK,和 20 个 MAPK(MAPK Group, 2002; Šamajová et al., 2013)。该信号通路的一个最大特点是,一个 MAPK 可以对不同的信号作出响应,同时一个 MAPK 的激活可以磷酸化不同的下游因子而调控不同过程。MAPKKK 分为 MEKK1 类(拟南芥有 10 个 MEKK 成员)和 RAF 类,其中研究得比较透彻的是 CTR1(constitutive triple response 1)和 EDR1(enhanced disease resistance 1),它们都属于 RAF 类,一般不直接磷酸化 MAPKK,而是通过磷酸化其他底物,分别参与了乙烯信号(Kieber et al., 1993)和抗病反应(Frye et al., 1998)。最新研究表明,EDR1 可以与 MKK4 和 MKK5 互作,调控下游 MPK3 和 MPK6,参与植物先天免疫反应,但 EDR1 对于 MKK4、MKK5-MPK3 和 MPK6 不是起激活作用,而是抑制作用(Zhao et al., 2014),说明植物中的 MAPK 信号通路十分复杂,并不是单一的线性关系。

MAPKK 分为 A(MKK1、2、6),B(MKK3),C(MKK4、5),D(MKK7~10)4 个亚类,是植物生长发育和环境响应的重要组分。MAPK 分为 A~D 4 个亚类,A、B、C 均具有 TEY 激活基序,D 类具有 TDY 激活基序。目前的研究表明,MAPK 信号通路在植物的抗病反应FLS2途径(Meng et al., 2013)、气孔发育的 ERECTA-YODA 途径(Kim et al., 2012)、乙烯信号转导(Meng et al., 2013)、活性氧产生(Asai et al., 2002)和衰老(Meng et al., 2013)等方面都发挥了重要调节作用。MAPK 作为重要的信号中间传输者,与 RLK

等协作参与了众多胞外胞内信号的级联传递。相关内容将在以下的章节进行重点介绍。

1.2.5 SnRK 与 TOR

植物中的 SnRK(sucrose non-fermenting-1-related kinase)家族共有 38 个成员，分为 SnRK1、SnRK2、SnRK3 3 个亚家族，其中 SnRK1 与酵母(*Saccharomyces cerevisiae*)中的 SNF1 或人中的 AMPK 为同工酶，有 3 个成员，命名为 SnRK1.1(也称作 KINASE10，KIN10)、1.2(也称作 KIN11)和 1.3(也称作 KIN9)；SnRK2 和 SnRK3 共包括 35 个成员，为植物所特有(Robaglia et al., 2012)。SnRK1 调控植物对胁迫、黑暗和能量的感应，可能通过磷酸化某些转录因子并协调相关基因的表达(Baena-González et al., 2007; Baena-González and Sheen, 2008)，调控了种子萌发、糖代谢、开花时间等过程(Tsai et al., 2014)。糖不仅为生命活动提供了能源，也做为信号调控了植物的生长发育。糖代谢中的海藻糖-6-磷酸(trehalose-6-phosphate, T6P)可以直接作用于 SnRK1 并抑制其活性，从而调控下游基因的表达(Zhang et al., 2009; Debast et al., 2011; Nunes et al., 2013)。拟南芥 KIN10 和 KIN11 双突变体 *kin10 kin11* 的植株矮小瘦弱、开花延迟，*KIN10* 过表达的植株不能正常感知营养缺乏，表明 SnRK1 在协调植物能量与生长发育方面具有重要功能(Baena-González et al., 2007)。有意思的是，PP2C 通过互作调控了 SnRK1 活性(Rodrigues et al., 2013)。

SnRK2 是植物激素脱落酸(abscisic acid, ABA)信号途径的重要组分，其在拟南芥中有 10 个成员。ABA 与受体 PYR/PYL/RCAR(parabactin resistance 1/PYR like/regulatory component of ABA receptor)结合后，抑制 PP2C 活性，解除 PP2C 对 SnRK2 的抑制作用；SnRK2 被激活后磷酸化下游通道蛋白、转录因子等(Kulik et al., 2011; Wang et al., 2013)，我们将在后面重点介绍这一植物蛋白磷酸化的经典范例(见 3.1 节)。SnRK3 一般与钙结合蛋白 CBL 互作，受到 Ca^{2+} 的调控，常称为 CIPK，在 CBL-CIPK 一节已做介绍，不再赘述。

TOR(target of rapamycin)也是细胞内能量感知的重要受体，是细胞周期的重要调控元件。拟南芥中只有一个 TOR 成员，其缺失致死(Xiong et al., 2013)，因此对其研究一般采用 RNA 干扰(RNAi)的方法。研究表明，TOR 一方面可以磷酸化 S6K1，通过调控蛋白翻译影响细胞周期(Mahfouz et al., 2006; Schepetilnikov et al., 2013)，与动物或酵母中的作用机制比较类似。另一方面，TOR 可以直接磷酸化转录因子 E2Fa，调控细胞周期 S 期基因的表达，参与根尖分生组织干细胞状态的维持(Xiong et al., 2013)。该结果阐明了一种新的 TOR 参与细胞周期调控的机制，但是否为植物特有还有待于更多的研究。在 SnRK1 和 TOR 的研究过程中，Jen Sheen 实验室都通过基因表达芯片的方法对二者所参与的生物学过程进行细致分析，这也是全面解析某个重要蛋白质分子机制时通常采用的一种高通量手段(Baena-González et al., 2007; Xiong et al., 2013)。

1.2.6 GSK3

糖原合酶激酶 3(GSK3)也称为类 SHAGGY 激酶，是真核生物中普遍存在的一类蛋白激酶，动物和真菌中的研究表明，GSK3 在发育过程中具有十分重要的功能(Saidi et al.,

2012)。GSK3 最早发现于动物的胰岛素途径中,它可以磷酸化糖原合酶(Cohen et al., 1982)。其后的研究表明,GSK3 在多个信号转导途径中都发挥着重要作用。GSK3 在 Wnt 信号通路中处于核心地位——Wnt 信号通路对于胚胎前后轴的发育和干细胞的维持非常重要,其异常会导致各种各样肿瘤的产生,因此是很多药物的靶点。在没有 Wnt 的情况下,β-连环蛋白(β-catenin)会被 CK1α 和 GSK 磷酸化,然后由 APC 复合体降解,下游基因不表达;当 Wnt 结合其受体 Fz 和 LRP 后,CK1γ 可以磷酸化并激活 LRP 蛋白,招募 APC 复合体中的支架蛋白 Axin,而 CK1ε 通过磷酸化 DVL 蛋白,抑制 GSK3 发挥作用,整体上保护 β-catenin 不被磷酸化降解,从而进入核中发挥功能。GSK3 与 CK1 通过磷酸化同一个蛋白上的不同位点,形成一种相互协作的调控模式。

GSK3 在植物激素油菜素内酯(brassinosteroid, BR)信号通路中的作用模式与动物 Wnt 信号通路中的作用模式十分相似。GSK3 作为 BR 信号途径的负调控因子,最早是在筛选 BR 不敏感的突变体时获得的,命名为 BIN2(BR-insensitive 2)(Li et al., 2001, 2002)。没有外源 BR 或者信号通路没有激活的情况下,BIN2 通过磷酸化 BZR1 使其在细胞质中与 14-3-3 结合,进一步被降解;当 BR 与受体 BRI1-BAK1 结合后,通过磷酸化并激活另一个蛋白激酶 BSK1,BSK1 进一步磷酸化并激活蛋白磷酸酶 BSU1,BSU1 可以去磷酸化 BIN2 并降低其活性,这样 BZR1 在 PP2A(type 2A protein phosphatase)的去磷酸化作用下,以非磷酸化的形式在核中稳定积累,激活下游基因的表达(Wang et al., 2012)。

作为信号中间传递者,GSK3 参与了多种信号通路。BIN2 不仅是 BR 信号通路的重要组分,也参与了其他调控过程,如通过磷酸化 SPCH(speechless)(Gudesblat et al., 2012) 或者 YODA(Kim et al., 2012)调控气孔发育。此外,BIN2 可以通过磷酸化 ARF(auxin response factor)家族成员 ARF2、ARF7、ARF19 参与生长素信号,调控侧根发育等(Vert et al., 2008; Cho et al., 2013)。最近的实验表明,BIN2 还可以通过磷酸化 EGL3(enhancer of glabra 3)和 TTG1(transparent testa glabra 1)调控根毛起始(Cheng et al., 2014)。

1.2.7 CK1

CK1(又称 CKI)是一种丝氨酸/苏氨酸蛋白激酶。CK 可以在体外磷酸化酪蛋白,早期曾被称为酪蛋白激酶(casein kinase),但后来的研究发现其对酪蛋白的磷酸化无任何体内证据和生理意义,因此一般称作 CK。CK 广泛存在于酵母、植物与哺乳动物等几乎所有真核生物中。根据理化特性的不同,可以分为 CK1 与 CK2。

CK1 在真核生物中的结构十分保守,其 N 端(约 300 个氨基酸)是高度保守的激酶活性功能域,C 端(13~200 个氨基酸)则具有多样性,可能与其底物特异性和亚细胞定位有关。CK1 可以对多种底物蛋白磷酸化(现在已发现的底物超过 50 种),因此功能多样。

植物中 CK1 的研究最初主要从生化角度来分离纯化并证明其具有磷酸化酪蛋白的活性(Murra et al., 1978; Neeraj et al., 1989),随着分子生物学的兴起和遗传学研究,目前已先后在大豆下胚轴(Murra et al., 1978)、花椰菜花序(Murra et al., 1978)、小麦胚芽(Rychlik et al., 1980)、玉米幼苗(Dobrowolska et al., 1987)、甘蓝(Klimcza et al., 1993)、翼豆(Mu Khopadhya et al., 1997)、拟南芥(Klimcza et al., 1995; Lee et al., 2005)、玉米胚乳(Babatsikos et al., 2000; Christos et al., 2001)、水稻(Liu et al., 2003)、烟草(Lee et al.,

2005),以及芝麻(Kim et al., 2010)中分离纯化了 CK1 或者克隆了 *CK1* 基因。分析表明拟南芥基因组中编码了 17 个 CK1,水稻中有 15 个(Tan et al., 2013)。

在 CK1 的生理功能方面,近几年来取得了比较多的成果。2003 年,Hongwei Xue 实验室曾经报道水稻中 *OsCKI1* 的表达受油菜素内酯及脱落酸诱导,调控了细胞延伸、细胞分裂及其相关信号转导途径,*OsCKI1* 表达下调对影响水稻根部发育(Liu et al., 2003)。其后的研究发现 *OsCKI1* 的 357 位氨基酸变异会导致一个 *japonica* 亚型的 *Koshihikari*(异亮氨酸)与一个 *indica* 亚型的 *Habataki*(赖氨酸)之间的杂种劣势(Yamamoto et al., 2010)。最近研究表明 *OsCKI1* 是一个低温相关的调控位点(low temperature growth 1, *LTG1*),参与了水稻在低温下的生长调控(Lu et al., 2014)。2010 年,Hongwei Xue 实验室发现,水稻 CK1 成员,EL1(early flowering 1),定位于细胞核中,通过磷酸化 SLR1 负调控 GA 信号,影响水稻的开花时间、茎杆伸长等诸多过程(Dai and Xue, 2010)。最新的研究表明 EL1 是一个重要的开花相关的数量性状位点(quantitative trait locus, QTL),还可以磷酸化开花途径中的重要转录因子 Ghd7(grain number, plant height, and heading date 7)(Hori et al., 2013; Kwon et al., 2014)。此外,水稻 CK1 激酶结构域的晶体结构也获得了解析,对其并且发现 CK1 与脂肪酸代谢密切相关(Park et al., 2012; Do and Park, 2012; Park et al., 2012)。

从烟草(*Nicotiana tabacum*)中分离的胞间连丝结合蛋白激酶(plasmadesmata associated protein kinase, PAPK)是一个 CK1 成员,可以在体外磷酸化烟草花叶病毒(TMV)运动蛋白(MP)的 C 端,从而影响其功能(Lee et al., 2005)。PAPK 在拟南芥中的同工酶 CKL6 可以在体内与皮层微管结合,在体外磷酸化微管蛋白。CKL6 的异常表达或者不具有激酶活性的突变体形式会导致皮层微管组织的改变,以及细胞的极性扩张,表明 CKL6 在细胞极性生长与形状建成中起重要作用(Gili et al., 2008)。进一步的研究表明,CKL6 定位于类似晚期内吞体的囊泡上,可能参与了叶肉细胞的极性建成(Gili et al., 2010),暗示时空特异性的亚细胞定位,以及与底物的相互作用可能是 CK1 底物众多而又参与具体生理功能的一种方式(Lee et al., 2009)。

对拟南芥基因组编码的 14 个低分子量 CK1 成员进行了亚细胞定位研究,根据定位情况可以将拟南芥 CK1 分成 3 类:第一类在细胞外围具有十分显著的点刻状分布;第二类在细胞核内具有很强信号,但也存在于细胞质中;第三类在细胞质中的信号十分强而在细胞核中很弱(Lee et al., 2005)。最新的研究表明,两个可定位于细胞核中的 CK1 成员,CK1.3 与 CK1.4 可以磷酸化蓝光受体隐花色素 2(cryptochrome2, CRY2),参与光形态建成和开花的调控(Tan et al., 2013)。

酵母双杂交实验证明芝麻的一个 CK1 成员,SeCKI,可以与一个 bHLH 转录因子结合(体外实验证明可将其磷酸化),共同调控了种子中微粒体油酸去饱合酶(SeFAD2)的表达,从而影响芝麻种子中的脂肪酸代谢(Kim et al., 2010)。棉花(*Gossypium hirsutum*)中与拟南芥 *AtCKL2* 同源的 *GhCKI* 是一个重要的雄性育性相关基因,*GhCKI* 受高温诱导在花药中高表达,其过表达不论在棉花中还是在拟南芥中都会导致雄性不育(Min et al., 2013)。

尽管对 CK1 功能的报道不断增多,研究结果也表明 CK1 在多个生长发育过程和信号途径中具有重要作用,但相关的机制研究还比较少。此外,动物中 CK1 与 GSK3 都是重要的信号中间传输者,其磷酸化位点具有相互导引(priming)作用,一个激酶的磷酸化可能会促进另一个激酶的磷酸化,相较于植物中的 GSK3 而言,对于植物 CK1 作用机制还需要更加系统深入的研究。

1.2.8 CK2

与 CK1 类似,CK2 也是真核生物中普遍存在的一类丝氨酸/苏氨酸蛋白激酶。CK2 是一个四聚体(tetrameric holoenzyme),由 2 个 α 亚基和 2 个 β 亚基构成,但是其亚基也可以作为单体(monomer)独立发挥作用。近年来的研究表明,植物 CK2 通过磷酸化转录因子或相关调控蛋白,改变了这些底物蛋白的 DNA 结合能力、二聚化、稳定性、与其他蛋白质互作或者亚细胞定位等,参与调控了光信号转导、生命节律等过程(参见综述 Mulekar and Huq, 2013),以及细胞分裂(Moreno-Romero et al., 2011)、DNA 损伤修复(Moreno-Romero et al., 2012)、ABA 信号响应(Mulekar et al., 2012)和生长素对侧根发育的调控等(Moreno-Romero et al., 2008)。

拟南芥 CK2 的 α 亚基和 β 亚基均由 4 个基因编码,存在一定的冗余性。其中 CK2α4 定位于叶绿体中,可以磷酸化质体的 σ 因子 AtSIG6(Salinas et al., 2006; Schweer et al., 2010)。最近的研究表明,CK2α 亚基 α1、α2、α3 缺失的三突变体(*α1 α2 α3*)具有发育上的多种缺陷,遗传与生化实验证明 CK2α 亚基参与了开花时间、ABA 信号和抗盐反应的调控(Lu et al., 2011; Mulekar et al., 2012)。

CK2 是光信号转导中的重要调控组分,早期关于 CK2 的研究中,采用反义 RNA 方法抑制内源 *CK2* 表达导致转基因植株叶片变小,叶绿体代谢相关基因表达增强(Lee et al., 1999)。其后的研究表明,CK2 可以磷酸化光形态建成所需积累的重要转录因子 HY5(long hypocotyl 5)(Hardtke et al., 2000)和 HFR1(long hypocotyl in far-red 1)(Duek et al., 2004; Yang et al., 2005),抑制其与 COP1(constitutive photomorphogenic 1)互作和进一步的泛素化降解;同时,CK2 还可以磷酸化负调控光形态建成的转录因子 PIF1 (phytochrome interacting factor 1),促进其红光诱导的泛素化降解(Bu et al., 2011)。CK2 磷酸化对不同底物的调控结果可能与其磷酸化位点的特点(是否存在磷酸化的密码?)和其他互作组分有关。

植物 CK2 与动物 CK2 具有一些类似的功能,也是植物生命节律的关键组分,拟南芥 CK2α 的三突变体(*α1 α2 α3*)在短日照下开花大大延迟,该表型可被外源赤霉素或春化处理恢复(Lu et al., 2011; Mulekar et al., 2012)。CCA1 在 *α1 α2 α3* 突变体中大量积累,导致生物钟周期变长(Lu et al., 2011)。研究表明 CK2 通过磷酸化 CCA1(circadian clock associated 1)和 LHY(late elongated hypocotyl)调控了 CCA1 的活性(Sugano et al., 1998, 1999; Daniel et al., 2004; Portoles and Mas, 2010)。利用水稻的研究获得了类似的结果,一个开花时间相关的 QTL,heading date 6(*Hd6*),编码了 CK2 的 α 亚基,通过影响 Hd1 (heading date 1;拟南芥中 CO 的同源蛋白之一)的活性,参与了开花调控。此外,Hd6 (heading date 6)可以磷酸化 CCA 和 OsLHY,调控水稻的节律(Ogiso et al., 2010)。拟南芥和小麦中 CK2 可以磷酸化重组起始因子 eIF2α、eIF2β、eIF3c、eIF4B、eIF5 和组蛋白

去乙酰化酶 2B(HAT2B)(Dennis and Browning, 2009; Dennis et al., 2009), 调控 eIF2β 的稳定性,以及 eIF3c 与 eIF1、eIF2 和 eIF5 的互作,调控蛋白质的翻译(Dennis and Browning, 2009)。从目前的研究结果来看,CK2 是一种重要的看家激酶(housekeeping kinase),几乎在所有生命活动中都发挥了重要。至今已发现了 CK2 的 37 种底物(Mulekar and Huq, 2013),涉及了植物的整个生长发育过程。考虑到 CK2 具有特异性的磷酸化位点,对底物的磷酸化并不是持续进行,其呈现的动态变化可能与底物本身的构象状态对磷酸化位点暴露的影响有关。

1.3 蛋白磷酸酶

根据蛋白质的一级序列与催化机制差异,蛋白磷酸酶可以分为 4 大类:磷酸化蛋白磷酸酶(phosphoprotein phosphatase, PPP)、金属离子依赖的蛋白磷酸酶(metal dependent protein phosphatase/protein phosphatase 2C, PPM/PP2C)、磷酸化酪氨酸磷酸酶(phosphotyrosine phosphatase, PTP)与双性蛋白磷酸酶(dual specificity phosphatase, DSP)(Kerk et al., 2002; Farkas et al., 2007)。PPP 与 PPM 主要以磷酸化的丝氨酸/苏氨酸为底物,而 PTP 主要以磷酸化的酪氨酸为底物,而 DSP 既可以作用于丝氨酸/苏氨酸,又可以作用于酪氨酸。目前发现植物具有除 PTP 外的 3 大类磷酸酶,下面分别进行介绍。

1.3.1 PPP 类蛋白磷酸酶

PPP 类丝氨酸/苏氨酸蛋白磷酸酶包括 7 类:PP1(protein phosphatase 1)、PP2(PP2A)、PP3(PP2B)、PP4、PP5、PP6、PP7,在拟南芥中有 26 个基因编码 PPP 类蛋白磷酸酶(Farkas et al., 2007)。这些蛋白磷酸酶在结构上具有同源性,其活性可被 microcystin 和 okadaic acid 抑制。与动物相比,植物没有 PP3 类,但是有其他特有的种类如 PPKL(protein phosphatases with kelch-like repeat domain)与 SLP(shewanella-like protein phosphatase)。

植物 PP1 家族包括 9 个成员,主要定位于胞质与细胞核中。PP1 对底物的调控一般需要一个中间蛋白来介导,该中间蛋白往往具有 PP1 所能识别的 RVxF 基序(Heroes et al., 2013),其与 PP1 的互作一般抑制 PP1 的磷酸酶活性或者招募底物能力。对分离出的一些 PP1 互作蛋白的分析推测其可能影响了细胞周期、蓝光信号转导、根伸长或胚胎发育等(Templeton et al., 2011; Takemiya et al., 2013; Ogawa et al., 2011),但是相关调控机制仍有待深入研究。

从进化上看,PP2A、PP4 与 PP6 可能起源于同一祖先分子。其蛋白 C 端均具有一个 YFL 基序,可能受到亮氨酸甲基化的调控(Sents et al., 2013)。PP2A 的全酶是一个三聚体,由 A、B、C 3 个亚基构成,其中 A 亚基起支架(scaffolding)作用,有 3 个成员,分别为 A1(roots curl in naphthylphthalamic acid 1, RCN1)、A2、A3;B 亚基起调控(regulatory)作用,有 17 个成员,分为 B、B′、B″ 3 个亚类;C 亚基为催化(catalytic)亚基,有 5 个成员。PP2A 在植物中具有多样功能,由于其底物众多,起到了重要调控作用。酵母双杂交与体外 pull down 实验均证明 PP2A 的 A 亚基 RCN1 可以直接与蓝光受体向光素 PHOT2 互作,减弱蓝光所诱导的 PHOT2 磷酸化,从而降低植物向光性和气孔开放敏感性(Tseng et al., 2010)。PP2A 可以与 PINOID 共同调控 PIN 蛋白的磷酸化水平，影响其极性定位，从而参与生长素极性运输的调控(Michniewicz et al., 2007; Ballesteros et al., 2013; Li et al., 2011)；此外，通过对 BR 信号转导中重要组分 BRI1(BR insensitive 1)和 BZR1(brassinazole resistant 1)/BES1(BRI1-EMS-suppressor 1，也称作 BZR2)去磷酸化并影响 BRI1 的稳定性和 BZR1/BES1 的核定位，从而调控 BR 信号(Tang et al., 2011)；或通过对 ACC 合酶(1-aminocyclopropane-1-caboxylate synthase, ACS)的去磷酸化影响乙烯合成(Skottke et al., 2011)。

对于 PP2A 的上游调控机制，目前也取得了一些进展。通过遗传筛选 BR 受体突变体 bri1-5 的抑制子时获得了 sbi1(suppressor of bri 1)，分析表明 SBI1 是一个亮氨酸碳端甲基转移酶(leucine carboxylmethyltransferase)，它可以直接与 PP2A 的 C 亚基互作，并对其 C 端的 YFL 基序进行甲基化，从而影响 PP2A 全酶的活性；此外，磷脂酸 PA(phosphatidic acid)可以直接与 A 亚基 RCN1 结合，调控 PP2A 活性，从而参与生长素极性运输(Gao et al., 2013)与 BR 信号(Wu et al., 2013)的调控。

植物中 PP4 尚无明确的功能报道。PP6 可以磷酸化 ABI5(ABA insensitive 5)影响 ABA 反应(Dai et al., 2012)，或者磷酸化 PIN 蛋白调控生长素的极性运输(Dai et al., 2013)。

PP5 的 N 端具有一个保守的 TPR(tetratricopeptide)结构域，C 端是一个催化结构域。在拟南芥与番茄(Lycopersicon esculentum)中，都只有一个基因编码 PP5，为 PPAP5，但均存在两个可变剪切(alternative splicing)版本，编码的蛋白分别为 62kDa 和 55kDa，前者定位于内质网中，后者定位于细胞核和胞质中(de la Fuente van Bentem et al., 2003)。研究表明 PP5 参与了抗病反应(de la Fuente van Bentem et al., 2005; Iki et al., 2012)、高温耐受性(Park et al., 2011)、光信号转导(Ryu et al., 2005)，以及质体信号转导等(Barajas-Lopez Jde et al., 2013)。有意思的是 PPAP5 可以与 PhyA 互作并对其去磷酸化，从而稳定 PhyA 以增强红光信号(Ryu et al., 2005)。

与动物 PP7 相比，植物 PP7 不具有与 Ca^{2+} 结合的 EF 结构域，但可以通过与钙调蛋白(camodulin)结合受到 Ca^{2+} 的调控。遗传研究表明 PP7 可能参与了光信号转导(Moller et al., 2003; Genoud et al., 2008; Sun et al., 2012)。

植物拥有特有的 PPP、PPKL 及 SLP 磷酸酶。PPKL 的 N 端是一个 Kelch 重复基序结构域，可能负责与其他蛋白质的相互作用。拟南芥基因组编码了 4 个 PPKL 成员：分别为 BSU1(BRI1 suppressor 1)、BSL1(BSU like 1)、BSL2 和 BSL3，在功能上有所类似，参与了 BR 信号转导(Mora-Garcia et al., 2004; Kim et al., 2009, 2013)。BSU1 最早在筛选 BRI1 的抑制子时获得，起初被认为可能通过对 BZR1 去磷酸化而参与 BR 信号转导(Mora-Garcia et al., 2004)，后来证明其主要是去磷酸化 BIN2(Kim et al., 2009)，具体在后面(见 3.2 节)介绍。

最近的研究发现 PPKL 是控制水稻籽粒长度乃至产量的关键基因。qGL3/PPKL1/GL3.1(LOC_Os03g44500)是一个调控水稻籽粒长度的重要数量性状位点，其编码了一个 PPKL 家族的成员，是籽粒长度的负调控因子(Qi et al., 2012; Zhang et al., 2012)。在水稻大粒生态型 WY3(Waiyin-3)中 OsPPKL1 的磷酸酶活性下降，该突变版本的过表达会促进籽粒变长。通过酵母双杂交方法筛选到了一个与 OsPPKL1 互作的蛋白 Cyclin-T1;3，进一步的实验表明 OsPPKL1 通过去磷酸化 Cyclin-T1;3 而影响细胞分裂，因此调控籽粒长度。有趣的是，水稻中有 3 个 PPKL1 成员(PPKL1、2、3)，其中 PPKL3 与 PPKL1 比较同源，对细胞分裂或细胞伸长有负调控作用；相反，PPKL2 与拟南芥 BSU1 同源性较高，对细胞分裂或细胞伸长有正调控作用(Zhang et al., 2012)。

植物 SLP 磷酸酶最早是通过生物信息学分析寻找 *Shewanella* 细菌中蛋白磷酸酶的同源蛋白时获得的，拟南芥中有两个成员，分别为 AtSLP1 和 AtSLP2，分别定位于叶绿体和细胞质中(Uhrig et al., 2011)。其具体功能尚不清楚。

早期曾有观点认为磷酸酶的数目少于蛋白激酶，按照某个底物存在可逆磷酸化的调控，预测应该由一个激酶——磷酸酶对(pair)来共同调控底物的磷酸化状态，所以磷酸酶可能作为看家酶(housekeeping enzyme)，不具有特定性。从目前的结果来看，蛋白磷酸化其实也具有特异性，推测植物的蛋白磷酸酶与激酶都具有一定的底物选择性与特异性，参与了特定发育过程的调控。

1.3.2 PPM/PP2C 类蛋白磷酸酶

植物的 PP2C 也是一个比较大的家族,拟南芥基因组中共有约 80 个基因编码了 PP2C 成员，水稻基因组中则有 78 个基因编码了 PP2C(Kerk et al., 2002; Xue et al., 2008)。根据蛋白序列上的差异，可以将 PP2C 分为 A-J 共 10 个亚进化分枝，此外还有 6 个独立的成员，该类蛋白磷酸酶的活性受到金属离子 Ca^{2+} 与 Mn^{2+} 的调控。从现在的研究结果来看，人们对于 A 类 PP2C 了解得比较清楚，特别是 ABI1(ABA insensitive 1)、ABI2、HAB1(hypersensitive to ABA1)等，它们是脱落酸(abscisic acid, ABA)信号的重要组分，可以与 PYR/RCAR 家族的蛋白质形成共受体，介导 ABA 信号转导。此部分将在 ABA 信号部分做详细阐述。

近期的研究表明，HAI(highly ABA-induced 1)也参与了 ABA 信号转导(Bhaskara et al., 2012)。水稻 PP2C 类 XB15 通过对受体激酶 XA21 的去磷酸化，负调控其介导的先天免疫反应(Park et al., 2008)。XB15 与拟南芥 POL(poltergeist)同源性较高，遗传分析表明 POL 处于 CLVATA1 的下游，参与 CLV 信号途径并调控顶端分生组织的建成(Yu et al., 2000, 2003)。

1.3.3 DSP 类蛋白磷酸酶

植物基因组共编码了 22 个 DSP 类的蛋白磷酸酶，其中一类是 MAPK 蛋白磷酸酶，包括 5 个成员：DsPTP1(dual-specificity protein tyrosine phosphatase 1)，MKP2(MAPK phosphatase 2)，IBR5(indole-3-butyric acid response 5)，PHS1(propyzamide hypersensitive 1)及 MKP1。它们都可以与 MAPK 互作并调控其活性(参见综述 Bartels et al., 2010)。这些成员主要参与了植物对逆境信号的响应，将在后面 MAPK 信号通路部分做详细阐述。

蛋白磷酸酶还参与了表面遗传调控。Detlef Weigel 实验室构建了一个巧妙的报告基因转基因株系 *35S::miLUC in 35S::LUC*，由于引入了针对 *LUC* 的 miRNA，该株系不会发出荧光，以此可以筛选调控基因沉默的因子。通过重测序方法，作者快速克隆到了发生突变的基因，一个蛋白磷酸酶 CPL1(C-terminal domain phosphatase-like 1)(Manavella et al., 2012)。遗传与生化实验表明，CPL1 通过与小 RNA 加工重要的蛋白 HYL1(hyponastic leaves 1)互作并对其去磷酸化，从而调控其对小 RNA 加工的过程 (Manavella et al., 2012)。这个结果表明蛋白磷酸酶参与了表观遗传过程，该研究所采用的新的测序技术也大大提高了遗传学研究的效率。

2. 蛋白磷酸化在植物生长发育中的功能

做为一种广泛存在的蛋白质翻译后修饰，可逆磷酸化几乎参与了植物生长发育的每一个方面，以及植物与环境间的物质、能源、信息交流。其形式与调控机制多样，下面将对植物中一些磷酸化参与的重要发育或环境响应过程进行介绍。

2.1 CLAVATA 通路在茎顶端分生组织维持中的功能

植物的胚后发育由顶端分生组织完成，根据它们的位置可以分为根顶端分生组织 (root apical meristem, RAM)和茎顶端分生组织(shoot apical meristem, SAM)。拟南芥的 SAM 的细胞分为 3 层，从外表皮向内依次为 L1、L2 和 L3。L1 是 SAM 最外层的一个细胞层，L2 是紧挨 L1 里面的一层细胞，两层细胞组成 SAM 的被膜(tunica)，L3 是 SAM 的主体部分(corpus)。SAM 包括很多不同种类的细胞，分为中心区(central zone, CZ)，周边区(peripheral zone, PZ)和肋状区(rib zone, RZ)(Fletcher, 2002)。叶、茎和花都是由 SAM 分化而来的，因此其起始、发育和维持对于植物的正常生长发育非常关键(Evans and Barton, 1997)。SAM 的主要生理活动包括顶端干细胞的维持，植物地上组织器官的分化。目前的研究主要集中在生长素(auxin)的极性运输和时空特异分布、不同转录因子的相互作用、不同植物激素的协调作用及类受体蛋白激酶途径等(Bowman and Floyd, 2008)。

由于 SAM 可以分化成各种组织，维持干细胞的分化和再生之间的平衡就十分重要。过多的干细胞分化会导致干细胞数目下降，而干细胞分裂过于旺盛则会导致分生组织过于庞大，这些都会影响 SAM 的正常形态及器官发生。目前的研究表明植物主要依靠 CLV(CLAVATA)与 WUS(WUSCHEL)之间的反馈机制完成对分生组织的精确调控，使分生组织干细胞维持在一定的数量，保持 SAM 的完整性(参见综述 Clark, 2001; Carles and Fletcher, 2003)。CLV-WUS 通路是拟南芥正向遗传学兴起时植物发育领域的重大发现之一，在 20 世纪 90 年代即形成了一个相对完整的分子模型，作为 SAM 维持的核心分子枢纽，众多的体内信号(如生长素、细胞分裂素等)都参与了 CLV-WUS 通路，这里我们主要介绍蛋白磷酸化在其中的功能。

拟南芥的 CLV 信号的接收元件包括 CLV1、CLV2 和 CLV3。*CLV1* 编码一个 LRR-RLK，包含一个胞外配体结合区(21 个 LRR)、一个跨膜疏水和一个胞内丝氨酸/苏氨酸激酶区(Clark et al., 1997)；*CLV2* 编码一个与 CLV1 作用密切相关的无激酶域的 LRR 蛋白(Jeong et al., 1999)，包含 20 个 LRR、一个跨膜区；*CLV3* 编码一个 96 氨基酸的多肽，N端有一个分泌到胞外的信号肽(Fletcher et al., 1999; Lenhard and Laux, 2003)，主要在 L1 和 L2 层表达。成熟的 CLV3 包含 12 或 13 个氨基酸，翻译后加工和糖基化修饰对其与 CLV1 的亲和性和生物学功能十分重要，未成熟的 CLV3 对分生组织的限制作用较弱(Kondo et al., 2006; Ohyama et al., 2009)。*CLV1* 在 CZ 的 L3 层表达，*clv1* 突变体中，SAM 和花原基与野生型相比都有一定程度的增大，分生组织细胞的数目增多，导致植物地上组织器官发育延迟；*clv2* 和 *clv3* 突变也 SAM 和花原基增大，和 *clv1* 突变体中这些组织的发育相似(Fletcher et al., 1999; Jeong et al., 1999)。*CLV2* 在不同器官中都有表达，可能还具有其他功能。研究表明 *clv* 突变体分生组织的增大可能是由于 CZ 区细胞的过度增生导致的，因为 *clv* 突变体分生组织细胞的形态大小没有发生变化(Laufs et al., 1998)。

CLV1 可以和 CLV2 形成异源二聚体，其胞外区域结合 CLV3 后使 CLV1 磷酸化而激活，进一步激活下游 MAPK 通路(Ogawa et al., 2008; Betsuyaku et al., 2011)，从而限制 SAM 细胞数目增多，促进器官发生。这一过程可被 WUS 转录因子所拮抗，以保证一定数量的分生组织细胞(Schoof et al., 2000)。*WUS* 基因在胚 16 细胞期就开始表达，随着发育过程，WUS 主要在活跃的茎顶端分生区下面的一小部分细胞中，即 SAM 的组织中心(organizing center)，维持干细胞的全能性(Mayer et al., 1998)。*WUS* 过表达导致植物分生组织增大，出现类似于 *clv* 突变体的表型，*CLV3* 的过表达植株表现出分生组织细胞减少，和 *wus* 突变体类似(Laux et al., 1996; Mayer et al., 1998; Brand et al., 2000)。CLV 信号减弱会导致 *WUS* 的表达升高并进一步诱导 *CLV3* 的表达，从而激活 CLV 信号通路来限制 WUS 的作用。这样构成了一个 CLV-WUS 的反馈回路，共同精确调节分生组织的大小(Mayer et al., 1998)。但是 CLV 信号如何传递到细胞核中并进一步调控 WUS 表达分子机制并不清楚。

利用 CLV1 抗体通过对花椰菜蛋白提取液的柱层析发现 CLV1 出现在一个大约 450kDa 的复合体中(Trotochaud et al., 1999)。该复合体含有一个 PP2C 类蛋白磷酸酶 KAPP 和一个 Rop(ρ-related protein)，可能通过激活下游 MAPK 通路的将信号转递到细胞核中(Betsuyaku et al., 2011)。KAPP 在整个茎顶端分生组织都表达，其过表达导致花原基增多，与 *clv* 突变体表型类似。KAPP 能直接与 CLV1 被磷酸化的激酶域结合，降低 CLV1 的磷酸化状态，减弱 CLV 通路的信号强度(Williams et al., 1997; Stone et al., 1998)，是 CLV 信号的负调节子。此外，KAPP 也能被 CLV1 磷酸化(Trotochaud et al., 1999)。

水稻和玉米中也存在类似的 CLV 通路。水稻 FON1(floral organ number 1)编码一个类似 CLV1 的类受体激酶(Suzaki et al., 2004a)，其突变体 *fon1* 表现出花原基增多；FON2、FON4 类似于 CLV3，是一个具有 CLE 结构域的分泌蛋白(Suzaki et al., 2006)，*fon2*、*fon4* 突变体没有明显表型，可能是由于基因冗余性造成的；FOS1(FON2 spare 1)是一个 CLE 蛋白，与 FON2、FON4 有相同的作用(Suzaki et al., 2009)；FCP1(FON2-like CLE protein 1)和 FCP2 与 FON2 的结构非常相似。这些 CLV 蛋白与 WOX4(wuschel-related homeobox 4)构成反馈途径，调节水稻 SAM 的大小(Ohmori et al., 2013)。玉米 TD1(thick tassel dwarf 1)和 FEA2(fasciated ear 2)分别类似于 CLV1 和 CLV2，调节玉米花分生组织的大小(Taguchi-Shiobara et al., 2001; Bommert et al., 2005)。最近发现一个 Gα 蛋白 CT2(compact plant 2)通过与 FEA2 结合，参与玉米的 CLV 调控途径。突变体 *ct2* 的 SAM 增大，花原

基数目增多，类似 *fea2* 突变体的表型(Bommert et al., 2013)。单次跨膜的 RLK 作为 G 蛋白耦联受体增进了人们对植物 G 蛋白的认识，但其是否具有普遍性尚需进一步研究。

除 CLV 通路外，与 CLV 相关的 BAM 受体也在 SAM 中发挥作用。*bam1*、*bam2* 和 *bam3* 的单突变体没有明显表型，双突变体和三突变体 SAM 变小，甚至停止生长(DeYoung et al., 2006)。这与 CLV 途径在 SAM 调控上的作用正好相反。BAM 可与一个 CLE 多肽相结合，但是亲和力不高(Guo et al., 2010)。另外，RPK2(receptor-like protein kinase 2，也称作 TOADSTOOL2，TOAD2)通过同源二聚体也可以识别 CLV3 信号，限制 SAM 的扩增，但 RPK2 与 CLV1 和 CLV2 均不能形成二聚体，且 *rpk2* 突变体与 *clv1* 和 *clv2* 突变体有叠加作用，表明 RPK2 与 CLV 信号途径是并行的(Kinoshita et al., 2010)。

ERECTA 可能通过响应细胞分裂素参与调控 SAM，*erecta* 突变体中 *WUS* 的表达升高，导致 SAM 增大和花原基增多(Torii et al., 1996; Uchida et al., 2013; Mandel et al., 2014)。

2.2 蛋白磷酸化在根顶端分生组织维持中的功能

拟南芥的根可以分为 4 个发育区：根冠、分生区、伸长区和成熟区。根顶端分生区细胞(RAM)位于根尖，是根组织的来源；静止中心(quiescent center, QC)区细胞位于根顶端中心区域，具有缓慢的分裂能力，和 SAM 中心区的细胞类似。静止中心对干细胞的维持很重要，其缺失导致周围的干细胞产生分化(vandenBerg et al., 1997; Sarkar et al., 2007)。静止中心附近是一些分裂缓慢的细胞，包括根冠柱起始细胞(columella initial)、表皮-侧生根冠起始细胞(epidermal-lateral root cap initial)、皮层-内皮层起始细胞(cortical-endodermal initial)和维管组织起始细胞(vascular initial)。根顶端分生组织(RAM)产生的细胞经过分裂、分化和伸长，直至形成整个地下组织，这和茎尖分生组织相似。根的生长发育受到各种激素(生长素、乙烯、细胞分裂素和脱落酸)信号途径中转录因子的调控，包括生长素依赖或非依赖转录因子的调控，如 PLT(plethora)、SHR(shortroot)、SCR(scarecrow)和 WOX5(wuschel-related homebox 5)等(Osmont et al., 2007; Scheres, 2007; Sablowski, 2011; Petricka et al., 2012)。除激素外，类受体蛋白激酶也对 RAM 有重要调节作用，根中有一个类似 SAM 的 CLAVATA 途径，由 CLV2 和一个 CLV3 的同源信号蛋白 CLE19 参与(Fiers et al., 2005a)。拟南芥根中过表达 CLE19 促进 RAM 细胞大量分化，导致 RAM 干细胞库细胞数目减少。利用 CLE19 对 RAM 的抑制作用，筛选到了 *SOL1* 和 *SOL2*(*CRN*)两个基因，它们参与了对 CLE19 的识别(Casamitjana-Martínez et al., 2003)。SOL2 是一个类受体蛋白激酶，缺少一个明显的胞外域，SOL2/CRN 可以与 CLV2 结合，而且是 CLV2 定位到细胞膜上所必需的，二者形成的复合体可以识别和传递根部的 CLE 信号，对 RAM 干细胞库细胞数目的正常维持十分重要(Fiers et al., 2005b; Miwa et al., 2008; Muller et al., 2008; Bleckmann et al., 2010)。RPK1(receptor-like protein kinase 1)、RPK2 对根部位置信息的传递起着重要作用，影响胚胎期根的发育(Nodine et al., 2007)。*RPK2* 在 RAM、SAM 及花原基中都有表达，其突变体的根对 CLV3 的抑制效应减弱，与 *clv2* 和 *sol2* 的突变体类似，表明 RPK2 可能与 CLV2 和 SOL2 共同在 CLE 信号对根的发育调控发挥重要作用(Kinoshita et al., 2010)。

ACR4(arabidopsis crinkly 4)编码一个 CRINKLY4 家族的 RLK(De Smet et al., 2008)，通过识别一个与 CLV3 十分同源的 CLE40 信号，限制 WOX5 的作用，对 RAM 的细胞进行调控(Stahl et al., 2009)。在球型胚期，CLE40 在整个胚中都表达并随着胚发育逐渐特异地在胚底部积累，这些底部的细胞随后发育为根的分生组织和维管组织。*cle40* 突变体的根比野生型短且根尖的形态发生变化，根部干细胞的分化出现延迟(Hobe et al., 2003)。WOX5 在 RAM 的作用相当于 SAM 的 WUS，维持着 RAM 的平衡(Sarkar et al., 2007)。*wox5* 突变体中 RAM 不能维持，过量表达 *WOX5* 则导致 RAM 不能正常分化，出现分生组织膨大(Sarkar et al., 2007)。*cle40* 突变体中 *WOX5* 的表达区域扩大，*acr4* 突变体中 *WOX5* 的表达与 *cle40* 突变体相同，有一定程度的增大；此外，CLE40 能促进 *ACR4* 的表达(Stahl et al., 2009)。

侧根发生需要在特定部位形成一个分生组织，ACR4 在其中发挥着双重作用，一方面促进侧根的形成，另一方面限制侧根起始分裂的数目。*acr4* 突变体中新生侧根之间的空间很小，甚至是相邻在一起(De Smet et al., 2008)。ACR4 可能与另一个有激酶活性的 RLK 结合共同介导相关信号的传递，并且独立于 CLV 途径(Stahl et al., 2009)。

2.3 蛋白磷酸化与叶表皮细胞发育

叶是植物重要的营养器官，不仅是光合作用的主要场所，也在植物感受环境变化和响应过程中起重要作用。叶是从茎顶端分生组织的周边区分化而来的，通过建立近-远轴、基-顶轴和中-侧轴的极性发育过程，最终成为成熟叶。叶片内分部着用于支持叶片伸展和输导水分及营养物质的叶脉，叶表皮以一定的间隔排布着大量气孔，负责植物与大气间的气体交换，对光合作用和水分利用十分重要(Raven, 2002)。植物通过调节气孔的开闭、大小和数目来调节对外界的气体交换，从而适应生存环境(Hetherington and Woodward, 2003)。

气孔由两个保卫细胞构成，其分化最初由表皮原细胞(protodermal cell)分化成拟分生母细胞(meristemoid mother cell, MMC)，拟分生母细胞经过一次不对称分裂产生拟分生细胞(meristemoid, M)和其姐妹细胞(stomatal lineage ground cell, SLGC)，拟分生细胞再分化为保卫母细胞(guard mother cell, GMC)，保卫母细胞经过一次对称分裂产生两个保卫细胞(guard cell, GC)(参见综述 Pillitteri and Torii, 2012)。正常情况下，叶片表面的两个气孔不会相邻，由普通的铺列细胞(pavement cell)间隔，呈现出固定间隔和密度的模式，这是发育过程中由一系列的不对称细胞分裂产生的，其分裂过程和细胞间的信号交流受到严格控制，包括一些转录因子的参与(Casson and Gray, 2008)，也受到膜上 EPF(epidermal patterning factor)类多肽小分子和膜上受体介导的磷酸化通路调控(参见综述 Pillitteri and Torii, 2012)。这些膜上受体包括 TMM(too many mouths)(Nadeau and Sack, 2002)、ER(ERECTA)(Shpak et al., 2005)和 GSO1-GSO2(GASSHO1-GASSHO2)(Tsuwamoto et al., 2008a)等。TMM 是一个 LRR-RLP(receptor-like protein)家族的类受体蛋白，含有 10 个 LRR 结构域，没有胞内激酶域，是最早分离的调控气孔发育的类受体蛋白(Yang and Sack, 1995)。*tmm* 突变体的子叶和叶片上出现大量气孔，但是下胚轴和茎上却没有气孔，表明其在不同器官中行使不同的作用(Geisler et al., 1998)，这可能与其参与识别的信号有关(Abrash and Bergmann, 2010)。*TMM* 在茎顶端表皮细胞中表达，在根中没有表达，其在叶中的表达最早出现在叶原基中，随着叶发育，表达 *TMM* 的细胞逐渐减少，至叶片完全伸展开时，*TMM* 表达消失(Nadeau and Sack, 2002)。TMM 通过和 ERECTA 家族的蛋白相互结合介导气孔发生(Lee et al., 2012)。ERECTA 属于 LRR-RLK 家族，有 3 个成员，ER、ERL1 和 ERL2，表现出部分功能冗余性和叠加效应，对气孔发生有抑制作用(Shpak et al., 2005)。在 *er erl1 erl2* 三突变中，叶片上出现大量气孔，与 *tmm* 突变体类似。ER 在整个原表皮层都有表达，主要抑制早期的不对称分裂，使细胞不能起始分化。ERL1 抑制 GMC 的分化和不对称间隔分裂(Shpak et al., 2005)。此外，富含半胱氨酸的短肽 EPF1 也是气孔发育重要的调控因子，其过量表达使气孔发育停止，*epf1* 突变体则出现一定程度的气孔增加(Hara et al., 2007)，表明 EPF1 可能作为一个细胞间信号抑制周围细胞分化为气孔。EPF2 则限制拟分生母细胞的不对称分裂，其突变体的气孔数目明显增多，但没有导致气孔的聚集(Hara et al., 2009; Hunt and Gray, 2009)。此外，EPFL9(STOMAGEN)则对气孔的发育起促进作用，其过表达出现大量的气孔聚集，类似于 ER 和 TMM 的缺失突变体(Kondo et al., 2010; Sugano et al., 2010)。有趣的是，*EPFL9* 在叶肉细胞中表达，而不是叶表皮细胞，说明气孔发育不只是表皮细胞之间信号的交流，其邻近的叶肉细胞对气孔形成也十分重要。

　　实验发现，EPF 能与 ER/ERL1/ERL2 结合，而与 TMM 结合的能力较弱(Lee et al., 2012)，表明 EPF 可能通过 ERECTA 家族来响应并传递下游信号，以抑制周围细胞分化为气孔。ER 和 TMM 的信号可能激活了下游的 MAPK 通路，包括 YODA(MAPKKK)、MKK4、MKK5(MAPKK)和 MPK3、MPK6(MAPK)等。总结目前了解的气孔发育调控机制(图版 II 图 1)：在拟分生母细胞中，bHLH 类转录因子 SCRM、SCRM2 和 SPCH 蛋白可以形成二聚体 SCRM-SPCH 或 SCRM2-SPCH，激活下游基因表达，促进拟分生母细胞向拟分生细胞转化；在这个过程中，*EPF2* 也被 SCRM-SPCH 或 SCRM2-SPCH 调控表达，形成一种反馈调控机制。EPF2 在拟分生母细胞加工成熟后分泌到周围细胞中，结合受体 ER、ERL1、ERL2，以一种尚不清楚的机制激活 YODA，进一步通过 MKK4、MKK5(可能也包括 MKK7、MKK9)激活 MPK3、MPK6，MPK3 和 MPK6 可以再磷酸化 SCRM、SCRM2、SPCH 蛋白，抑制 SCRM-SPCH 或 SCRM2-SPCH 的功能，从而抑制其向拟分生细胞分化。在拟分生母细胞中，EPFL9 可以竞争 EPF2，保证其向拟分生细胞分化。拟分生细胞进一步由 SCRM-MUTE 或 SCRM2-MUTE 介导向保卫母细胞分化。在此过程中，EPF1 通过相同的信号转导途径抑制 SCRM-MUTE 或 SCRM2-MUTE 的功能，抑制周边拟分生细胞向保卫母细胞转化。最后由 YODA – MKK7、9 – MPK3、6 介导促进 SCRM-FAMA 或 SCRM2-FAMA，促进保卫母细胞一分为二，成为两个保卫细胞，形成气孔。其中 SCRM、SCRM2 与 SPCH、MUTE、FAMA 均可形成二聚体，前两者抑制气孔发育，FAMA 则促进气孔发育。除细胞间的信号交流外，最新的研究表明油菜素内酯可以通过下游激酶 BIN2 磷酸化 SPCH(Gudesblat et al., 2012)或 YODA(Kim et al., 2012)，参与到气孔发育调控。

　　在玉米(*Zea mays*)中发现了 PAN1(PANGLOSS 1)，其属于 LRR-RLK 家族类受体蛋白激酶，可以调节 SMC(subsidiary mother cell)到 GMC 的转变(Cartwright et al., 2009)。

PAN1 定位在 SMC 中与 GMC 相接触的部位，其胞内激酶域没有活性。Rho 家族 G 蛋白可以与 PAN1 相互作用，因此有可能共同介导下游信号传递(Humphries et al., 2011)。在拟南芥中，是否由小 G 蛋白介导 ERECTA 到 YODA 之间的信号转导，有什么样的具体机制，尚需进一步的研究。

2.4 受体激酶介导的蛋白磷酸化与花发育

拟南芥的花包括 4 种器官：萼片、花瓣、雄蕊和心皮，从花分生组织分化开始经过一系列的发育进程完成。花分生组织由数层具有分生能力的细胞组成，可以分化出各个花器官。花分生组织和 SAM 类似，细胞可以分为 3 层，从外表皮向内依次为 L1、L2 和 L3。花分生组织的起始和维持受很多转录因子(LFY、AP1 等)和激酶(CLV1、CLV2、CLV3 和 PINOID 等)的调控(Zik and Irish, 2003; Christensen et al., 2000)。和茎分生组织一样，花分生组织的维持也由 CLV-WUS 途径调控，并受到多种因子的影响。*clv1* 和 *clv3* 突变体都表现出花分生组织增大(Clark et al., 1993, 1995)。*CLV1* 的同源基因 *FON1* 和 *TD1* 参与了水稻和玉米花分生组织的维持(Suzaki et al., 2004b; Bommert et al., 2005)。

根据发育形态特征，拟南芥雄蕊花药发育可以分为 14 个时期，包含一系列特化体细胞的形成和雄配子的产生，最终形成成熟的花粉(Goldberg et al., 1993; Sanders et al., 1999)。L1、L2 和 L3 层的细胞分裂分化开始了花药的发育，到第 5 时期时，花药的细胞层从外到内依次是：表皮、药室内壁、中层、绒毡层和小孢子母细胞。减数分裂完成是在第 7 时期，第 10 和 11 时期时绒毡层完成降解。花药的发育包含了细胞分裂、细胞分化、细胞与细胞之间信号的传递和细胞凋亡等，因此大量的调控因子参与了其进程，研究表明蛋白磷酸化，特别是受体激酶对雄蕊的发育至关重要。

拟南芥的 EMS1/EXS(excess microsporocytes 1/extra sporogenous cell)是一个 LRR-RLK，由包含了 21 个 LRR 构成的胞外结构域、跨膜区和丝氨酸/苏氨酸激酶域构成。EMS1/EXS 主要调控雄蕊中 L2 层细胞的分裂分化，其突变体表现出雄性不育，没有花粉粒，具体表现为雄孢子体过多，但是缺少绒毡层和中层细胞(Canales et al., 2002; Zhao et al., 2002)。进一步的细胞学标记证明 *ems1/exs* 的突变体中雄孢子体的数目和野生型雄孢子体与绒毡层细胞数目之和相近，表明过多的雄孢子体可能是由绒毡层的前体而来的。*ems1/exs* 突变体的雄孢子体是正常的，说明 EMS1/EXS 可能不参与早期细胞的分裂活动。*EMS1/EXS* 在花分生组织、花药、胚珠及种子中都被转录，但是 *ems1/exs* 突变体没有表现出胚珠发育上的缺陷，说明 EMS1/EXS 在胚珠的发育过程中是不必要的。分泌小肽 TPD1(tapetum determinant 1)可以与 EMS1/EXS 的 LRR 配体结合域结合，激活 EMS1/EXS 的激酶活性(Yang et al., 2003; Jia et al., 2008)，决定雄蕊细胞的分化。*tpd1* 突变体的表型与 *ems1/exs* 突变体相似，缺少绒毡层细胞。

SERK1(somatic embryogenesis receptor kinase 1)和 *SERK2* 编码两个同源性很高的 LRR-RLK，共同决定绒毡层细胞的分化过程(Albrecht et al., 2005; Colcombet et al., 2005)。*SERK1* 和 *SERK2* 在雄蕊发育的第 6 期之前在花粉室内遍在表达，之后只在绒毡层细胞中表达。*serk1* 和 *serk2* 双突变表现出与 *tpd1* 和 *ems1/exs* 突变体类似的表型，绒毡

层细胞不能正常分化。

水稻中也发现了 *EMS1/EXS* 和 *TPD1* 的同源基因，*MSP1*(multiple sporocytes 1)和 *TPL1*，它们也调控水稻的雄蕊发育。但是 MSP1 只和 TPL1A 结合，且 *TPL1* 在 *EMS1/EXS* 不表达的区区域也表达，表明其可能参与了其他一些调控途径(Nonomura et al., 2003; Yang et al., 2003; Zhao et al., 2008; Zhang et al., 2011)。

BAM1 和 BAM2 也参与了雄蕊发育的调控，与 EMS1/EXS 类似的影响了细胞分化。*bam1* 和 *bam2* 双突变在雄蕊早期不能正常形成药室内壁、中间层和绒毡层细胞，表明其参与了细胞命运决定及细胞间的信号交流(Hord et al., 2006)。

RPK2(receptor-like protein kinase 2)在中间层和绒毡层细胞分化、花药开裂和花粉成熟上起着与 EMS1/EXS-SERK 调控途径相反的作用(Mizuno et al., 2007)。ERECTA 家族通过 MAPK 途径也参与了雄配子的发育调控，*er erl1 erl2* 三突变体雄性不育，缺少 1 个或所有 4 个药瓣(Hord et al., 2008)。此外，FERONIA(Yang et al., 2010; Chae and Lord, 2011)可以调控花粉管是否能够接受雄配子，在生殖隔离中起重要作用，FERONIA 特异的定位在助细胞的细胞膜上限制花粉管生长和破裂，促进雄配子释放与雌配子结合(Escobar-Restrepo et al., 2007; Berger, 2009)。两个 RLCK 类的激酶 LIP1 和 LIP2 主要在花粉管中表达，对花粉管的伸长有一定的指引作用。LIP1 和 LIP2 锚定在花粉管细胞尖端的细胞膜上，其缺失导致母本的 AtLURE1 信号不能被识别，花粉管也不能朝向 AtLURE1 的方向生长(Liu et al., 2013)。MLPK(M-locus protein kinase)可以结合多肽信号 SCR(S-locus cysteine rich protein)或 SP11(S-locus protein 11)(Murase et al., 2004; Swanson et al., 2004; Kakita et al., 2007)，对植物自交不亲和很重要。LePRK1 和 LePRK2 主要在番茄花粉和萌发中的花粉管中表达(Muschietti et al., 1998)，参与花粉与柱头的识别。尽管类受体激酶及其介导的磷酸化在花发育及受精过程中具有十分重要的功能，相关分子机制仍有待进一步阐明。

2.5 蛋白磷酸化与种子发育

种子发育主要包括胚和胚乳的发育，是一个复杂且高度系统化的进程，通过胚胎发生产生了初步的植物器官，并建立了极性。胚胎极性包括顶端-基部轴和辐射轴，其中顶端-基部轴极性在合子中就建立了，合子经过一次不对称分裂产生一个小的顶细胞和一个比较大的基细胞(Mansfield and Briarty, 1991)，顶端-基部轴的建立形成了后来的 SAM 和 RAM。辐射轴对不同细胞层的排列十分重要，从最外面的表皮细胞到中部的导管组织的正常发育都需要辐射轴极性的建立。拟南芥的胚胎发生依次经历合子期、合子伸长期、1 细胞期、2 细胞期、8 细胞期、16 细胞期、球形胚期、心形胚期、鱼雷形胚期和成熟胚期等(West and Harada, 1993; Jenik et al., 2007; Lau et al., 2012)。

早期胚发育受到多种信号途径的调控，如生长素信号(Moller and Weijers, 2009)、*WOX* 等。蛋白磷酸化在拟南芥早期合子对母体信号响应，以及细胞不对称分裂等过程起重要调控作用(Nodine et al., 2011)。利用基因芯片对水稻种子发育早期 RLK 的表达进行了研究，结果表明主要以 RLCK 和 LRR 家族为主，且其表达量随着胚胎发生时期的转换而改变(Gao and Xue, 2012)。

SSP(short suspensor)是最早发现的在胚发育中行使功能的RLCK，其可以激活下游的YODA类MAPK通路，调控合子的第一次不对称分裂(Bayer et al., 2009)，其突变体中合子第一次分裂产生的基细胞不能正常伸长。*SSP*在花粉中转录，但该蛋白不在花粉中翻译，受精后可以在合子中观察到SSP蛋白。最近鉴定了一个特异在极核和胚乳中表达的多肽ESF1(embryo surrounding factor 1)，其可能是SSP上游的信号因子。ESF1与玉米的MEG同源，是一个富半胱氨酸的多肽(Costa et al., 2014)。*ESF1*在受精后开始表达，*esf1_RNAi*能正常受精，但成熟种子体积变小，胚形态发育异常。此外，*esf1_RNAi*的胚的胚柄细胞数量减少，胚柄标记基因*WOX5*、*ARF13*和*IAA10*的表达紊乱，细胞的正常分化被破坏。由于ESF1不能与SSP直接相互作用，推断可能存在其他的RLK接收母本的ESF1信号，这些RLK将信号传递给SSP，再激活YODA(Costa et al., 2014)。

RPK1和TOAD2/RPK2通过识别细胞间的信号来调控胞间交流，在早期球形胚期的辐射轴建立和基部极分化中发挥重要作用。*rpk1*和*toad2*双突变在球形胚期停止生长，表现出伞状胚(Nodine et al., 2007)。RPK1和TOAD2接收何种信号和下游信号传递链的组成目前还不清楚，推测RPK1与ABA途径有一定联系(Osakabe et al., 2010; Lee et al., 2011)。ACR4除了影响侧根起始外，对胚表层细胞的分化也十分重要。在胚中表达*ACR4*的反义序列导致胚产生不同程度的畸形(Tanaka et al., 2002)。ALE2也调控胚的外层细胞分化，其突变体与*cr4*突变体类似，二者可能共同调节表皮细胞的分化(Tanaka et al., 2007)。ALE2和ACR4与ALE1的双突变植株的原表皮细胞层不能正常分化，且原表皮细胞层发育的重要转录因子ATML1也不能正常表达。鉴于ALE1是一个分泌型蛋白酶，主要在胚乳中表达，在早期胚中也有微弱的表达，因此ALE2和ACR4可能识别一个从胚乳中传来的信号进一步调节胚原表皮层的正常发育。

GASSHO1(GSO1)和GSO2是LRRⅪ家族的LRR-RLK，主要在果荚、花芽、种子和根中表达，与早期心形胚期表皮发育有关(Tsuwamoto et al., 2008)。RLK7也是一个LRR-RLK，在种子成熟的时候表达量逐渐升高，其突变体种子萌发推迟，对氧化刺激响应敏感(Pitorre et al., 2010)。尽管众多受体激酶和胞质类受体激酶参与了种子发育的调控，除个别知道其与MAPK通路相关外，相关下游信号通路尚需进一步的研究。

3. 蛋白磷酸化在植物激素信号转导中的作用

植物激素(生长素、乙烯、赤霉素、脱落酸、细胞分裂素、油菜素内酯、茉莉酸、水杨酸、独角金内酯及多肽类激素等)广泛参与了植物生长发育与环境响应的调控，目前的研究表明植物激素在核内的信号感知主要由泛素化途径构成(参见综述Vierstra, 2009)，而胞内或膜上感知系统则主要由蛋白磷酸化完成，其中乙烯和细胞分裂素的受体为组氨酸激酶。下文将对蛋白磷酸化在植物激素信号转导中的作用进行介绍。

3.1 PP2C-SnRK2 介导的脱落酸信号转导

脱落酸(ABA)在植物生长发育以及环境应答方面起重要作用，是一种重要的植物激素。经过近些年来的系统研究，已经清楚 ABA 信号转导的核心为可逆磷酸化反应。回顾 ABA 的研究历史，从早期的植物生理学研究，到20世纪八九十年代兴起的拟南芥正向遗传学研究，继而的生物化学(biochemistry)与化学遗传学(chemical genetics)筛选，结构生物学(structural biology)、磷酸化组学(phospho-proteomics)和系统生物学(system biology)手段，一直到基于晶体结构的拮抗剂(antagonist)的设计与应用——ABA 的研究历程伴随着现代分子生物学的发展(图版Ⅱ图2)。

通过正向遗传学筛选对 ABA 敏感性发生变化的突变体，获得了 abi(abscisic acid insensitive) 1~5 五个突变体(Koornneef et al., 1984)，其中 ABI1(Meyer et al., 1994; Leung et al., 1994, 1997)与 ABI2(Leung et al., 1994, 1997; Rodriguez et al., 1998)是属于 PP2C 家族的磷酸酶，ABI3(Finkelstein et al., 1994; Parcy et al., 1994)、ABI4(Finkelstein et al., 1998)、ABI5(Finkelstein et al., 2000)是转录因子。ABI1 在 ABA 信号转导中起负调控作用(Gosti et al., 1999)。

2002 年，Mustilli 等采用红外线照相机检测叶片温度差异的方法从拟南芥中筛选出了一个叶片温度较低的 AAPK 同源基因的突变体，其气孔持续开放，因而命名为 open stomata 1(ost1)，OST1 在 ROS 和 Ca^{2+} 上游起作用调控气孔开度，可能正调控 ABA 信号 (Mustilli et al., 2002; Yoshida et al., 2002)。OST1 属于 SnRK2 激酶家族，拟南芥基因组编码了10个 SnRK2 成员，OST1 为 SnRK2.6。通过分析，证明 SnRK2.2、2.3 通过磷酸化激活 ABA 信号通路中的转录因子 ABF1、ABF2 及 ABI5，介导 ABA 信号转导(Fujii et al., 2007)。*snrk2.1 snrk2.2 snrk2.3 snrk2.4 snrk2.5 snrk2.6 snrk2.7 snrk2.8 snrk2.9 snrk2.10* 十突变体的表型十分严重，气孔不能关闭，保水能力极差，只能在高湿度的环境生存，不能结实，整个植株非常矮小，此外，*snrk2.1 snrk2.3 snrk2.4 snrk2.5 snrk2.6 snrk2.7 snrk2.8 snrk2.9 snrk2.10 snrk2.2*(+/-)杂合体的种子不能正常成熟(Fujii et al., 2011)。

PP2C 与 SnRK2 一个是磷酸酶，一个是激酶，它们之间有什么样的关系？又是如何协同调控 ABA 信号的？2009 年，两个课题组分别采用化学遗传学(Sean Cutler 实验室通过一种能模拟 ABA 部分功能的小分子 pyrabactin 作正常遗传学筛选，获得了 *pyr1* 突变体)和生物化学(Erwin Grill 实验室通过酵母双杂交的方法，以 PP2C 作钓饵筛选到的 RCAR)的方法鉴定出了 ABA 的受体蛋白 PYR1(pyrabactin resistance 1)/PYL(PYR1 like)/RCAR(regulatory component of ABA receptor)(Park et al., 2009; Ma et al., 2009; Santiago et al., 2009)。PYR1/PYL/RCAR 是 START 家族成员，可以直接与 A 类 PP2C(ABI1、ABI2、HAB 等)互作并抑制其活性，从而激活 ABA 信号响应。PP2C 则通过直接调控 SnRK2 类的成员参与 ABA 响应(Umezawa et al., 2009)。Jiankang Zhu 实验室利用原生质体系统，以一个 ABA 响应基因 *RD29B* 的启动子驱动荧光素酶 *RD29B-LUC* 作为报告基因(reporter)，当 ABF2 被激活时便可以检测到荧光，进一步检测了 PYR1/PYL/RCAR、PP2C、SnRK2.2、SnRK2.3、SnRK2.6 各个组分对 ABF2 的激活作用及上下游关系,结合体外磷酸化实验 PYR1/PYL/RCAR 和 PP2C 对 SnRK2 磷酸化 ABF2 的作用,证实了这些组分构成了一个 PYR1/PYL/RCAR-PP2C-SnRK2.2(2.3、2.6)-ABF2 的信号通路,精细介导 ABA 对下游基因表达的调控(Fujii et al., 2010)。

进一步结构生物学的研究为深入阐明 ABA 作用机制提供了重要线索。多个研究组对 PYR1/PYL/RCAR 的单个蛋白或与 PP2C 共同的晶体结构作了解析(Melcher et al., 2009; Miyazono et al., 2009; Nishimura et al., 2009; Santiago et al., 2009; Yin P et al., 2009),分析表明 PYR1/PYL/RCAR 具有一个与 ABA 结合的凹槽(cave),与 ABA 结合后诱导凹槽邻近的一个类似门(gate)的基序闭合,同时凹槽另一侧的基序也会闭合,形成一个门栓(latch)似的结构,将 ABA 裹在凹槽内部,形成一个可与 PP2C 互作的表面;与 PP2C 结合后可抑制其磷酸酶的活性。此前的结果表明 PP2C 对于 SnRK2 的抑制作用为去磷酸化,由于酶与底物一般是瞬时作用,构象高度动态,不易结晶,Eric Xu 实验室设计了一个 linker 进一步解析了 HAB1(hypersensitive to ABA 1)-SnRK2.6 融合蛋白的结构,发现 SnRK2 的激活环(activation loop)可以插入到 PP2C 的活性中心,而 PP2C 与 ABA 互作的色氨酸可以直接插入到 SnRK2 激酶区与底物结合的大凹槽中,抑制 SnRK2 的活性。这表明 PP2C 不仅可以通过去磷酸化大大降低 SnRK2 的活性,还可以通过直接结合完全阻断 SnRK2 的活性(Soon et al., 2012)。其他结构生物学的研究表明不同的 PYR1/PYL/RCAR 与 PP2C 的作用机制有所不同(参见最新综述 Ng et al., 2014)。

磷酸化组学与系统生物学的研究为全面了解 ABA 的作用机制提供了丰富线索。2013 年,Jiankang Zhu 实验室通过定量磷酸化组学的方法,高通量地分析了 SnRK2 家族蛋白激酶的底物,通过对比野生型(WT)与 *snrk2.2 snrk2.3 snrk2.6* 三突变体对 ABA 的响应,鉴定出了 58 个 SnRK2 的底物,它们参与了植物生长发育调控的各个方面,包括开花时间、核酸结合、miRNA 与表观遗传、信号转导、叶绿体功能等(Wang et al., 2013)。最近,Peter McCourt 实验室通过系统生物学的方法,对受 ABA 诱导的基因所编码蛋白之间的相互作用分析发现,围绕 PYR/PYL/RCAR-PP2C-SnRK2-ABF2/SLAC1 的核心信号传递链,拟南芥中 138 个蛋白质通过 500 多对相互作用构成了 ABA 信号反应网络(Lumba et al., 2014)。也发现更多的 PP2C 成员介导了 ABA 响应,且 PP2C 不仅可以与 SnRK2 家族成员相互作用,还可以与 MAPK、SnRK3 和 CIPK 类激酶相互作用;下游转录因子 ABF 更是与众多其他家族的转录因子存在相互作用,这些结果从分子水平解释了 ABA 参与植物几乎所有生命活动调控的机制(图版Ⅲ图 1)。此外,有证据表明,磷脂分子 PA 可以结合 PP2C,影响其向细胞核内的转运(Mishra et al., 2006)。

Jiankang Zhu 实验室利用带有组氨酸(His)标签的 PYR1 和带有生物素(biotin)标签的 HAB1 筛选能够促进二者互作的化学小分子,获得了一种激活剂(angonist)AM1,它可以部分模拟 ABA 的作用,增强植物保水和抗旱能力,具有潜在的农业应用价值(Cao et al., 2013)。ABA 受体与 ABA 互作的信息显示 ABA 的 3′环上 CH 位于与 PP2C 结合的表面,据此设计了该位置更长的一些 ABA 的类似物(analog),获得了一个拮抗剂(antogonist; Takeuchi et al., 2014),可以有效地阻断 PYL-PP2C 之间的互作。这些工作为今后基于结构生物学设计激活剂或抑制剂提供了思路。

3.2 蛋白磷酸化在油菜素内酯信号转导中的重要作用

油菜素甾醇(brassinosteroid, BR)最初是从花粉里分离出来的(Grove et al., 1979)，参与了植物光形态建成、种子萌发、细胞伸长和分裂、气孔发育、育性及抗逆性等(Li and Chory, 1999; Bishop and Koncz, 2002)。BR 的受体 BRI1(BR-insensitive 1)属于 LRR-RLK 家族，有 24 个 LRR 结构域(BR 结合区在第 21 和 22 个 LRR 结构域之间)(Kinoshita et al., 2005)，胞内有丝氨酸/苏氨酸激酶域。BRI1 在 BR 存在下磷酸化 BKI1(BRI1 kinase inhibitor 1)(Wang and Chory, 2006)，使 BKI1 解离，然后 BRI1 通过激酶域与 BAK1(BRI1-associated receptor kinase 1，也是 LRR-RLK 家族成员)形成异源二聚体。BR 能促进 BRI1 和 BAK1 之间的相互作用，并且增加二者的磷酸化状态(Wang et al., 2005; Wang et al., 2008)。

磷酸化激活的 BRI1 能够磷酸化 BSK1(BR signaling kinase 1)(Tang et al., 2008)，BSK1 再与 BSU1 结合，导致 BIN2 去磷酸化而失活(Li and Nam, 2002; Kim et al., 2009)，从而转录因子 BZR1 不被 BIN2 磷酸化降解(Wang et al., 2002)，进入核中促进转录(图版 Ⅲ图 2)(Kim and Wang, 2010; Gudesblat and Russinova, 2011)。近些年来对于 BR 信号通路的研究获得了很多重要结果，对其信号转导机制的了解相对清楚，在此不再赘述，具体可参考相关综述(Kim and Wang, 2010)。

3.3 蛋白磷酸化在乙烯信号通路中的作用

乙烯(ethylene, ETH)调控了植物生长发育、果实成熟、叶片衰老及抗逆反应等，蛋白磷酸化在其合成及信号转导中发挥了重要作用。根据外源乙烯处理后的三重反应(triple response，包括下胚轴与主根变短、下胚轴变粗、顶端弯钩过度闭合等)，遗传筛选乙烯不敏感或组成型三重反应的突变体，从拟南芥中克隆到了参与乙烯合成及信号转导的关键因子，并结合分子生物学、生物化学、细胞生物学的手段，基本上了解了乙烯从合成到发挥作用的过程(参见综述 Merchante et al., 2013)。

拟南芥中共有 5 个乙烯受体成员：ETR1(ethyleneresponse 1)(Chang et al., 1993)、ETR2(Sakai et al., 1998)、ERS1(ethylene response sensor 1)(Hua et al., 1995)、ERS2(Hua et al., 1998; Hua and Meyerowitz, 1998)、EIN4(ethylene insensitive 4)(Hua et al., 1998; Hua and Meyerowitz, 1998)，它们定位在内质网上(Hua and Meyerowitz, 1998)，蛋白结构比较类似，其 N 端为跨膜结构域，负责结合乙烯分子；中间为 GAF 结构域，负责不同受体之间相互作用形成二聚体；C 端与细菌中二元组分系统的组氨酸激酶同源，但是仅 ETR1 与 ERS1 具有组氨酸激酶(histidine kinase)活性，而且其激酶活性并不是乙烯信号传递所必需的。一般认为 C 端与重要的乙烯组分 CTR1(constitutive triple response 1)(Kieber et al., 1993)结合，并在乙烯激活受体时抑制 CTR1 活性。有研究表明 ETR1 的组氨酸激酶活性对乙烯信号传导有一定的影响，但具体机制尚不清楚(Hall et al., 2012)。

乙烯与受体 ETR1 等在内质网上结合后，ETR1 可以通过由 CTR1 和 EIN2 介导的信号转导激活重要转录因子 EIN3/EIL，启动下游基因表达。最近，有 3 个研究组同时报道了 EIN2 从内质网上剪切入核的结果(Ju et al., 2012; Qiao et al., 2012; Wen et al., 2012)，加深了对乙烯信号转导的认识。在乙烯不结合 ETR1 等受体时，ETR1 与 CTR1、EIN2 在内质网上结合，CTR1 磷酸化 EIN2 的 S645 和 S924，保护其不被剪切并稳定的存在于内质网上，下游乙烯响应不被激活；乙烯与受体结合后，受体通过一种未知机制抑制 CTR1 的活性，导致 CTR1 不能持续磷酸化 EIN2，可能通过某个磷酸酶去磷酸化 EIN2，其 C 端被剪切下来，进入到细胞核内与转录因子 EIN3(ethylene insensitive 3)结合，促进 ERF(ethylene response factor)类转录因子的表达，进一步激活下游响应(Ju et al., 2012; Qiao et al., 2012; Wen et al., 2012)(图版Ⅳ图 1)。在这个过程中，其他一些蛋白质修饰类的因子也参与了对乙烯信号的调控，如 Joseph Ecker 实验室通过酵母双杂交方法筛选到的两个 F-box 类 E3 泛素连接酶 ETP1(EIN2 targeting protein 1)与 ETP2：在没有乙烯的情况下，ETP1 与 ETP2 通过泛素化促进 EIN2 的降解，当乙烯与受体结合激活信号通路后，它们会促进 EIN2 的积累(Qiao et al., 2009)。此外，而 EBF1(EIN3-binding F box protein 1)与 EBF2 可以降解 EIN3(Guo et al., 2003; Potuschak et al., 2003)。

酵母中的 PP2A 与 CK1 共同调控粘连蛋白(cohesin)组分 REC8 的磷酸化水平，磷酸化会诱导 REC8 被裂解酶(separase)剪切(Ishiguro et al., 2010; Katis et al., 2010)，参与到细胞分裂中姐妹染色单体的分离。而 CTR1 对于 EIN2 的磷酸化，对其剪切来说是一种保护机制，与 REC8 并不一样。磷酸化调控 EIN2 剪切的机制需要更加深入的研究。

3.4　二元组分系统与细胞分裂素信号通路

细胞分裂素(cytokinin)也是一种重要的植物激素，在愈伤组织再生、叶片扩展和种子萌发、叶片衰老、顶端优势和生命节律等方面起重要调控作用(参见综述 Hwang et al., 2012)。在细胞分裂素信号转导的研究过程中，与传统的遗传学方法不同，Kakimoto 实验室通过激活标签构建 T-DNA 突变体的方法，发现了 CKI1(cytokinin-independent 1)(Kakimoto et al., 2001)。CKI1 是细胞分裂素的受体，它是一个杂合型组氨酸激酶家族的成员，既具有组氨酸激酶结构域，又具有响应调控结构域，构成了一个磷酸传递系统(phosphorelay system)，也称为二元组分系统(two-component signaling system)。二元组分系统是原核生物中一种十分古老的磷酸化系统，可以将磷酸基团转移至底物的组氨酸上，该系统一般包括至少两个成员——组氨酸激酶响应子(histidine kinase sensor)和响应调控子(response regulator)，后者一般具有保守的组氨酸和天冬氨酸，可以被磷酸化，负责传递磷酸基团。

拟南芥有 4 个细胞分裂素受体(均为组氨酸激酶)：CKI1、AHK2(*arabidopsis* His kinase 2)(Hwang and Sheen, 2001; Hwang et al., 2002)、AHK3、AHK4/CRE1(cytokinin reponsive 1)(Inoue et al., 2001)/WOL(woodenleg)，其中 AHK2、3、4 可以感知细胞分裂素信号；组氨酸磷酸传递蛋白为 AHP(*arabidopsis* His phosphotransfer protein)(Suzuki et al., 1998)；响应调控子蛋白为 ARR(*arabidopsis* response regulator)(Sakai et al., 2000, 2001)。细胞分裂素与受体 AHK 结合后，AHK 的 His 通过自磷酸化，继而将磷酸基团传

递到天冬氨酸(Asp)后继续传递到 AHP1~5 上，AHP 可以进入细胞核，将磷酸基团最终转移到 ARR 蛋白上，其中 B 类的 ARR(ARR1、2、10~14、18~21)可以激活下游响应基因的表达，参与细胞分裂、生长素信号途径等；而 A 类 ARR(ARR3~9、15~17)主要做为一种反馈抑制机制，负调控 B 类 ARR 的功能(图版Ⅳ图2)。最近的研究表明，细胞分裂素可以直接调控生长素输出蛋白 PIN1 的胞内囊泡循环，调控生长素极性运输，但细胞分裂素对 PIN1 的影响依赖于 AHK 类受体和下游的 B 类 ARR，表明这是一个间接的过程(Marhavý et al., 2011, 2014)。

3.5 生长素极性运输的磷酸化调控

生长素是最早发现的植物激素，参与了植物生长发育各个过程的调节。生长素有一个特有的现象即极性运输，其不仅介导生长素在特定细胞、组织及发育阶段时生长素的定位，也参与了植物向性、胚胎发育、器官发育等过程的调控作用。生长素的极性运输涉及 AUX1 类输入载体、ABCB 类运输载体 PGP1(P-glycoprotein 1)和 PIN(PIN formed)类输出载体，其中 PIN 蛋白在细胞内极性分布，对于生长素极性运输调控起关键作用(图版Ⅴ图1)。

PIN 蛋白在拟南芥中有 8 个成员，其中 PIN1、PIN2、PIN3、PIN4、PIN7 定位于细胞膜上，与生长素向胞外的极性运输相关；而 PIN5、PIN6 和 PIN8 定位于内质网中，与生长素的胞内平衡有关，其中 PIN6 的功能比较特殊，与蜜腺(nectar)发育相关(Bender et al., 2013)。PIN 蛋白在细胞内通过囊泡运输在质膜与胞质内膜间动态循环，其极性定位由 PINOID、WAG1(wavy root growth 1)和 WAG2 3 个 AGC 类蛋白激酶(Christensen et al., 2000; Benjamins et al., 2001; Friml et al., 2004; Sukumar et al., 2009)与 PP2A 磷酸酶(Zhou et al., 2004; Michniewicz et al., 2007)共同调控其磷酸化水平来实现。PID 是 AGC 家族的蛋白激酶，WAG1 和 WAG2 是其同源蛋白(Santner et al., 2006)，它们都定位在细胞膜上，通过磷酸化 PIN 蛋白一个保守基序 TPRXS(N/S)中的丝氨酸，调控其极性定位(Huang et al., 2009; Dhonukshe et al., 2010)。PID 缺失突变体 pid 中 PIN 蛋白定位于基部(basal)或者根方向(rootward)，而在 PID 过表达或 pp2aa1,3 突变体中 PIN 蛋白定位于顶部(apical)或茎端方向(shootward) (Friml et al., 2004; Michniewicz et al., 2007)。研究表明磷酸化状态与 PIN 蛋白极性定位密切相关，磷酸化的 PIN 定位于顶部，而非磷酸化的 PIN 则定位于基部(Michniewicz et al., 2007)，其调控主要是由于磷酸化的 PIN 可以被招募到 GNOM 介导的囊泡运输途径中(Kleine-Vehn et al., 2009)。

植物细胞内 PIN 蛋白(包括 PIN1、PIN2、PIN4)的磷酸化水平一般较低，令其定位于基部，介导生长素的向基运输，维持植物的向重性生长及正常发育(Michniewicz et al., 2007)。PIN1 蛋白的 Ser337 与 Thr340 的磷酸化对其极性定位和生物学功能十分重要，但是体外磷酸化实验中虽然 PID 可以磷酸化 PIN1 的亲水环(hydrophilic loop)，却并不能直接磷酸化包含这两个位点的短肽(30 个氨基酸)，表明 PIN1 蛋白可能存在其他的蛋白激酶(也不能排除是由于短肽丧失了与 PID 的互作位点所致) (Zhang et al., 2010)。最近，Schwechheimer 课题组发现，AGC1 类的 D6PK 在生长素极性运输中具有十分重要的功能。

D6PK 具有 4 个同源蛋白，分别为 D6PK、D6PKL1、D6PKL2 和 D6PKL3，其多突变体 *d6pk d6pkl1 d6pkl2*、*d6pk d6pkl1 d6pkl3* 和 *d6pk d6pkl1 d6pkl2 d6pkl3* 表现为植株矮小、侧枝减少、侧根发生受阻和向性减弱等多种生长素缺陷表型，这与其突变体中生长素的极性运输受阻和分布异常有关(Zourelidou et al., 2009; Willige et al., 2013)。D6PK 极性定位于细胞基部(Barbosa et al., 2014)，可以直接磷酸化细胞膜上的 PIN1、PIN2、PIN3、PIN4 和 PIN7 蛋白的胞内亲水环(Zourelidou et al., 2009)，促进生长素极性运输。与 PID 介导的磷酸化作用同时影响 PIN 的极性分布和转运生长素活性不同，D6PK 介导的磷酸化并不影响 PIN 的极性分布，而是影响其转运生长素的能力(Zourelidou et al., 2014)。PIN 的磷酸化可以发生在多个位点，包括 PIN1 的 Ser231、252、271、290 和 PIN3 的 Ser215、226、243、262、283，而 PID(偏好磷酸化 PIN1 的 Ser231、252、290 和 PIN3 的 Ser226、243、283)和 D6PK(偏好磷酸化 PIN1 的 Ser271 和 PIN3 的 Ser215、262)的磷酸化位点偏好性有所差异，由此产生了不同的作用(Zourelidou et al., 2014)。

PID 的活性受到钙调蛋白 TOUCH3(TCH3)(Benjamins et al., 2003)、PDK1(Zegzouti et al., 2006)、NPY 类蛋白(Cheng et al., 2007, 2008)的调控。此外，Ca^{2+} 相关的蛋白激酶 CRK5 也可以磷酸化 PIN2 影响生长素的极性运输，参与向重性反应和侧根发生(Rigó et al., 2013)。

3.6 类受体激酶介导的磷酸化与植物多肽类激素

植物的多肽类激素对细胞间的信息交流十分重要，如 CLV3、CLE、系统素、PSK、SCR 和 ENOD40 等(Matsubayashi and Sakagami, 2006a; Matsubayashi, 2011)。CLV3 和 CLE 对分生组织的调控已经在前面讲述，可以由 CLAVATA1、CLAVATA2 等相应的受体识别。植物的叶片和茎在伤害或昆虫取食的情况下会产生系统素(systemin)，导致细胞内钙浓度增加，MAPK 通路激活和茉莉酸的前体释放等。系统素的受体是 LRR-RLK 家族的 SR160，其氨基酸序列与 BRI1 有很高的同源性，而且和番茄 BR 受体 Cu-3(tBRI1) 相同(Montoya et al., 2002; Scheer and Ryan, 2002)。SR160 可以与系统素直接结合，但相关结合位点目前还不知道，此外，SR160 如何将信号传递到下游目前仍不清楚(Scheer et al., 2003)。

PSK(phytosulfokine)促进植物细胞在较低的细胞量时进行分裂和生长，影响细胞间的信息交流；也促进木质部管状分子(tracheary element, TE)的分化、胚胎发生和花粉萌发(Matsubayashi et al., 1999; Chen et al., 2000; Hanai et al., 2000)。PSK 前体是一个大约 80 个氨基酸的蛋白，N 端是一个信号肽；其受体 PSKR1 也是一个 LRR-RLK。过表达 *PSKR1* 可以促进植物生长，而抑制 *PSKR1* 的表达植物虽然能够成活和增殖，但是最终形成的愈伤组织较小。

此外，FERONIA(FER)可以促进植物对生长素的反应，但是对 ABA 途径产生抑制作用，*fer* 突变体和 *gef1 gef4 gef10* 表现出对 ABA 超敏感。FER 通过与 GEF1、GEF4 和 GEF10 相互作用激活 GTPase ROP11/ARAC10，以增强 ABI2 活性，从而抑制 ABA 信号(Yu et al., 2012)。

4. 蛋白磷酸化在植物-环境相互作用中的功能

植物营固着生活,对于环境变化的及时感知和响应就极其重要。目前的研究表明,蛋白磷酸化广泛参与了植物向重力性、向光性、营养吸收、抗盐反应、抗旱反应、抗病反应等过程。其中植物的向重力性主要由生长素的极性运输来介导,已在前面论述(见3.5节),下文就其他方面做一简要介绍。

4.1 磷酸化与植物的向光性

蓝光在植物向光性中起关键作用。蓝光受体向光素(phototropin 1)是 AGC 类的蛋白激酶,是向光性反应中的重要光受体;光敏色素(phytochrome, Phy)和隐花色素(cryptochrome, Cry)也参与了向光性反应(参见综述 Christie, 2012; Goyal et al., 2013)。1988 年,Briggs 实验室从豌豆黄化苗的上胚轴中分离出一种受蓝光诱导发生磷酸化的位于细胞膜上的蛋白(Gallagher et al., 1988)。他们进而利用拟南芥筛选下胚轴丧失向光性(nonphototropic hypocotyl, nph)的突变体 nph1~4,发现在野生型背景下有一个 120kDa 左右的蛋白发生强烈磷酸化,而在 nph1 的 4 个等位突变体中该磷酸化消失,暗示 NPH1 可能编码了这个 120kDa 左右的蛋白(Liscum et al., 1995)。通过图位克隆,他们发现 NPH1 编码了一个蛋白激酶,其 N 端具有两个 LOV(light, oxygen, or voltage)结构域,其中 LOV2 结构域可以结合吸光辅基 FMN(flavin mononucleotide),而其 C 端是一个激酶结构域,NPH1 定位于细胞膜上,可以发生自磷酸化,将其命名为向光素(phototropin, Phot)(Huala et al., 1997; Christie et al., 1999)。

Phot 广泛存在于植物中,不仅介导植物的向光生长,还介导光调控植物的子叶舒展、叶片运动、气孔运动和叶绿体运动等(Christie, 2012)。拟南芥中有两个 Phot,分别为 Phot1 和 Phot2,其中 Phot1 对光更敏感,主要介导光对下胚轴伸长的抑制,而 Phot2 主要介导强光条件下叶绿体快速的避光运动,二者均参与了光诱导的气孔开放(Kagawa et al., 2001; Jarillo et al., 2001)。对 Phot1 的结构解析发现,黑暗下 LOV2 可以与该结构域靠近 C 端的 Jα 基序互作,接受蓝光后 LOV2 发生一定的构象变化,这种互作被释放,从而发生自磷酸化,因此如果 Jα 基序突变后,LOV2 不接受蓝光也可以发生自磷酸化(Harper et al., 2003)。Phot1 的自磷酸化发生在 LOV1 和 LOV2 之间的区域,对于其介导植物向光性是必需的(Salomon et al., 2003)。最新研究表明,PP2A 的 A 亚基 RCN1 可以对 Phot2 去磷酸化降低其活性,但是对 Phot1 没有影响,表明二者的磷酸化调控并不相同(Tseng and Briggs, 2010)。Phot1 主要通过与 NPH3 和 RPT2(root phototropism 2)等 NPH3 家族的蛋白以及 PKS(phytochrome kinase substrate)互作,调控植物的向性运动(Christie, 2012)。

生长素极性运输在植物向光运动中起关键作用,光抑制 PIN 蛋白的磷酸化,影响其极性定位,从而改变生长素的分布(Blakeslee et al., 2004; Ding et al., 2011)。PIN 蛋白的磷酸化受到 AGC 家族的蛋白激酶 PID 和 D6PK，以及蛋白磷酸酶 PP2A 的共同调控（具体请参考前面章节的论述，见 3.5 节）。PIN3 在向光性方面具有重要作用，*pin3* 突变体的向光性不敏感。观察发现暗时 PIN3 均匀地定位于细胞各侧，而光可以抑制 *PINOID* 的表达，推测在单侧光照射时，向光一侧的 *PINOID* 的表达量快速下调，导致 PIN3 极性定位于细胞内侧，使生长素向背离光的方向运输（Ding et al., 2011），从而产生向光性。此外，*NPH4* 编码为 ARF7 蛋白（也称 MSG1 或 TIR5），*MSG2*（Massugu2）编码了 IAA19 蛋白，均为生长素响应中的重要组分，为向光性所必需，这也进一步证明了生长素在植物向光性中的重要功能（Christie, 2012）。目前，Phot 通过何种机制影响生长素的极性运输尚不清楚，有待于更加深入地研究。

4.2 SOS 途径与抗盐反应

细胞内外的离子平衡，特别是 Na^+ 和 K^+ 浓度的相对平衡，具有非常重要的意义。Jiankang Zhu 实验室通过对拟南芥盐过度敏感（salt overlay sensitive, *sos*）突变体的筛选，鉴定了植物抗盐途径的关键因子，建立了植物抗盐的 SOS 通路（Wu et al., 1996; Liu et al., 1998）。SOS 通路主要包括 SOS1、SOS2、SOS3 三个组分：SOS1 是一个质膜定位的 Na^+/H^+ 逆向转运蛋白（Na^+/H^+ antiporter）；SOS2 是一个 CIPK（CIPK24）（Guo et al., 2001）（可以参考前面的论述，见 1.2.2 节）；SOS3 是一个 CBL（CBL4）（Liu et al., 1998），可以被 *N*-肉豆蔻酰化修饰（*N*-myristoylation）而定位到质膜上（Ishitani et al., 2000; Batistič et al., 2008）。研究表明 SOS3 可与 SOS2 结合，并将其招募到质膜上，在 Ca^{2+} 的作用下被激活，磷酸化并激活 SOS1，启动离子转运（Qiu et al., 2002; Quintero et al., 2002, 2011）。该途径在水稻、杨树等多个植物物种中都得到了验证（参见综述 Mahajan et al., 2008）。

SOS3 的同源蛋白，SCaBP8（SOS3 like calcium bind protein 8）/CBL10 也可以作为 SOS2 的互作蛋白激活 SOS 通路（Kim et al., 2007; Quan et al., 2007），而 SOS2 可以磷酸化 SCaBP8，以形成稳定的复合体（Lin et al., 2009）。在 SOS3-SOS2 的底物层面，它们还可以磷酸化众多的膜上运输蛋白和其他底物，如质膜上 H^+/Ca^{2+} 逆向转运蛋白（Cheng et al., 2004）、液泡膜上 H^+-ATP 酶（Qiu et al., 2004）和 Na^+/H^+ 逆向转运蛋白（Batelli et al., 2007）、PP2C（Ohta et al., 2003）、NDK2（Verslues et al., 2007）、过氧化氢酶 CAT2 与 CAT3，以及生命节律蛋白 GIGANTEA（Kim et al., 2013）等，参与多种细胞活动和生长发育过程的调控。有趣的是，SOS2 磷酸化 GIGANTEA 影响植物开花时间，揭示了植物在逆境环境下协调营养生长和生殖生长的机制（Kim et al., 2013）。

最近的研究表明，14-3-3 蛋白可以与 SOS2 结合并抑制其激酶活性，而且这种结合会在非盐胁迫条件下降低，表明 14-3-3 的作用可以做为 SOS 途径的一种减敏机制来调控植物的正常生长（Zhou et al., 2014）。

4.3 蛋白磷酸化与抗病反应

一些微生物严重影响植物生长发育，植物相应地进化出了完善的机制抵御有害病菌的入侵，研究表明植物一般通过 RLK 识别有害病菌，进一步通过磷酸化级联反应启动相应信号，产生抗病反应(参见综述 Tena et al., 2011; Böhm et al., 2014; Han et al., 2014; Ntoukakis et al., 2014)。目前了解较多的是 FLS2 介导的抗病机制。FLS2 属于 LRR-RLK 家族，可以识别病原菌鞭毛蛋白 flg22，其与 flg22 结合后启动下游信号通路(Gómez-Gómez and Boller, 2000)。另外一个受体激酶 BAK1 可与 FLS2 结合，共同介导 flg22 激发的反应，在 bak1 突变体中由 flg22 引发的抗病反应减弱(Chinchilla et al., 2007; Heese et al., 2007)。研究表明 BAK1 的 Tyr610 的自磷酸化对于抗病非常重要，该位点突变后导致很多抗病基因不能正常表达(Oh et al., 2010)。虽然 BAK1 也参与激素 BR 的信号转导，但是 BR 信号途径与 FLS2 抗病途径之间的关系尚不能肯定(Albrecht et al., 2012; Belkhadir et al., 2012)。FLS2/BAK1 激活后，磷酸化下游蛋白激酶 BIK1 或通过 MAPK 信号通路，激活活性氧(ROS)途径、乙烯途径、各种次生代谢、茉莉酸途径等，启动免疫反应，或将信号转移到核中，激活转录因子调控相关基因表达(Tena et al., 2011)。在此过程中，BAK1 可以磷酸化两个带有 U-box 的 E3 泛素连接酶——PUB12 和 PUB13，促使二者与 FLS2 的结合并泛素化降解 FLS2，作为一种减敏机制调控抗病反应(Lu et al., 2011)。最近的研究还发现 BIK1 可以直接磷酸化 NADPH 氧化酶(RbohD)，调控 ROS 产生(Li et al., 2014; Kadota et al., 2014)。

除 flg22 外，EF-Tu(elongation factor Tu)是细菌体内含量很多的蛋白之一，也可以作为 PAMP(pathogen-associated molecular patterns)被拟南芥等识别，进一步引起 PTI(PAMP-triggered immunity)(Kunze et al., 2004)。EF-Tu 的受体是 EFR(EF-Tu receptor)，与 FLS2 一样属于 LRRXII类受体激酶。efr 突变体容易感病，但用 EFR 转化烟草后，转基因植株对 EF-Tu 产生反应，表明相关下游应答机制在不同物种间是保守的，可以用来提高植物抗性(Zipfel et al., 2006)。该过程中 EFR 自身酪氨酸 Tyr836 位点的磷酸化对其激活十分重要，最近研究表明一种常见的植物致病细菌假单孢菌(*Pseudomonas syringae*)可以分泌出一种效应蛋白(effector)HopAO1，这是一种酪氨酸磷酸酶，能够对 EFR 进行去磷酸化，降低其敏感性(Macho et al., 2014)。植物与病原菌之间的这种"军备竞赛"(arm races)在抗病反应途径中体现在各个层面，如假单胞菌中另外一种效应蛋白 AvrPphB，它是一种蛋白裂解酶，侵染植物后，可以对植物 ETI(effector-triggered immunity)途径中重要的胞质受体激酶 PBS1 进行剪切，激活 PBS1 并进而引起 RPS5 介导的 ETI 免疫反应(Shao et al., 2003); AvrPphB 也可以对 PBS1 的同源蛋白，BIK1、PBL1、PBL2 进行剪切，由于这 3 个胞质受体激酶是 FLS2 介导的 PTI 通路中重要的组分，所以对 PTI 是一种抑制作用(Zhang et al., 2010)。此外，番茄(*Solanum lycopersicum*)中的研究表明，AvrPtoB 是一个 E3 泛素连接酶，可以通过对宿主体内的 Fen 激酶进行泛素化降解，Fen 是与 Pto 高度同源的一个激酶，在植物的免疫反应中具有重要作用，通过这样一种机制，番茄假单胞菌可以较为顺利地入侵植物(Rosebrock et al., 2007); 有意思的是，Pto 可以磷酸化 AvrPtoB，抑制其泛素连接酶活性，形成一种抵御机制(Ntoukakis et al., 2009)。

植物在进化中也产生了一种非常"高明"的保护机制，即顶端分生组织等处一般不易被感染。研究表明负责茎顶端分生组织(SAM)维持的多肽 CLV3 也可以通过激活 FLS2

途径保护 SAM 不受病菌侵害(Lee et al., 2011a)。此外，也有其他一些磷酸化途径参与抗病反应(参见综述 Tena et al., 2011)：PEPR1(proline extensin-like receptor kinase 1)和 PEPR2 通过与 PEP1 和 PEP2 结合，感受细胞壁的物理变化，参与伤害和 PAMP 响应(Silva and Goring, 2002; Yamaguchi et al., 2010)。水稻 Xa21 能够增强黄单胞菌属(*Xanthomonas*)的抗性(Song et al., 1995)，其原因可能是识别并与 Ax21 结合，激活了 Xa21 的磷酸化活性(Lee et al., 2009)。番茄 Pto 识别假单胞杆菌(*Pseudomonas syringae*)表达的 AvrPto 和 AvrPtoB 蛋白(Scofield et al., 1996; Kim et al., 2002)，自身构象变化而激活，进一步磷酸化另一蛋白激酶 Pti1，发生超敏应答(hypersensitive response)(Zhou et al., 1995)。最近还发现植物凝血素受体激酶(lectin receptor kinase)DORN1 和 DORN2 可以通过结合胞外 ATP 对外界伤害作出应急反应，如胞内钙浓度增加等(Choi et al., 2014)。

4.4 其他

基于不同逆境处理下的基因表达数据表明，RLK 在逆境应答过程中有重要作用(Vij et al., 2008; Lehti-Shiu et al., 2009; Marshall et al., 2012)。ARCK1 属于 CRR-RLK(cysteine-rich repeat RLK)，在逆境诱导下表达上调，其突变体 *arck1* 在幼苗期表现出对渗透压和 ABA 超敏感。ARCK1 能够与 CRK36 结合，且 CRK36 在体外能够磷酸化 ARCK1，*CRK36* 的 RNAi 转基因株系表现出对 ABA 和渗透压更高的敏感性，表明 CRK36 可能和 ARCK1 共同抑制 ABA 和渗透压信号的传递(Tanaka et al., 2012)。*RPK1* 过表达的植株对 ABA 响应提高，减少了水分散失，植株对逆境的抵抗能力增强；*rpk1* 突变体则水分散失增加，对 ABA 敏感性下降(Osakabe et al., 2010)。水稻 *OsSIK1* 在高盐、干旱和 H_2O_2 处理下表达增高，过表达 *OsSIK1* 可以增强植株抗逆性(Ouyang et al., 2010)。此外，WAK(cell wall-associated receptor kinase)和 WAKL(WAK-like kinase)家族可以被多种离子所诱导，WAK 可以促进 Zn^{2+} 的吸收并减少有害离子的伤害(Sivaguru et al., 2003; Hou et al., 2005)。

磷吸收对农业生产很重要，增加磷的有效利用能大幅度增加粮食产量及农作物的有效种植范围。最近发现，一个 RLCK 家族成员 PSTOL1(phosphorus-starvation tolerance 1)可以增加植物的磷利用效率，过表达 *PSTOL1* 明显增加水稻在低磷土壤中的产量，其原因可能是 *PSTOL1* 促进了早期根的生长(Gamuyao et al., 2012; Lopez-Arredondo et al., 2014)。

SYMRK [nodulation receptor kinase(NORK)/symbiosis receptor-like kinase]在苜蓿、荷花和豌豆中参与早期共生信号的识别，其突变体不能形成根瘤(Endre et al., 2002; Stracke et al., 2002)。NFR1(nod factor receptor 1)和 NFR5 使植物能识别共生菌，与受体肽结合后介导下游信号(Radutoiu et al., 2003)。另外，Ca^{2+} 可以通过激活 DMI3(CCaMK 成员, calcium and calmodulin-dependent protein kinase)参与根瘤发育(Levy et al., 2004)。

5. 总结与展望

植物蛋白磷酸化的研究伴随植物分子生物学的发展，特别是近 20 年来取得了丰硕成果。磷酸化修饰在植物生长发育中具有重要作用，参与了胞间信号交流和"植物-环境"相互作用等多个过程，对各个"激酶/磷酸酶-底物-表型"关系的明确和相关机制的深入了解，有助于人们从分子水平上认识生命过程。近年来发展的结构生物学、化学遗传学和磷酸化组学等技术为蛋白磷酸化的深入研究提供了更全面的手段。

在蛋白磷酸化的研究中，要清楚一个蛋白激酶或蛋白磷酸酶的作用机制，证明酶与底物的关系十分重要，在这个过程中，往往需要通过一系列的体外实验(如 yeast two hybrid、GST pull down、in vitro kinase assay 甚至晶体结构分析)来证明酶与底物的直接互作或修饰关系，为可能的分子互作提供一个模型。进一步要想证明体内这种情况是否发生，以及这种互作关系的生理功能，就需要根据体外实验得到的线索，比如互作的结构域、基序及修饰的位点(某个氨基酸)对酶或底物作相应的突变或重组，进行体内验证。体内、体外实验的相互佐证(图版 V 图 2)，才能够有效地证明蛋白磷酸化的生理功能。

从目前蛋白磷酸组的质谱数据来看，细胞内全蛋白的相当大一部分都受到磷酸化修饰，且蛋白激酶或磷酸化酶的数量远远不是"一酶一底物"这种一对一的关系。由于许多蛋白激酶具有类似的生化性质，表明一个蛋白激酶或蛋白磷酸酶可能负责体内许多底物的修饰，从而参与不同生理过程的调控。从这点看，质谱技术的发展为全面研究某个激酶或磷酸酶的功能及机制提供了新手段，即不仅可以直接鉴定某个特定底物的磷酸化位点，也可以高通量地在全蛋白质组的基础上鉴定可能的底物(如 Jiankang Zhu 实验室对 SnRK2 的研究, Wang et al., 2013)。此外，生物信息学和系统生物学研究让人们从整体上了解某个途径成为可能，如前面已述的 Peter McCourt 实验室对 ABA 信号途径的系统生物学研究(Lumba et al., 2014)。

结构生物学研究不仅可以帮助阐明受体作用的重要位点，也为设计同源类似物(analog)提供了借鉴，如在 PP2C 调控 SnRK2 的作用机制方面提供了普通生化手段不可能发现的有效信息。质谱技术及系统生物学、结果生物学的发展为进一步研究蛋白激酶或蛋白磷酸酶的功能提供了新的技术手段和思路，也是未来植物蛋白磷酸化研究的一个新方向。

致谢

植物蛋白磷酸化是一个很大的领域，新成果层出不穷。我们力图通过对该领域的重要发现作一系统性总结，以为研究者提供一份翔实的"全景"报告。但由于篇幅和时间所限，特别是能力有限，不能一一列举，对本文未提及的研究成果，作者表示诚挚的歉意。

参 考 文 献

Abrash E B, Bergmann D C. 2010. Regional specification of stomatal production by the putative ligand CHALLAH. Development, 137(3): 447-455.

Afzal A J, Wood A J, Lightfoot D A. 2008. Plant receptor-like serine threonine kinases: roles in signaling and plant defense. Mol Plant Microbe Interact, 21(5): 507-517.

Aichinger E, Kornet N, Friedrich T, et al. 2012. Plant stem cell niches. Annu Rev Plant Biol, 63: 615-636.

Albrecht C, Boutrot F, Segonzac C, et al. 2012. Brassinosteroids inhibit pathogen-associated molecular pattern-triggered immune signaling independent of the receptor kinase BAK1. Proc Natl Acad Sci USA, 109(1): 303-308.

Albrecht C, Russinova E, Hecht V, et al. 2005. The *Arabidopsis thaliana* SOMATIC EMBRYOGENESIS RECEPTOR-LIKE KINASES1 and 2 control male sporogenesis. Plant Cell, 17(12): 3337-3349.

Anthony R G, Henriques R, Helfer A, et al. 2004. A protein kinase target of a PDK1 signalling pathway is involved in root hair growth in *Arabidopsis*. EMBO J, 23(3): 572-581.

Asai T, Tena G, Plotnikova J, et al. 2002. MAP kinase signalling cascade in *Arabidopsis* innate immunity. Nature, 415(6875): 977-983.

Babatsikos C, Yupsanis T. 2000. Separation and purification of both CK I and CK II casein kinases in developing maize endosperm phosphorylation of native HMG proteins. J. Plant Physiol, 156(4): 492-503.

Baena-González E, Rolland F, Thevelein J M, et al. 2007. A central integrator of transcription networks in plant stress and energy signalling. Nature, 448(7156): 938-942.

Ballesteros I, Domínguez T, Sauer M, et al. 2013. Specialized functions of the PP2A subfamily II catalytic subunits PP2A-C3 and PP2A-C4 in the distribution of auxin fluxes and development in *Arabidopsis*. Plant J, 73(5): 862-872.

Barbosa I C, Zourelidou M, Willige B C, et al. 2014. D6 PROTEIN KINASE activates auxin transport-dependent growth and PIN-FORMED phosphorylation at the plasma membrane. Dev Cell, 29(6): 674-685. Bartels S, González Besteiro M A, Lang D, et al. 2010. Emerging functions for plant MAP kinase phosphatases. Trends Plant Sci, 15(6): 322-329.

Batelli G, Verslues P E, Agius F, et al. 2007. SOS2 promotes salt tolerance in part by interacting with the vacuolar H+-ATPase and upregulating its transport activity. Mol Cell Biol, 27(22): 7781-7790.

Bayer M, Nawy T, Giglione C, et al. 2009. Paternal Control of Embryonic Patterning in *Arabidopsis thaliana*. Science, 323(5920): 1485-1488.

Belkhadir Y, Jaillais Y, Epple P, et al. 2012. Brassinosteroids modulate the efficiency of plant immune responses to microbe-associated molecular patterns. Proc Natl Acad Sci USA, 109(1): 297-302.

Belkhadir Y, Wang X, Chory J. 2006. *Arabidopsis* brassinosteroid signaling pathway. Sci STKE, 2006(364): cm5.

Bender R L, Fekete M L, Klinkenberg P M, et al. 2013. PIN6 is required for nectary auxin response and short stamen development. Plant J, 74(6): 893-904.

Benjamins R, Ampudia C S, Hooykaas P J, et al. 2003. PINOID-mediated signaling involves calcium-binding proteins. Plant Physiol, 132(3): 1623-1630.

Benjamins R, Quint A, Weijers D, et al. 2001. The PINOID protein kinase regulates organ development in *Arabidopsis* by enhancing polar auxin transport. Development, 128(20): 4057-4067.

Berger F. 2009. Reproductive biology: Receptor-like kinases orchestrate love songs in plants. Curr Biol, 19(15): R647-R649.

Betsuyaku S, Takahashi F, Kinoshita A, et al. 2011. Mitogen-Activated Protein Kinase Regulated by the CLAVATA Receptors Contributes to Shoot Apical Meristem Homeostasis. Plant Cell Physiol, 52(1): 14-29.

Bhaskara G B, Nguyen T T, Verslues P E. 2012. Unique drought resistance functions of the highly ABA-induced clade A protein phosphatase 2Cs. Plant Physiol, 160(1): 379-395.

Bishop G J, Koncz C. 2002. Brassinosteroids and plant steroid hormone signaling. Plant Cell, 14(Suppl): S97-S110.

Blakeslee J J, Bandyopadhyay A, PeerW A, et al. 2004. Relocalization of the PIN1 auxin efflux facilitator plays a role in phototropic responses. Plant Physiol, 134(1): 28-31.

Bleckmann A, Weidtkamp-Peters S, Seidel C A, et al. 2010. Stem cell signaling in *Arabidopsis* requires CRN to localize CLV2 to the plasma membrane. Plant Physiol, 152(1): 166-176.

Bögre L, Okrész L, Henriques R, et al. 2003. Growth signalling pathways in *Arabidopsis* and the AGC protein kinases. Trends Plant Sci, 8(9): 424-431.

Böhm H, Albert I, Fan L, et al. 2014. Immune receptor complexes at plant cell surface. Curr Opin Plant Biol, 20C: 47-54

Bommert P, Je B I, Goldshmidt A, et al. 2013. The maize G alpha gene *COMPACT PLANT2* functions in CLAVATA signalling to control shoot meristem size. Nature, 502(7472): 555-558.

Bommert P, Lunde C, Nardmann J, et al. 2005. Thick tassel dwarf1 encodes a putative maize ortholog of the *Arabidopsis* CLAVATA1 leucine-rich repeat receptor-like kinase. Development, 132(6): 1235-1245.

Boudsocq M, Willmann M R, McCormack M, et al. 2010. Differential innate immune signalling via Ca^{2+} sensor protein kinases. Nature, 464(7287): 418-422.

Bowman J L, Floyd S K. 2008. Patterning and polarity in seed plant shoots. Annu Rev Plant Biol, 59: 67-88.

Brand U, Fletcher J C, Hobe M, et al. 2000. Dependence of stem cell fate in *Arabidopsis* on a feedback loop regulated by CLV3 activity. Science, 289(5479): 617-619.

Brandt B, Brodsky D E, Xue S, et al. 2012. Reconstitution of abscisic acid activation of SLAC1 anion channel by CPK6 and OST1 kinases and branched ABI1 PP2C phosphatase action. Proc Natl Acad Sci USA, 109(26): 10593-10598.

Bu Q, Zhu L, Dennis M D, et al. 2011. Phosphorylation by CK2 enhances the rapid light-induced degradation of phytochrome interacting factor 1 in *Arabidopsis*. J Biol Chem, 286(14): 12066-12074.

Camehl I, Drzewiecki C, Vadassery J, et al. 2011. The OXI1 kinase pathway mediates Piriformospora indica-induced growth promotion in *Arabidopsis*. PLoS Pathog, 7(5):e1002051.

Canales C, Bhatt A M, Scott R, et al. 2002. EXS, a putative LRR receptor kinase, regulates male germline cell number and tapetal identity and promotes seed development in *Arabidopsis*. Curr Biol, 12(20): 1718-1727.

Cao M, Liu X, Zhang Y, et al. 2013. An ABA mimicking ligand that reduces water loss and promotes drought resistance in plants. Cell Res, 23(8): 1043-1054.

Carles C C, Fletcher J C. 2003. Shoot apical meristem maintenance: the art of a dynamic balance. Trends Plant Sci, 8(8): 394-401.

Cartwright H N, Humphries J A, Smith L G. 2009. PAN1: A Receptor-Like Protein That Promotes Polarization of an Asymmetric Cell Division in Maize. Science, 323(5914): 649-651.

Casamitjana-Martínez E, Hofhuis H F, Xu J, et al. 2003. Root-Specific CLE19 Overexpression and the sol1/2 Suppressors Implicate a CLV-like Pathway in the Control of *Arabidopsis* Root Meristem Maintenance. Curr Biol, 13(16): 1435-1441.

Casson S, Gray J E. 2008. Influence of environmental factors on stomatal development. New Phytol, 178(1): 9-23.

Chae K, Lord E M. 2011. Pollen tube growth and guidance: roles of small, secreted proteins. Ann Bot, 108(4): 627-636.

Chang C, Kwok SF, Bleecker AB, et al. 1993. *Arabidopsis* ethylene-response gene *ETR1*: similarity of product to two component regulators. Science, 262(5133): 539-544.

Chen Y F, Matsubayashi Y, Sakagami Y. 2000. Peptide growth factor phytosulfokine-alpha contributes to the pollen population effect. Planta, 211(5): 752-755.

Cheng N H, Pittman J K, Zhu J K, et al. 2004. The protein kinase SOS2 activates the *Arabidopsis* H^+/Ca^{2+} antiporter CAX1 to integrate calcium transport and salt tolerance. J Biol Chem, 279(4): 2922-2926.

Cheng Y, Qin G, Dai X, et al. 2007. NPY1, a BTB-NPH3-like protein, plays a critical role in auxin-regulated organogenesis in *Arabidopsis*. Proc Natl Acad Sci USA, 104(47): 18825-18829.

Cheng Y, Qin G, Dai X, et al. 2008. *NPY* genes and AGC kinases define two key steps in auxin-mediated organogenesis in *Arabidopsis*. Proc Natl Acad Sci USA, 105(52): 21017-21022.

Cheng Y, Zhu W, Chen Y, et al. 2014. Brassinosteroids control root epidermal cell fate via direct regulation of a MYB-bHLH-WD40 complex by GSK3-like kinases. eLife. 3: e02525.

Cheong Y H, Pandey G K, Grant J J, et al. 2007. Two calcineurin B-like calcium sensors, interacting with protein kinase CIPK23, regulate leaf transpiration and root potassium uptake in *Arabidopsis*. Plant J, 52(2): 223-239.

Chinchilla D, Zipfel C, Robatzek S, et al. 2007. A flagellin-induced complex of the receptor FLS2 and BAK1 initiates plant defence. Nature, 448(7152): 497-500.

Cho H, Ryu H, Rho S, et al. 2014. A secreted peptide acts on BIN2-mediated phosphorylation of ARFs to potentiate auxin response during lateral root development. Nat Cell Biol, 16(1): 66-76.

Choi J, Tanaka K, Cao Y R, et al. 2014. Identification of a Plant Receptor for Extracellular ATP. Science, 343(6168): 290-294.

Christensen S K, Dagenais N, Chory J, et al. 2000. Regulation of auxin response by the protein kinase PINOID. Cell, 100(4): 469-478.

Christie J M, Salomon M, Nozue K, et al. 1999. LOV (light, oxygen, or voltage) domains of the blue-light photoreceptor phototropin (nph1): binding sites for thechromophore flavin mononucleotide. Proc Natl Acad Sci USA, 96(15): 8779-8783.

Christie J M. Phototropin blue-light receptors. 2007. Annu Rev Plant Biol, 58: 21-45.

Christos B, Traianos Y. Similarities and differences in the properties of multiple isoforms of maize endosperm casein kinase I (CK-I). 2001. J Plant Physiol, 158(2): 167-175.

Clark K L, Larsen P B, Wang X X, et al. 1998. Association of the *Arabidopsis* CTR1 Raf-like kinase with the ETR1 and ERS ethylene receptors. Proc Natl Acad Sci USA, 95(9): 5401-5406.

Clark S E, Running M P, Meyerowitz E M. 1993. CLAVATA1, a regulator of meristem and flower development in *Arabidopsis*. Development, 119(2): 397-418.

Clark S E, Running M P, Meyerowitz E M. 1995. CLAVATA3 is a specific regulator of shoot and floral meristem development affecting the same processes as CLAVATA1. Development, 121(7): 2057-2067.

Clark S E, Williams R W, Meyerowitz E M. 1997. The CLAVATA1 gene encodes a putative receptor kinase that controls shoot and floral meristem size in *Arabidopsis*. Cell, 89(4): 575-585.

Clark S E. 2001. Cell signalling at the shoot meristem. Nat Rev Mol Cell Biol, 2(4): 276-284.

Colcombet J, Boisson-Dernier A, Ros-Palau R, et al. 2005. *Arabidopsis* SOMATIC EMBRYOGENESIS RECEPTOR KINASES1 and 2 are essential for tapetum development and microspore maturation. Plant Cell, 17(12): 3350-3361.

Costa L M, Marshall E, Tesfaye M, et al. 2014. Central Cell-Derived Peptides Regulate Early Embryo Patterning in Flowering Plants. Science, 344(6180): 168-172.

Cutler S R, Rodriguez P L, Finkelstein R R, et al. 2010. Abscisic acid: emergence of a core signaling network. Annu Rev Plant Biol, 61: 651-679.

Dai C, Xue H W. 2010. Rice early flowering1, a CKI, phosphorylates DELLA protein SLR1 to negatively regulate gibberellin signalling. EMBO J, 29(11): 1916-1927.

Dai M, Xue Q, Mccray T, et al. 2013. The PP6 phosphatase regulates ABI5 phosphorylation and abscisic acid signaling in *Arabidopsis*. Plant Cell, 25(2): 517-534.

Dai M, Zhang C, Kania U, et al. 2012. A PP6-type phosphatase holoenzyme directly regulates PIN phosphorylation and auxin efflux in *Arabidopsis*. Plant Cell, 24(6): 2497-2514.

Daniel X, Sugano S, Tobin E M. 2004. CK2 phosphorylation of CCA1 is necessary for its circadian oscillator function in *Arabidopsis*. Proc Natl Acad Sci USA USA, 101(9): 3292-3297.

de la Fuente van Bentem S, Vossen J H, Vermeer JE, et al. 2003. The subcellular localization of plant protein phosphatase 5 isoforms is determined by alternative splicing. Plant Physiol, 133(2): 702-712.

de la Fuente van Bentem S, Vossen JH, de Vries KJ, et al. 2005. Heat shock protein 90 and its co-chaperone protein phosphatase 5 interact with distinct regions of the tomato I-2 disease resistance protein. Plant J, 43(2): 284-298.

De Smet I, Vassileva V, De Rybel B, et al. 2008. Receptor-like kinase ACR4 restricts formative cell divisions in the *Arabidopsis* root. Science, 322(5901): 594-597.

Debast S, Nunes-Nesi A, Hajirezaei M R, et al. 2011. Altering trehalose-6-phosphate content in transgenic potato tubers affects tuber growth and alters responsiveness to hormones during sprouting. Plant Physiol, 156(4): 1754-1771.

Dennis M D, Browning K S. 2009. Differential phosphorylation of plant translation initiation factors by *Arabidopsis thaliana* CK2 holoenzymes. J Biol Chem, 284(31): 20602-20614.

Dennis M D, Person M D, Browning K S. 2009. Phosphorylation of plant translation initiation factors by CK2 enhances the in vitro interaction of multifactor complex components. J Biol Chem, 284(31): 20615-20628.

DeYoung B J, Bickle K L, Schrage K J, et al. 2006. The CLAVATA1-related BAM1, BAM2 and BAM3 receptor kinase-like proteins are required for meristem function in *Arabidopsis*. Plant J, 45(1): 1-16.

Dhonukshe P, Huang F, Galvan-Ampudia C S, et al. 2010. Plasma membrane-bound AGC3 kinases phosphorylate PIN auxin carriers at TPRXS(N/S) motifs to direct apical PIN recycling. Development, 137(19): 3245-3255.

Ding Z, Galván-Ampudia C S, Demarsy E, et al. 2011. Light-mediated polarization of the PIN3 auxin transporter for the phototropic response in *Arabidopsis*. Nat Cell Biol, 13(4): 447-452.

Do K H, Park H H. 2012. Crystallization and preliminary X-ray crystallographic studies of casein kinase I-like protein from rice. Acta Crystallogr Sect F Struct Biol Cryst Commun, 68(Pt 3):298-300.

Dobrowolska G, Meggio F, Pinna L A. 1987. Characterization of multiple forms of maize seedling kinases reminiscent of animal casein kinases S (type 1) and TS (type 2). Biochimica et Biophysica Acta (BBA) - Mol Cell Res, 931: 188-195.

Duek P D, Elmer M V, Van Oosten V R, et al. 2004. The degradation of HFR1, a putative bHLH class transcription factor involved in light signaling, is regulated by phosphorylation and requires COP1. Curr Biol, 14(24): 2296-2301.

Endre G, Kereszt A, Kevei Z, et al. 2002. A receptor kinase gene regulating symbiotic nodule development. Nature, 417(6892): 962-966.

Escobar-Restrepo J M, Huck N, Kessler S, et al. 2007. The FERONIA receptor-like kinase mediates male-female interactions during pollen tube reception. Science, 317(5838): 656-660.

Evans M M S, Barton M K. 1997. Genetics of angiosperm shoot apical meristem development. Annu Rev Plant Biol Plant Mol Biol, 48: 673-701.

Fiers M, Golemiec E, Xu J, et al. 2005. The 14-amino acid CLV3, CLE19, and CLE40 peptides trigger consumption of the root meristem in *Arabidopsis* through a CLAVATA2-dependent pathway. Plant Cell, 17(9): 2542-2553.

Finkelstein R R, Lynch T J. 2000. The *Arabidopsis* abscisic acid response gene ABI5 encodes a basic leucine zipper transcription factor. Plant Cell, 12(4): 599-609.

Finkelstein R R, Wang M L, Lynch T J, et al. 1998. The *Arabidopsis* abscisic acid response locus ABI4 encodes an APETALA 2 domain protein. Plant Cell, 10(6): 1043-1054.

Finkelstein R R. 1994. Mutations at two new *Arabidopsis* ABA response loci are similar to the abi3 mutations. Plant J, 5: 765-771.

Fletcher J C. 2002. Shoot and floral meristem maintenance in *Arabidopsis*. Annu Rev Plant Biol, 53: 45-66.

Fletcher L C, Brand U, Running M P, et al. 1999. Signaling of cell fate decisions by CLAVATA3 in *Arabidopsis* shoot meristems. Science, 283(5409): 1911-1914.

Forzani C, Carreri A, de la Fuente van Bentem S, et al. 2011. The *Arabidopsis* protein kinase Pto-interacting 1-4 is a common target of the oxidative signal-inducible 1 andmitogen-activated protein kinases. FEBS J, 278(7): 1126-1136.

Friml J, Yang X, Michniewicz M, et al. 2004. A PINOID-dependent binary switch in apical-basal PIN polar targeting directs auxin efflux. Science, 306(5697): 862-865.

Fujii H, Chinnusamy V, Rodrigues A, et al. 2009. In vitro reconstitution of an abscisic acid signalling pathway. Nature, 462(7273): 660-664.

Fujii H, Verslues P E, Zhu J K. 2007. Identification of two protein kinases required for abscisic acid regulation of seed germination, root growth, and gene expression in *Arabidopsis*. Plant Cell, 19(2): 485-494.

Fujii H, Verslues PE, Zhu JK. 2011. *Arabidopsis* decuple mutant reveals the importance of SnRK2 kinases in osmotic stress responses in vivo. Proc Natl Acad Sci USA, 108(4): 1717-1722.

Gallagher S, Short T W, Ray P M, et al. 1988. Light-mediated changes in two proteins found associated with plasma membrane fractions from pea stem sections. Proc Natl Acad Sci USA, 85(21): 8003-8007.

Galvan-Ampudia C S, Offringa R. 2007. Plant evolution: AGC kinases tell the auxin tale. Trends Plant Sci, 12(12): 541-547.

Gamuyao R, Chin J H, Pariasca-Tanaka J, et al. 2012. The protein kinase Pstol1 from traditional rice confers tolerance of phosphorus deficiency. Nature, 488(7412): 535-539.

Gao H B, Chu Y J, Xue H W. 2013. Phosphatidic Acid(PA) Binds PP2AA1 to Regulate PP2A Activity and PIN1 Polar Localization. Mol Plant, 6(5): 1692-1702.

Gao L L, Xue H W. 2012. Global analysis of expression profiles of rice receptor-like kinase genes. Mol Plant, 5(1): 143-153.

Gao X, Chen X, Lin W, et al. 2013. Bifurcation of *Arabidopsis* NLR immune signaling via Ca^{2+}-dependent protein kinases. PLoS Pathog, 9(1): e1003127.

Garcia A V, Al-Yousif M, Hirt H. 2012. Role of AGC kinases in plant growth and stress responses. Cell Mol Life Sci, 69(19): 3259-3267.

Ge X C, Chang F, Ma H. 2010. Signaling and transcriptional control of reproductive development in *Arabidopsis*. Curr Biol, 20(22): R988-R997.

Geiger D, Maierhofer T, Al-Rasheid K A, et al. 2011. Stomatal closure by fast abscisic acid signaling is mediated by the guard cell anion channel SLAH3 and the receptor RCAR1. Sci Signal, 4(173): ra32.

Geiger D, Scherzer S, Mumm P, et al. 2010. Guard cell anion channel SLAC1 is regulated by CDPK protein kinases with distinct Ca^{2+} affinities. Proc Natl Acad Sci USA, 107(17): 8023-8028.

Geisler M, Yang M, Sack F D. 1998. Divergent regulation of stomatal initiation and patterning in organ and suborgan regions of the *Arabidopsis* mutants too many mouths and four lips. Planta, 205(4): 522-530.

Genoud T, Santa Cruz M T, Kulisic T, et al. 2008. The protein phosphatase 7 regulates phytochrome signaling in *Arabidopsis*. PLoS One, 3(7): e2699.

Gili B N, Cui W, Kim D J, et al. 2008. *Arabidopsis* Casein Kinase 1-Like 6 contains a microtubule binding domain and affects the organization of cortical microtubules. Plant Physiol, 148(4): 1897-1907.

Gili B N, Yang Y D, Lee J Y. 2010. Partitioning of casein kinase 1-like 6 to late endosome-like vesicles. Protoplasma, 240(1-4): 45-56.

Goldberg R B, Beals T P, Sanders P M. 1993. Anther development: basic principles and practical applications. Plant Cell, 5(10): 1217-1229.

Gómez-Gómez L, Boller T. 2000. FLS2: An LRR receptor-like kinase involved in the perception of the bacterial elicitor flagellin in *Arabidopsis*. Mol Cell, 5(6): 1003-1011.

Gosti F, Beaudoin N, Serizet C, et al. 1999. ABI1 protein phosphatase 2C is a negative regulator of abscisic acid signaling. Plant Cell, 11(10): 1897-1910.

Goyal A, Szarzynska B, Fankhauser C. 2013. Phototropism: at the crossroads of light-signaling pathways. Trends Plant Sci, 18(7): 393-401.

Grove M D, Spencer G F, Rohwedder W K, et al. 1979. Brassinolide, a plant growth-promoting steroid isolated from Brassica-Napus Pollen. Nature, 281(5728): 216-217.

Gudesblat G E, Russinova E. 2011. Plants grow on brassinosteroids. Curr Opin Plant Biol, 14(5): 530-537.

Gudesblat G E, Schneider-Pizoń J, Betti C, et al. 2012. SPEECHLESS integrates brassinosteroid and stomata signalling pathways. Nat Cell Biol, 14(5): 548-554.

Guo H, Ecker J R. 2003. Plant responses to ethylene gas are mediated by SCF (EBF1/EBF2)-dependent proteolysis of EIN3 transcription factor. Cell, 115(6): 667-677.

Guo Y F, Han L Q, Hymes M, et al. 2010. CLAVATA2 forms a distinct CLE-binding receptor complex regulating *Arabidopsis* stem cell specification. Plant J, 63(6): 889-900.

Guo Y, Halfter U, Ishitani M, et al. 2001. Molecular characterization of functional domains in the protein kinase SOS2 that is required for plant salt tolerance. Plant Cell, 13(6): 1383-1400.

Hall B P, Shakeel S N, Amir M, et al. 2012. Histidine kinase activity of the ethylene receptor ETR1 facilitates the ethylene response in *Arabidopsis*. Plant Physiol, 159(2): 682-695.

Hamel L P, Sheen J, Séguin A. 2014. Ancient signals: comparative genomics of green plant CDPKs. Trends Plant Sci, 19(2): 79-89.

Han Z, Sun Y, Chai J. 2014. Structural insight into the activation of plant receptor kinases. Curr Opin Plant Biol, 20C: 55-63.

Hanai H, Matsuno T, Yamamoto M, et al. 2000. A secreted peptide growth factor, phytosulfokine, acting as a stimulatory factor of carrot somatic embryo formation. Plant Cell Physiol, 41(1): 27-32.

Hara K, Kajita R, Torii K U, et al. 2007. The secretory peptide gene EPF1 enforces the stomatal one-cell-spacing rule. Genes Dev, 21(14): 1720-1725.

Hara K, Yokoo T, Kajita R, et al. 2009. Epidermal cell density is autoregulated via a secretory peptide, EPIDERMAL PATTERNING FACTOR 2 in *Arabidopsis* leaves. Plant Cell Physiol, 50(6): 1019-1031.

Hardtke C S, Gohda K, Osterlund M T, et al. 2000. HY5 stability and activity in *Arabidopsis* is regulated by phosphorylation in its COP1 binding domain. EMBO Journal, 19(18): 4997-5006.

Harper S M, Neil L C, Gardner K H. 2003. Structural basis of a phototropin light switch. Science, 301(5639):1541-1544.

Heese A, Hann D R, Gimenez-Ibanez S, et al. 2007. The receptor-like kinase SERK3/BAK1 is a central regulator of innate immunity in plants. Proc Natl Acad Sci USA, 104(29): 12217-12222.

Hematy K, Hofte H. 2008. Novel receptor kinases involved in growth regulation. Curr Opin Plant Biol, 11(3): 321-328.

Henriques R, Magyar Z, Monardes A, et al. 2010. *Arabidopsis* S6 kinase mutants display chromosome instability and altered RBR1-E2F pathway activity. EMBO J, 29(17): 2979-2993.

Heroes E, Lesage B, Görnemann J, et al. 2013. The PP1 binding code: a molecular-lego strategy that governs specificity. FEBS J, 280(2): 584-595.

Hetherington A M, Woodward F I. 2003. The role of stomata in sensing and driving environmental change. Nature, 424(6951): 901-908.

Hobe M, Muller R, Grunewald M, et al. 2003. Loss of CLE40, a protein functionally equivalent to the stem cell restricting signal CLV3, enhances root waving in *Arabidopsis*. Dev Genes Evol, 213(8): 371-381.

Hord C L H, Sun Y J, Pillitteri L J, et al. 2008. Regulation of *Arabidopsis* early anther development by the Mitogen-Activated Protein Kinases, MPK3 and MPK6, and the ERECTA and Related Receptor-Like Kinases. Mol Plant, 1(4): 645-658.

Hord C L, Chen C, Deyoung B J, et al. 2006. The BAM1/BAM2 receptor-like kinases are important regulators of *Arabidopsis* early anther development. Plant Cell, 18(7): 1667-1680.

Hori K, Ogiso-Tanaka E, Matsubara K, et al. 2013. Hd16, a gene for casein kinase I, is involved in the control of rice flowering time by modulating the day-length response. Plant J, 76(1): 36-46.

Hou X, Tong H, Selby J, et al. 2005. Involvement of a Cell Wall-Associated Kinase, WAKL4, in *Arabidopsis* Mineral Responses. Plant Physiol, 139(4): 1704-1716.

Hua J, Chang C, Sun Q, et al. 1995. Ethylene insensitivity conferred by *Arabidopsis* ERS gene. Science, 269(5231):1712-1714.

Hua J, Meyerowitz E M. 1998. Ethylene responses are negatively regulated by a receptor gene family in *Arabidopsis thaliana*. Cell, 94(2): 261-271.

Hua J, Sakai H, Nourizadeh S, et al. 1998. EIN4 and ERS2 are members of the putative ethylene receptor gene family in *Arabidopsis*. Plant Cell, 10(8): 1321-1332.

Huala E, Oeller P W, Liscum E, et al. 1997. *Arabidopsis* NPH1: a protein kinase with a putative redox-sensing domain. Science, 278(5346): 2120-2123.

Huang F, Zago M K, Abas L, et al. 2010. Phosphorylation of conserved PIN motifs directs *Arabidopsis* PIN1 polarity and auxin transport. Plant Cell, 22(4): 1129-1142.

Humphries J A, Vejlupkova Z, Luo A, et al. 2011. ROP GTPases act with the receptor-like protein PAN1 to polarize asymmetric cell division in maize. Plant Cell, 23(6): 2273-2284.

Hunt L, Gray J E. 2009. The signaling peptide EPF2 controls asymmetric cell divisions during stomatal development. Curr Biol, 19(10): 864-869.

Hwang I, Chen H C, Sheen J. 2002. Two-component signal transduction pathways in *Arabidopsis*. Plant Physiol, 129(2): 500-515.

Hwang I, Sheen J, Müller B. 2012. Cytokinin signaling networks. Annu Rev Plant Biol, 63: 353-380.

Hwang I, Sheen J. 2001. Two-component circuitry in *Arabidopsis* cytokinin signal transduction. Nature, 413(6854):383-389.

Iki T, Yoshikawa M, Meshi T, et al. 2012. Cyclophilin 40 facilitates HSP90-mediated RISC assembly in plants. EMBO J, 31(2): 267-278.

Inoue T, Higuchi M, Hashimoto Y, et al. 2001. Identification of CRE1 as a cytokinin receptor from *Arabidopsis*. Nature, 409(6823): 1060-1063.

Ishiguro T, Tanaka K, Sakuno T, et al. 2010. Shugoshin-PP2A counteracts casein-kinase-1-dependent cleavage of Rec8 by separase. Nat Cell Biol, 12(5): 500-506.

Jaillais Y, Hothorn M, Belkhadir Y, et al. 2011. Tyrosine phosphorylation controls brassinosteroid receptor activation by triggering membrane release of its kinase inhibitor. Genes Dev, 25(3): 232-237.

Jarillo J A, Gabrys H, Capel J, et al. 2001. Phototropin-related NPL1 controls chloroplast relocation induced by blue light. Nature, 410(6831): 952-954.

Jenik P D, Gillmor C S, Lukowitz W. 2007. Embryonic patterning in *Arabidopsis thaliana*. Annu Rev Cell Dev Biol, 23: 207-236.

Jeong S, Trotochaud A E, Clark S E. 1999. The *Arabidopsis* CLAVATA2 gene encodes a receptor-like protein required for the stability of the CLAVATA1 receptor-like kinase. Plant Cell, 11(10): 1925-1933.

Jia G, Liu X, Owen H A, et al. 2008. Signaling of cell fate determination by the TPD1 small protein and EMS1 receptor kinase. Proc Natl Acad Sci USA, 105(6): 2220-2225.

Ju C, Yoon G M, Shemansky J M, et al. 2012. CTR1 phosphorylates the central regulator EIN2 to control ethylene hormone signaling from the ER membrane to the nucleus in *Arabidopsis*. Proc Natl Acad Sci USA, 109(47): 19486-19491.

Kadota Y, Sklenar J, Derbyshire P, et al. 2014. Direct regulation of the NADPH oxidase RBOHD by the PRR-associated kinase BIK1 during plant immunity. Mol Cell, 54(1):43-55.

Kagawa T, Sakai T, Suetsugu N, et al. 2001. *Arabidopsis* NPL1: a phototropin homolog controlling the chloroplast high-light avoidance response. Science, 291(5511): 2138-2141.

Kakimoto T. 1996. CKI1, a histidine kinase homolog implicated in cytokinin signal transduction. Science, 274(5289): 982-985.

Kakita M, Murase K, Iwano M, et al. 2007. Two distinct forms of M-locus protein kinase localize to the plasma membrane and interact directly with S-locus receptor kinase to transduce self-incompatibility signaling in *Brassica rapa*. Plant Cell,19(12): 3961-3973.

Katis V L, Lipp J J, Imre R, et al. 2010. Rec8 phosphorylation by casein kinase 1 and Cdc7-Dbf4 kinase regulates cohesin cleavage by separase during meiosis. Dev Cell, 18(3): 397-409.

Kieber J J, Rothenberg M, Roman G, et al. 1993. CTR1, a negative regulator of the ethylene response pathway in *Arabidopsis*, encodes a member of Raf family of protein kinases. Cell, 72(3): 427-551.

Kim B G, Waadt R, Cheong Y H, et al. 2007. The calcium sensor CBL10 mediates salt tolerance by regulating ion homeostasis in *Arabidopsis*. Plant J, 52(3): 473-484.

Kim M J, Go Y S, Lee S B, et al. 2010. Seed-expressed casein kinase I acts as a positive regulator of the SeFAD2 promoter via phosphorylation of the SebHLH transcription factor. Plant Mol Biol, 73(4-5): 425-437.

Kim T W, Guan S, Burlingame A L, et al. 2011. The CDG1 kinase mediates brassinosteroid signal transduction from BRI1 receptor kinase to BSU1 phosphatase and GSK3-like kinase BIN2. Mol Cell, 43(4): 561-571.

Kim T W, Guan S, Sun Y, et al. 2009. Brassinosteroid signal transduction from cell-surface receptor kinases to nuclear transcription factors. Nat Cell Biol, 11(10): 1254-1260.

Kim T W, Michniewicz M, Bergmann D C, et al. 2012. Brassinosteroid regulates stomatal development by GSK3-mediated inhibition of a MAPK pathway. Nature, 482(7385): 419-422.

Kim T W, Wang Z Y. 2010. Brassinosteroid Signal Transduction from Receptor Kinases to Transcription Factors. Annu Rev Plant Biol, 61: 681-704.

Kim W Y, Ali Z, Park H J, et al. 2013. Release of SOS2 kinase from sequestration with GIGANTEA determines salt tolerance in *Arabidopsis*. Nat Commun, 4:1352.

Kim Y J, Lin N C, Martin G B. 2002. Two distinct Pseudomonas effector proteins interact with the Pto kinase and activate plant immunity. Cell, 109(5): 589-598.

Kinoshita A, Betsuyaku S, Osakabe Y, et al. 2010. RPK2 is an essential receptor-like kinase that transmits the CLV3 signal in *Arabidopsis*. Development, 137(22): 3911-3920.

Kinoshita T, Cano-Delgado A, Seto H, et al. 2005. Binding of brassinosteroids to the extracellular domain of plant receptor kinase BRI1. Nature, 433(7022): 167-171.

Kleine-Vehn J, Huang F, Naramoto S, et al. 2009. PIN auxin efflux carrier polarity is regulated by PINOID kinase-mediated recruitment into GNOM-independent trafficking in *Arabidopsis*. Plant Cell, 21(12): 3839-3849.

Klimcza L J, Cashmore A R. 1993. Purification and characterization of casein kinase I from broccoli. Biochemical J, 93(Pt 1): 283-288

Klimcza LJ, Farini D, Lin C, et al. 1995. Multiple isoforms of *Arabidopsis* casein kinase I combine conserved catalytic domains with variable carboxyl terminal extensions. Plant Physiol, 109(2): 687-696.

Kobayashi M, Ohura I, Kawakita K, et al. 2007. Calcium-dependent protein kinases regulate the production of reactive oxygen species by potato NADPH oxidase. Plant Cell, 19(3): 1065-1080.

Kondo T, Kajita R, Miyazaki A, et al. 2010. Stomatal density is controlled by a mesophyll-derived signaling molecule. Plant Cell Physiol, 51(1): 1-8.

Kondo T, Sawa S, Kinoshita A, et al. 2006. A plant peptide encoded by CLV3 identified by in situ MALDI-TOF MS analysis. Science, 313(5788): 845-848.

Koornneef M, Reuling G, Karssen C M. 1984. The isolation and characterization of abscisic acid-insensitive mutants of *Arabidopsis thaliana*. Physiol Plant, 61(3): 377-383.

Kudla J, Xu Q, Harter K, et al. 1999. Genes for calcineurin B-like proteins in *Arabidopsis* are differentially regulated by stress signals. Proc Natl Acad Sci USA, 96(8): 4718-4723.

Kulik A, Wawer I, Krzywińska E, et al. 2011. SnRK2 protein kinases-key regulators of plant response to abiotic stresses. OMICS, 15(12): 859-872.

Kunze G, Zipfel C, Robatzek S, et al. 2004. The N terminus of bacterial elongation factor Tu elicits innate immunity in *Arabidopsis* plants. Plant Cell, 16(12): 3496-3507.

Kwon C T, Yoo S C, Koo B H, et al. 2014. Natural variation in Early flowering1 contributes to early flowering in japonica rice under long days. Plant Cell Environ, 37(1): 101-112.

Lau S, Slane D, Herud O, et al. 2012. Early embryogenesis in flowering plants: setting up the basic body pattern. Annu Rev Plant Biol, 63: 483-506.

Laufs P, Grandjean O, Jonak C, et al. 1998. Cellular parameters of the shoot apical meristem in *Arabidopsis*. Plant Cell, 10(8): 1375-1389.

Laux T, Mayer K F X, Berger J, et al. 1996. The WUSCHEL gene is required for shoot and floral meristem integrity in *Arabidopsis*. Development, 122(1): 87-96.

Lee H, Chah O K, Sheen J. 2011a. Stem-cell-triggered immunity through CLV3p-FLS2 signalling. Nature, 473(7347): 376-U559.

Lee I C, Hong S W, Whang S S, et al. 2011b. Age-dependent action of an ABA-inducible receptor kinase, RPK1, as a positive regulator of senescence in *Arabidopsis* leaves. Plant Cell Physiol, 52(4): 651-662.

Lee J S, Kuroha T, Hnilova M, et al. 2012. Direct interaction of ligand-receptor pairs specifying stomatal patterning. Genes Dev, 26(2): 126-136.

Lee J Y, Taoka K, Yoo B C, et al. 2005. Plasmodesmal-associated protein kinase in tobacco and *Arabidopsis* recognizes a subset of non-cell-autonomous proteins. Plant Cell, 17(10): 2817-2831.

Lee J Y. 2009. Versatile casein kinase 1: Multiple locations and functions. Plant Signaling & Behavior, 4(7): 652-654.

Lee S C, Lan W Z, Kim B G, et al. 2007. A protein phosphorylation/dephosphorylation network regulates a plant potassium channel. Proc Natl Acad Sci USA, 104(40): 15959-15964.

Lee S-W, Han S-W, Sririyanum M, et al. 2009. A Type I secreted, sulfated peptide triggers XA21-mediated innate immunity. Science, 326(5954): 850-853.

Lehti-Shiu M D, Zou C, Hanada K, et al. 2009. Evolutionary history and stress regulation of plant receptor-like kinase/pelle genes. Plant Physiol, 150(1): 12-26.

Lenhard M, Laux T. 2003. Stem cell homeostasis in the *Arabidopsis* shoot meristem is regulated by intercellular movement of CLAVATA3 and its sequestration by CLAVATA1. Development, 130(14): 3163-3173.

Leung J, Bouvier-Durand M, Morris P C, et al. 1994. *Arabidopsis* ABA response gene ABI1: features of a calcium-modulated protein phosphatase. Science, 264(5164): 1448-1452.

Leung J, Merlot S, Giraudat J. 1997. The *Arabidopsis* ABSCISIC ACIDINSENSITIVE2 (ABI2) and ABI1 genes encode homologous protein phosphatases 2C involved in abscisic acid signal transduction. Plant Cell, 9(5): 759-771.

Levy J, Bres C, Geurts R, et al. 2004. A putative Ca^{2+} and calmodulin-dependent protein kinase required for bacterial and fungal symbioses. Science, 303(5662): 1361-1364.

Li H, Lin D, Dhonukshe P, et al. 2011. Phosphorylation switch modulates the interdigitated pattern of PIN1 localization and cell expansion in *Arabidopsis* leaf epidermis. Cell Res, 21(6): 970-978.

Li J M, Chory J. 1997. A putative leucine-rich repeat receptor kinase involved in brassinosteroid signal transduction. Cell, 90(5): 929-938.

Li J M, Chory J. 1999. Brassinosteroid actions in plants. J Exp Bot, 50(332): 275-282.

Li J M, Nam K H. 2002. Regulation of brassinosteroid signaling by a GSK3/SHAGGY-like kinase. Science, 295(5558): 1299-1301.

Li J, Nam KH, Vafeados D, Chory J. 2001. BIN2, a new brassinosteroid-insensitive locus in *Arabidopsis*. Plant Physiol, 127(1): 14-22.

Li J, Wen J Q, Lease K A, et al. 2002. BAK1, an *Arabidopsis* LRR receptor-like protein kinase, interacts with BRI1 and modulates brassinosteroid signaling. Cell, 110(2): 213-222.

Li L, Kim B G, Cheong Y H, et al. 2006. A Ca^{2+} signaling pathway regulates a K^+ channel for low-K response in *Arabidopsis*. Proc Natl Acad Sci USA, 103(33): 12625-12630.

Li L, Li M, Yu L, et al. 2014. The FLS2-associated kinase BIK1 directly phosphorylates the NADPH oxidase RbohD to control plant immunity. Cell Host Microbe, 15(3): 329-338.

Lin H, Yang Y, Quan R, et al. 2009. Phosphorylation of SOS3-LIKE CALCIUM BINDING PROTEIN8 by SOS2 protein kinase stabilizes their protein complex and regulates salt tolerance in *Arabidopsis*. Plant Cell, 21(5):1607-1619.

Lin W, Li B, Lu D, et al. 2014. Tyrosine phosphorylation of protein kinase complex BAK1/BIK1 mediates *Arabidopsis* innate immunity. Proc Natl Acad Sci USA, 111(9): 3632-3637.

Liscum E, Briggs W R. 1995. Mutations in the NPH1 locus of *Arabidopsis* disrupt the perception of phototropic stimuli. Plant Cell, 7(4): 473-485.

Liu J J, Zhong S, Guo X Y, et al. 2013. Membrane-bound RLCKs LIP1 and LIP2 are essential male factors controlling male-female attraction in *Arabidopsis*. Curr Biol, 23(11): 993-998.

Liu J, Zhu J K. 1998. A calcium sensor homolog required for plant salt tolerance. Science, 280(5371): 1943-1945.

Liu W, Xu Z H, Luo D, et al. 2003. OsCKI1, a rice casein kinase I, plays significant roles in auxin related root development and functions of plant hormones. Plant J, 36(2): 189-202.

Lopez-Arredondo D L, Leyva-Gonzalez M A, Gonzalez-Morales S I, et al. 2014. Phosphate nutrition: improving low-phosphate tolerance in crops. Annu Rev Plant Biol, 65: 95-123.

Lu D, Lin W, Gao X, et al. 2011. Direct ubiquitination of pattern recognition receptor FLS2 attenuates plant innate immunity. Science, 332(6036):1439-1442.

Lu G, Wu F Q, Wu W, et al. 2014. Rice LTG1 is involved in adaptive growth and fitness under low ambient temperature. Plant J, 78(3): 468-480.

Lu S X, Liu H, Knowles SM, Li J, Ma L, Tobin EM, Lin C. 2011. A role for protein kinase casein kinase2 α-subunits in the *Arabidopsis* circadian clock. Plant Physiol, 157(3): 1537-1545.

Luan S. 2009. The CBL-CIPK network in plant calcium signaling. Trends Plant Sci, 14: 37-42.

Ma H. 2005. Molecular genetic analyses of microsporogenesis and microgametogenesis in flowering plants. Annu Rev Plant Biol, 56: 393-434.

Ma Y, Szostkiewicz I, Korte A, et al. Regulators of PP2C phosphatase activity function as abscisic acid sensors. Science 2009; 324(5930): 1064-1068.

Macho A P, Schwessinger B, Ntoukakis V, et al. 2014. A bacterial tyrosine phosphatase inhibits plant pattern recognition receptor activation. Science, 343(6178): 1509-1512.

Mahfouz MM, Kim S, Delauney AJ, et al. 2006. *Arabidopsis* TARGET OF RAPAMYCIN interacts with RAPTOR, which regulates the activity of S6 kinase in response to osmotic stress signals. Plant Cell, 18(2): 477-490.

Manavella PA, Hagmann J, Ott F, et al. 2012. Fast-forward genetics identifies plant CPL phosphatases as regulators of miRNA processing factor HYL1. Cell, 151(4):859-870.

Mandel T, Moreau F, Kutsher Y, et al. 2014. The ERECTA receptor kinase regulates *Arabidopsis* shoot apical meristem size, phyllotaxy and floral meristem identity. Development, 141(4): 830-841.

Marshall A, Aalen R B, Audenaert D, et al. 2012. Tackling drought stress: RECEPTOR-LIKE KINASES present new approaches. Plant Cell, 24(6): 2262-2278.

Matsubayashi Y, Sakagami Y. 2006b. Peptide hormones in plants. Annu Rev Plant Biol, 57: 649-674.

Matsubayashi Y, Takagi L, Omura N, et al. 1999. The endogenous sulfated pentapeptide phytosulfokine-alpha stimulates tracheary element differentiation of isolated mesophyll cells of zinnia. Plant Physiol, 120(4): 1043-1048.

Matsubayashi Y. 2011. Post-translational modifications in secreted peptide hormones in plants. Plant Cell Physiol, 52(1): 5-13.

Mayer K F X, Schoof H, Haecker A, et al. 1998. Role of WUSCHEL in regulating stem cell fate in the *Arabidopsis* shoot meristem. Cell, 95(6): 805-815.

McCourt P, Creelman R. 2008. The ABA receptors-we report you decide. Curr Opin Plant Biol, 11(5): 474-478.

Meng X, Zhang S. 2013. MAPK cascades in plant disease resistance signaling. Annu Rev Phytopathol, 51: 245-266.

Merchante C, Alonso J M, Stepanova A N. 2013. Ethylene signaling: simple ligand, complex regulation. Curr Opin Plant Biol, 16(5): 554-560.

Meyer K, Leube M P, Grill E. 1994. A protein phosphatase 2C involved in ABA signal transduction in *Arabidopsis thaliana*. Science, 264(5164):1452-1455.

Michniewicz M, Zago M K, Abas L, et al. 2007. Antagonistic regulation of PIN phosphorylation by PP2A and PINOID directs auxin flux. Cell, 130(6): 1044-1056.

Min L, Zhu L, Tu L, et al. 2013. Cotton GhCKI disrupts normal male reproduction by delaying tapetum programmed cell death via inactivating starch synthase. Plant J, 75(5): 823-835.

Mishra G, Zhang W, Deng F, et al. 2006. A bifurcating pathway directs abscisic acid effects on stomatal closure and opening in *Arabidopsis*. Science, 312: 264-266.

Miwa H, Betsuyaku S, Iwamoto K, et al. 2008. The receptor-like kinase SOL2 mediates CLE signaling in *Arabidopsis*. Plant Cell Physiol, 49(11): 1752-1757.

Mizuno S, Osakabe Y, Maruyama K, et al. 2007. Receptor-like protein kinase 2 (RPK2) is a novel factor controlling anther development in *Arabidopsis thaliana*. Plant J, 50(5): 751-766.

Möller B, Weijers D. 2009. Auxin control of embryo patterning. Cold Spring Harb Perspect Biol, 1(5): a001545.

Møller S G, Kim Y S, Kunkel T, et al. 2003. PP7 is a positive regulator of blue light signaling in *Arabidopsis*. Plant Cell, 15(5): 1111-1119.

Montoya T, Nomura T, Farrar K, et al. 2002. Cloning the tomato curl3 gene highlights the putative dual role of the leucine-rich repeat receptor kinase tBRI1/SR160 in plant steroid hormone and peptide hormone signaling. Plant Cell, 14(12): 3163-3176.

Mora-García S, Vert G, Yin Y, et al. 2004. Nuclear protein phosphatases with Kelch repeat domains modulate the response to brassinosteroids in *Arabidopsis*. Genes Dev, 18(4): 448-460.

Moreno-Romero J, Armengot L, Mar Marquès-Bueno M, et al. 2012. CK2-defective *Arabidopsis* plants exhibit enhanced double-strand break repair rates and reduced survival after exposure to ionizing radiation. The Plant J, 71(4): 627-638.

Moreno-Romero J, Armengot L, Marquès-Bueno MM, et al. 2011. About the role of CK2 in plant signal transduction. Mol Cell Biochem, 356(1-2): 233-240.

Moreno-Romero J, Espunya MC, Platara M, et al. 2008. A role for protein kinase CK2 in plant development: evidence obtained using a dominant-negative mutant. The Plant J, 55(1): 118-130.

Mori I C, Murata Y, Yang Y, et al. 2006. CDPKs CPK6 and CPK3 function in ABA regulation of guard cell S-type anion- and Ca^{2+}-permeable channels and stomatal closure. PLoS Biol, 4(10): e327.

Mu Khopadhya Y K. 1997. Purification and characterization of a protein kinase from winged bean. Phytochemistry, 46(3): 461-467.

Mulekar J J, Huq E. 2013. Expanding roles of protein kinase CK2 in regulating plant growth and development. J Exp Bot, 65(11): 2883-2893.

Mulekar JJ, Bu Q, Chen F, Huq E. 2012. Casein kinase II α subunits affect multiple developmental and stress-responsive pathways in *Arabidopsis*. Plant J, 69(2): 343-354.

Muller R, Bleckmann A, Simon R. 2008. The receptor kinase CORYNE of *Arabidopsis* transmits the stem cell-limiting signal CLAVATA3 independently of CLAVATA1. Plant Cell, 20(4): 934-946.

Murase K, Shiba H, Iwano M, et al. 2004. A membrane-anchored protein kinase involved in Brassica self-incompatibility signaling. Science, 303(5663): 1516-1519.

Murra Y M G, Guil Foyl E T J, Key J L. 1978. Isolation and characterization of a chromatin associated protein kinase from soybean. Plant Physiol, 61(6): 1023-1030.

Murra Y M G, Guil Foyl E T J, Key J L. 1978. Isolation and preliminary characterization of a casein kinase from cauliflower nuclei. Plant Physiol, 62(3): 434-437.

Muschietti J, Eyal Y, McCormick S. 1998. Pollen tube localization implies a role in pollen-pistil interactions for the tomato receptor-like protein kinases LePRK1 and LePRK2. Plant Cell, 10(3): 319-330.

Mustilli AC, Merlot S, Vavasseur A, et al. 2002. *Arabidopsis* OST1 protein kinase mediates the regulation of stomatal aperture by abscisic acid and acts upstream of reactive oxygen species production. Plant Cell, 14(12): 3089-3099.

Myers C, Romanowsky S M, Barron Y D, et al. 2009. Calcium-dependent protein kinases regulate polarized tip growth in pollen tubes. Plant J, 59: 528-539.

Nadeau J A, Sack F D. 2002. Control of stomatal distribution on the *Arabidopsis* leaf surface. Science, 296(5573): 1697-1700.

Nam K H, Li J M. 2002. BRI1/BAK1, a receptor kinase pair mediating brassinosteroid signaling. Cell, 110(2): 203-212.

Neeraj D, Cashmore A R. 1989. Binding of a Pea Nuclear Protein to Promoters of Certain Photoregulated Genes 1s Modulated by Phosphorylation. Plant Cell, 1(11): 1 069-1077.

Ng L M, Melcher K, Teh B T, et al. 2014. Abscisic acid perception and signaling: structural mechanisms and applications. Acta Pharmacol Sin, 35(5): 567-584.

Nito K, Wong C C, Yates J R 3rd, et al. 2013. Tyrosine phosphorylation regulates the activity of phytochrome photoreceptors. Cell Rep, 3(6): 1970-1979.

Nodine M D, Bryan A C, Racolta A, et al. 2011. A few standing for many: embryo receptor-like kinases. Trends Plant Sci, 16(4): 211-217.

Nodine M D, Yadegari R, Tax F E. 2007. RPK1 and TOAD2 are two receptor-like kinases redundantly required for *Arabidopsis* embryonic pattern formation. Dev Cell, 12(6): 943-956.

Nonomura K I, Miyoshi K, Eiguchi M, et al. 2003. The MSP1 gene is necessary to restrict the number of cells entering into male and female sporogenesis and to initiate anther wall formation in rice. Plant Cell, 15(8): 1728-1739.

Ntoukakis V, Mucyn T S, Gimenez-Ibanez S, et al. 2009. Host inhibition of a bacterial virulence effector triggers immunity to infection. Science, 324(5928): 784-787.

Ntoukakis V, Saur I M L, Conlan B, et al. 2014. The changing of the guard: the Pto/Prf receptor complex of tomato and pathogen recognition. Curr Opin Plant Biol, 20C: 69-74.

Nunes C, O'Hara L E, Primavesi L F, et al. 2013. The trehalose 6-phosphate/SnRK1 signaling pathway primes growth recovery following relief of sink limitation. Plant Physiol, 162(3): 1720-1732.

Ogawa D, Abe K, Miyao A, et al. 2011. RSS1 regulates the cell cycle and maintains meristematic activity under stress conditions in rice. Nat Commun, 2: 278.

Ogawa M, Shinohara H, Sakagami Y, et al. 2008. *Arabidopsis* CLV3 peptide directly binds CLV1 ectodomain. Science, 319(5861): 294.

Ogiso E, Takahashi Y, Sasaki T, et al. 2010. The role of casein kinase II in flowering time regulation has diversified during evolution. Plant Physiol, 152(2): 808-820.

Oh C S, Martin G B. 2011. Effector-triggered immunity mediated by the Pto kinase. Trends Plant Sci, 16(3): 132-140.

Oh M H, Wang X, Wu X, et al. 2010. Autophosphorylation of Tyr-610 in the receptor kinase BAK1 plays a role in brassinosteroid signaling and basal defense gene expression. Proc Natl Acad Sci USA, 107(41): 17827-17832.

Ohmori Y, Tanaka W, Kojima M, et al. 2013. WUSCHEL-RELATED HOMEOBOX4 is involved in meristem maintenance and is negatively regulated by the CLE gene FCP1 in rice. Plant Cell, 25(1): 229-241.

Ohta M, Guo Y, Halfter U, et al. 2003. A novel domain in the protein kinase SOS2 mediates interaction with the protein phosphatase 2C ABI2. Proc Natl Acad Sci USA, 100(20): 11771-11776.

Ohyama K, Shinohara H, Ogawa-Ohnishi M, et al. 2009. A glycopeptide regulating stem cell fate in *Arabidopsis thaliana*. Nat Chem Biol, 5(8): 578-580.

Osakabe Y, Mizuno S, Tanaka H, et al. 2010. Overproduction of the membrane-bound receptor-like protein kinase 1, RPK1, enhances abiotic stress tolerance in *Arabidopsis*. J Biol Chem, 285(12): 9190-9201.

Osmont K S, Sibout R, Hardtke C S. 2007. Hidden branches: Developments in root system architecture. Annu Rev Plant Biol, 58: 93-113.

Ouyang S-Q, Liu Y-F, Liu P, et al. 2010. Receptor-like kinase OsSIK1 improves drought and salt stress tolerance in rice (*Oryza sativa*) plants. Plant J, 62(2): 316-329.

Parcy F, Valon C, Raynal M, et al. 1994. Regulation of gene expression programs during *Arabidopsis* seed development: roles of the ABI3 locus and of endogenous abscisic acid. Plant Cell, 6(11): 1567-1582.

Park C J, Peng Y, Chen X, et al. 2008. Rice XB15, a protein phosphatase 2C, negatively regulates cell death and XA21-mediated innate immunity. PLoS Biol, 6(9): e231.

Park H H. 2012. Casein kinase I-like protein linked to lipase in plant. Plant Signal Behav, 7(7): 719-721.

Park J H, Lee S Y, Kim W Y, et al. 2011. Heat-induced chaperone activity of serine/ threonine protein phosphatase 5 enhances thermotolerance in *Arabidopsis thaliana*. New Phytol, 191(3): 692-705.

Park S Y, Fung P, Nishimura N, et al. 2009. Abscisic acid inhibits type 2C protein phosphatases via the PYR/PYL family of START proteins. Science, 324(5930): 1068-1071.

Park Y I, Do K H, Kim I S, et al. 2012. Structural and functional studies of casein kinase I-like protein from rice. Plant Cell Physiol, 53(2):304-311.

Petricka J J, Winter C M, Benfey P N. 2012. Control of *Arabidopsis* root development. Annu Rev Plant Biol, 63: 563-590.

Pillitteri L J, Torii K U. 2012. Mechanisms of stomatal development. Annu Rev Plant Biol, 63: 591-614.

Pillitteri L J, Torii K U. 2012. Mechanisms of stomatal development. Annu Rev Plant Biol, 63:591-614.

Pitorre D, Llauro C, Jobet E, et al. 2010. RLK7, a leucine-rich repeat receptor-like kinase, is required for proper germination speed and tolerance to oxidative stress in *Arabidopsis thaliana*. Planta, 232(6): 1339-1353.

Potuschak T, Lechner E, Parmentier Y, et al. 2003. EIN3-dependent regulation of plant ethylene hormone signaling by two *Arabidopsis* F box proteins: EBF1 and EBF2. Cell, 115(6): 679-689.

Qi P, Lin Y S, Song X J, et al. 2012. The novel quantitative trait locus GL3.1 controls rice grain size and yield by regulating Cyclin-T1;3. Cell Res, 22(12): 1666-1680.

Qiao H, Chang K N, Yazaki J, et al. 2009. Interplay between ethylene, ETP1/ETP2 F-box proteins, and degradation of EIN2 triggers ethylene responses in *Arabidopsis*. Genes Dev, 23(4): 512-521.

Qiao H, Shen Z, Huang S S, et al. 2012. Processing and subcellular trafficking of ER-tethered EIN2 control response to ethylene gas. Science, 338(6105): 390-393.

Qiu Q S, Guo Y, Dietrich M A, et al. 2002. Regulation of SOS1, a plasma membrane Na^+/H^+ exchanger in *Arabidopsis thaliana*, by SOS2 and SOS3.Proc Natl Acad Sci USA, 99(12): 8436-8441.

Qiu Q S, Guo Y, Quintero F J, et al. 2004. Regulation of vacuolar Na^+/H^+ exchange in *Arabidopsis thaliana* by the salt-overly-sensitive (SOS) pathway. J Biol Chem, 279(1): 207-215.

Quan R, Lin H, Mendoza I, et al. 2007. SCABP8/CBL10, a putative calcium sensor, interacts with the protein kinase SOS2 to protect *Arabidopsis* shoots from salt stress. Plant Cell, 19(4): 1415-1431.

Quintero F J, Martinez-Atienza J, Villalta I, et al. 2011. Activation of the plasma membrane Na/H antiporter Salt-Overly-Sensitive 1 (SOS1) by phosphorylation of an auto-inhibitory C-terminal domain. Proc Natl Acad Sci USA, 108(6): 2611-2616.

Quintero F J, Ohta M, Shi H, et al. 2002. Reconstitution in yeast of the *Arabidopsis* SOS signaling pathway for Na^+ homeostasis. Proc Natl Acad Sci USA, 99(13): 9061-9066.

Radutoiu S, Madsen L H, Madsen E B, et al. 2003. Plant recognition of symbiotic bacteria requires two LysM receptor-like kinases. Nature, 425(6958): 585-592.

Raven J A. 2002. Selection pressures on stomatal evolution. New Phytol, 153(3): 371-386.

Rentel M C, Lecourieux D, Ouaked F, et al. 2004. OXI1 kinase is necessary for oxidative burst-mediated signalling in *Arabidopsis*. Nature, 427(6977): 858-861.

Rigó G, Ayaydin F, Tietz O, et al. 2013. Inactivation of plasma membrane-localized CDPK-RELATED KINASE5 decelerates PIN2 exocytosis and root gravitropic response in *Arabidopsis*. Plant Cell, 25(5): 1592-1608.

Robaglia C, Thomas M, Meyer C. 2012. Sensing nutrient and energy status by SnRK1 and TOR kinases. Curr Opin Plant Biol, 15(3): 301-307.

Rodrigues A, Adamo M, Crozet P, et al. 2013. ABI1 and PP2CA phosphatases are negative regulators of Snf1-related protein kinase1 signaling in *Arabidopsis*. Plant Cell, 25(10): 3871-3884.

Rodriguez M C, Petersen M, Mundy J. 2010. Mitogen-activated protein kinase signaling in plants. Annu Rev Plant Biol, 61: 621-649.

Rodriguez P L, Benning G, Grill E. 1998. ABI2, a second protein phosphatase 2C involved in abscisic acid signal transduction in *Arabidopsis*. FEBS Lett, 421(3): 185-190.

Rosebrock T R, Zeng L, Brady J J, et al. 2007. A bacterial E3 ubiquitin ligase targets a host protein kinase to disrupt plant immunity. Nature, 448(7151): 370-374.

Rychlik W and Zagorski W. 1980. Purification and characterization of adenosine-3′, 5′ 2 phosphate independent protein kinase from wheat germ. Eur J Biochem, 106(2): 653-659.

Ryu J S, Kim J I, Kunkel T, et al. 2005. Phytochrome-specific type 5 phosphatase controls light signal flux by enhancing phytochrome stability and affinity for a signal transducer. Cell, 120(3): 395-406.

Sablowski R. 2011. Plant stem cell niches: from signalling to execution. Curr Opin Plant Biol, 14(1): 4-9.

Saidi Y, Hearn T J, Coates J C. 2012. Function and evolution of 'green' GSK3/Shaggy-like kinases. Trends Plant Sci, 17(1): 39-46.

Sakai H, Aoyama T, Oka A. 2000. *Arabidopsis* ARR1 and ARR2 response regulators operate as transcriptional activators. Plant J. 24(6): 703-711

Sakai H, Honma T, Aoyama T, et al. 2001. ARR1, a transcription factor for genes immediately responsive to cytokinins. Science, 294(5546):1519-1521

Sakai H, Hua J, Chen QG, et al. 1998. ETR2 is an ETR1-like gene involved in ethylene signaling in *Arabidopsis*. Proc Natl Acad Sci USA, 95(10): 5812-5817.

Salinas P, Fuentes D, Vidal E, et al. 2006. An extensive survey of CK2 alpha and beta subunits in *Arabidopsis*: multiple isoforms exhibit differential subcellular localization. Plant Cell Physiol, 47(9): 1295-1308.

Salomon M, Knieb E, von Zeppelin T, et al. 2003. Mapping of low- and high-fluence autophosphorylation sites in phototropin 1. Biochemistry, 42(14): 4217-4225.

Sanders P M, Bui A Q, Weterings K, et al. 1999. Anther developmental defects in *Arabidopsis thaliana* male-sterile mutants. Sexual Plant Reprod, 11(6): 297-322.

Santiago J, Rodrigues A, Saez A, et al. 2009. Modulation of drought resistance by the abscisic acid receptor PYL5 through inhibition of clade A PP2Cs. Plant J, 60(4): 575-588.

Santner A A, Watson J C. 2006. The WAG1 and WAG2 protein kinases negatively regulate root waving in *Arabidopsis*. Plant J, 45(5): 752-764.

Santner A, Estelle M. 2009. Recent advances and emerging trends in plant hormone signalling. Nature, 459(7250): 1071-1078.

Sarkar A K, Luijten M, Miyashima S, et al. 2007. Conserved factors regulate signalling in *Arabidopsis thaliana* shoot and root stem cell organizers. Nature, 446(7137): 811-814.

Scheer J M, Pearce G, Ryan C A. 2003. Generation of systemin signaling in tobacco by transformation with the tomato systemin receptor kinase gene. Proc Natl Acad Sci USA, 100(17): 10114-10117.

Scheer J M, Ryan C A. 2002. The systemin receptor SR160 from Lycopersicon peruvianum is a member of the LRR receptor kinase family. Proc Natl Acad Sci USA, 99(14): 9585-9590.

Schepetilnikov M, Dimitrova M, Mancera-Martínez E, et al. 2013. TOR and S6K1 promote translation reinitiation of uORF-containing mRNAs via phosphorylation of eIF3h. EMBO J, 32(8): 1087-1102.

Scheres B. 2007. Stem-cell niches: nursery rhymes across kingdoms. Nat Rev Mol Cell Biol, 8(5): 345-354.

Schoof H, Lenhard M, Haecker A, et al. 2000. The stem cell population of *Arabidopsis* shoot meristems is maintained by a regulatory loop between the CLAVATA and WUSCHEL genes. Cell, 100(6): 635-644.

Schweer J, Türkeri H, Link B, et al. 2010. AtSIG6, a plastid sigma factor from *Arabidopsis*, reveals functional impact of cpCK2 phosphorylation. Plant J, 62(2): 192-202.

Scofield S R, Tobias C M, Rathjen J P, et al. 1996. Molecular basis of gene-for-gene specificity in bacterial speck disease of Tomato. Science, 274(5295): 2063-2065.

Sents W, Ivanova E, Lambrecht C, et al. 2013. The biogenesis of active protein phosphatase 2A holoenzymes: a tightly regulated process creating phosphatase specificity. FEBS J, 280(2): 644-661.

Shao F, Golstein, C, Ade J, et al. 2003. Cleavage of *Arabidopsis* PBS1 by a bacterial type III effector. Science, 301(5637): 1230-1233.

She J, Han Z F, Kim T W, et al. 2011. Structural insight into brassinosteroid perception by BRI1. Nature, 474(7352): 472-U496.

Shi J, Kim K N, Ritz O, et al. 1999. Novel protein kinases associated with calcineurin B like calcium sensors in *Arabidopsis*. Plant Cell, 11(12): 2393-2405.

Shiu S H, Bleecker A B. 2001a. Plant receptor-like kinase gene family: diversity, function, and signaling. Sci STKE, 2001(113): re22.

Shiu S H, Bleecker A B. 2001b. Receptor-like kinases from *Arabidopsis* form a monophyletic gene family related to animal receptor kinases. Proc Natl Acad Sci USA, 98(19): 10763-10768.

Shiu S H, Bleecker A B. 2003. Expansion of the receptor-like kinase/Pelle gene family and receptor-like proteins in *Arabidopsis*. Plant Physiol, 132(2): 530-543.

Shiu S H, Karlowski W M, Pan R, et al. 2004. Comparative analysis of the receptor-like kinase family in *Arabidopsis* and rice. Plant Cell, 16(5): 1220-1234.

Shpak E D, McAbee J M, Pillitteri L J, et al. 2005. Stomatal patterning and differentiation by synergistic interactions of receptor kinases. Science, 309(5732): 290-293.

Silva N F, Goring D R. 2002. The proline-rich, extensin-like receptor kinase-1 (PERK1) gene is rapidly induced by wounding. Plant Mol Biol, 50(4): 667-685.

Sivaguru M, Ezaki B, He Z H, et al. 2003. Aluminum-induced gene expression and protein localization of a cell wall-associated receptor kinase in *Arabidopsis*. Plant Physiol, 132(4): 2256-2266.

Skottke K R, Yoon G M, Kieber J J, et al. 2011. Protein Phosphatase 2A controls ethylene biosynthesis by differentially regulating the turnover of ACC Synthase isoforms. PLoS Genetics, 7(4): e1001370..

Song W Y, Wang G L, Chen L L, et al. 1995. A receptor kinase-like protein encoded by the rice disease resistance gene, Xa21. Science, 270(5243): 1804-1806.

Soon F F, Ng L M, Zhou X E, et al. 2012. Molecular mimicry regulates ABA signaling by SnRK2 kinases and PP2C phosphatases. Science, 335(6064): 85-88.

Stahl Y, Wink R H, Ingram G C, et al. 2009. A signaling module controlling the stem cell niche in *Arabidopsis* root meristems. Curr Biol, 19(11): 909-914.

Stone J M, Trotochaud A E, Walker J C, et al. 1998. Control of meristem development by CLAVATA1 receptor kinase and kinase-associated protein phosphatase interactions. Plant Physiol, 117(4): 1217-1225.

Stone J M, Walker J C. 1995. Plant Protein Kinase Families and Signal Transduction. Plant Physiol, 108(2): 451-457.

Stracke S, Kistner C, Yoshida S, et al. 2002. A plant receptor-like kinase required for both bacterial and fungal symbiosis. Nature, 417(6892): 959-962.

Stratmann J W. 2003. Long distance run in the wound response - jasmonic acid is pulling ahead. Trends Plant Sci, 8(6): 247-250.

Sugano S S, Shimada T, Imai Y, et al. 2010. Stomagen positively regulates stomatal density in *Arabidopsis*. Nature, 463(7278): 241-244.

Sugano S, Andronis C, Green R M, et al. 1998. Protein kinase CK2 interacts with and phosphorylates the *Arabidopsis* circadian clock-associated 1 protein. Proc Natl Acad Sci USA, 95(18): 11020-11025.

Sugano S, Andronis C, Ong M S, et al. 1999. The protein kinase CK2 is involved in regulation of circadian rhythms in *Arabidopsis*. Proc Natl Acad Sci USA, 96(22): 12362-12366.

Sukumar P, Edwards K S, Rahman A, et al. 2009. PINOID kinase regulates root gravitropism through modulation of PIN2-dependent basipetal auxin transport in *Arabidopsis*. Plant Physiol, 150(2): 722-735.

Sun X, Kang X, Ni M. 2012. Hypersensitive to red and blue 1 and its modification by protein phosphatase 7 are implicated in the control of *Arabidopsis* stomatal aperture. PLoS Genet, 8(5): e1002674.

Suzaki T, Ohneda M, Toriba T, et al. 2009. FON2 SPARE1 redundantly regulates floral meristem maintenance with FLORAL ORGAN NUMBER2 in rice. PLoS Genet, 5(10): e1000693.

Suzaki T, Sato M, Ashikari M, et al. 2004a. The gene FLORAL ORGAN NUMBER1 regulates floral meristem size in rice and encodes a leucine-rich repeat receptor kinase orthologous to *Arabidopsis* CLAVATA1. Development, 131(22): 5649-5657.

Suzaki T, Toriba T, Fujimoto M, et al. 2006. Conservation and diversification of meristem maintenance mechanism in Oryza sativa: Function of the FLORAL ORGAN NUMBER2 gene. Plant Cell Physiol, 47(12): 1591-1602.

Swanson R, Edlund A F, Preuss D. 2004. Species specificity in pollen-pistil interactions. Annu Rev Genet, 38: 793-818.

Taguchi-Shiobara F, Yuan Z, Hake S, et al. 2001. The fasciated ear2 gene encodes a leucine-rich repeat receptor-like protein that regulates shoot meristem proliferation in maize. Genes Dev, 15(20): 2755-2766.

Takemiya A, Yamauchi S, Yano T, et al. 2013. Identification of a regulatory subunit of protein phosphatase 1 which mediates blue light signaling for stomatal opening. Plant Cell Physiol, 54(1): 24-35.

Takeuchi J, Okamoto M, Akiyama T, et al. 2014. Designed abscisic acid analogs as antagonists of PYL-PP2C receptor interactions. Nat Chem Biol, 10(6): 477-482.

Tan S T, Dai C, Liu H T, et al. 2013. *Arabidopsis* casein kinase1 proteins CK1.3 and CK1.4 phosphorylate cryptochrome2 to regulate blue light signaling. Plant Cell, 25(7): 2618-2632.

Tanaka H, Osakabe Y, Katsura S, et al. 2012. Abiotic stress-inducible receptor-like kinases negatively control ABA signaling in *Arabidopsis*. Plant J, 70(4): 599-613.

Tanaka H, Watanabe M, Sasabe M, et al. 2007. Novel receptor-like kinase ALE2 controls shoot development by specifying epidermis in *Arabidopsis*. Development, 134(9): 1643-1652.

Tanaka H, Watanabe M, Watanabe D, et al. 2002. ACR4, a putative receptor kinase gene of *Arabidopsis thaliana*, that is expressed in the outer cell layers of embryos and plants, is involved in proper embryogenesis. Plant Cell Physiol, 43(4): 419-428.

Tang R J, Liu H, Yang Y, et al. Tonoplast calcium sensors CBL2 and CBL3 control plant growth and ion homeostasis through regulating V-ATPase activity in *Arabidopsis*. Cell Res, 2012, 22(12): 1650-1665.

Tang W, Kim T-W, Oses-Prieto J A, et al. 2008. BSKs mediate signal transduction from the receptor kinase BRI1 in *Arabidopsis*. Science, 321(5888): 557-560.

Tang W, Yuan M, Wang R, et al. 2011. PP2A activates brassinosteroid-responsive gene expression and plant growth by dephosphorylating BZR1. Nat Cell Biol, 13(2): 124-131.

Templeton G W, Nimick M, Morrice N, et al. 2011. Identification and characterization of AtI-2, an *Arabidopsis* homologue of an ancient protein phosphatase 1 (PP1) regulatory subunit. Biochem J, 435(1): 73-83.

Tena G, Boudsocq M, Sheen J. 2011. Protein kinase signaling networks in plant innate immunity. Curr Opin Plant Biol, 14(5):519-529.

Torii K U, Mitsukawa N, Oosumi T, et al. 1996. The *Arabidopsis* ERECTA gene encodes a putative receptor protein kinase with extracellular leucine-rich repeats. Plant Cell, 8(4): 735-746.

Trotochaud A E, Hao T, Wu G, et al. 1999. The CLAVATA1 receptor-like kinase requires CLAVATA3 for its assembly into a signaling complex that includes KAPP and a Rho-related protein. Plant Cell, 11(3): 393-405.

Tsai A Y, Gazzarrini S. 2014. Trehalose-6-phosphate and SnRK1 kinases in plant development and signaling: the emerging picture. Front Plant Sci, 5: 119.

Tseng T S, Briggs W R. 2010. The *Arabidopsis* rcn1-1 mutation impairs dephosphorylation of Phot2, resulting in enhanced blue light responses. Plant Cell, 22(2): 392-402.

Tsuwamoto R, Fukuoka H, Takahata Y. 2008. GASSHO1 and GASSHO2 encoding a putative leucine-rich repeat transmembrane-type receptor kinase are essential for the normal development of the epidermal surface in *Arabidopsis* embryos. Plant J, 54(1): 30-42.

Uchida N, Shimada M, Tasaka M. 2013. ERECTA-family receptor kinases regulate stem cell homeostasis via buffering its cytokinin responsiveness in the shoot apical meristem. Plant Cell Physiol, 54(3): 343-351.

Umezawa T, Sugiyama N, Mizoguchi M, et al. 2009. Type 2C protein phosphatases directly regulate abscisic acid-activated protein kinases in *Arabidopsis*. Proc Natl Acad Sci USA, 106(41): 17588-17593.

vandenBerg C, Willemsen V, Hendriks G, et al. 1997. Short-range control of cell differentiation in the *Arabidopsis* root meristem. Nature, 390(6657): 287-289.

Vert G, Walcher C L, Chory J, et al. 2008. Integration of auxin and brassinosteroid pathways by Auxin Response Factor 2. Proc Natl Acad Sci USA, 105(28): 9829-9834.

Vierstra R D. 2009. The ubiquitin-26S proteasome system at the nexus of plant biology. Nat Rev Mol Cell Biol, 10(6): 385-397.

Vij S, Giri J, Dansana P K, et al. 2008. The receptor-like cytoplasmic kinase (OsRLCK) gene family in rice: Organization, phylogenetic relationship, and expression during development and stress. Mol Plant, 1(5): 732-750.

Wang H C, Ngwenyama N, Liu Y D, et al. 2007. Stomatal development and patterning are regulated by environmentally responsive mitogen-activated protein kinases in *Arabidopsis*. Plant Cell, 19(1): 63-73.

Wang P, Xue L, Batelli G, et al. 2013. Quantitative phosphoproteomics identifies SnRK2 protein kinase substrates and reveals the effectors of abscisic acid action. Proc Natl Acad Sci USA, 110(27): 11205-11210.

Wang X F, Goshe M B, Soderblom E J, et al. 2005. Identification and functional analysis of in vivo phosphorylation sites of the *Arabidopsis* BRASSINOSTEROID-INSENSITIVE1 receptor kinase. Plant Cell, 17(6): 1685-1703.

Wang X F, Kota U, He K, et al. 2008. Sequential transphosphorylation of the BRI1/BAK1 receptor kinase complex impacts early events in brassinosteroid signaling. Dev Cell, 15(2): 220-235.

Wang X, Chory J. 2006. Brassinosteroids Regulate Dissociation of BKI1, a Negative Regulator of BRI1 Signaling, from the Plasma Membrane. Science, 313(5790): 1118-1122.

Wang Z Y, Bai M Y, Oh E, et al. 2012. Brassinosteroid signaling network and regulation of photomorphogenesis. Annu Rev Genet, 46: 701-724.

Wang Z Y, Nakano T, Gendron J, et al. 2002. Nuclear-localized BZR1 mediates brassinosteroid-induced growth and feedback suppression of brassinosteroid biosynthesis. Dev Cell, 2(4): 505-513.

Weigel D, Meyerowitz E M. 1994. The abcs of floral homeotic genes. Cell, 78(2): 203-209.

Wen X, Zhang C, Ji Y, et al. 2012. Activation of ethylene signaling is mediated by nuclear translocation of the cleaved EIN2 carboxyl terminus. Cell Res, 22(11): 1613-1616.

West M A L, Harada J J. 1993. Embryogenesis in higher plants - an overview. Plant Cell, 5(10): 1361-1369.

Williams R W, Wilson J M, Meyerowitz E M. 1997. A possible role for kinase-associated protein phosphatase in the *Arabidopsis* CLAVATA1 signaling pathway. Proc Natl Acad Sci USA, 94(19): 10467-10472.

Willige B C, Ahlers S, Zourelidou M, et al. 2013. D6PK AGCVIII kinases are required for auxin transport and phototropic hypocotyl bending in *Arabidopsis*. Plant Cell. 25(5): 1674-1688.

Wu P, Gao H B, Zhang L L, et al. 2013. Phosphatidic acid regulates BZR1 activity and brassinosteroid signal of *Arabidopsis*. Mol Plant, 7(2): 445-447.

Wu S J, Ding L, Zhu J K. 1996. SOS1, a genetic locus essential for salt tolerance and potassium acquisition. Plant Cell, 8(4): 617-627.

Xiong Y, McCormack M, Li L, et al. 2013. Glucose-TOR signalling reprograms the transcriptome and activates meristems. Nature, 496(7444): 181-186.

Xu J, Li H D, Chen L Q, et al. 2006. A protein kinase, interacting with two calcineurin B like proteins, regulates K^+ transporter AKT1 in *Arabidopsis*. Cell, 125(7): 1347-1360.

Xue T, Wang D, Zhang S, et al. 2008. Genome-wide and expression analysis of protein phosphatase 2C in rice and *Arabidopsis*. BMC Genomics, 9: 550.

Yamaguchi Y, Huffaker A, Bryan A C, et al. 2010. PEPR2 is a second receptor for the Pep1 and Pep2 reptides and contributes to defense responses in *Arabidopsis*. Plant Cell, 22(2): 508-522.

Yamamoto E, Takashi T, Morinaka Y, et al. 2010. Gain of deleterious function causes an autoimmune response and Bateson-Dobzhansky-Muller incompatibility in rice. Mol Genet Genomics, 283(4): 305-315.

Yang H P, Matsubayashi Y, Nakamura K, et al. 1999. Oryza sativa PSK gene encodes a precursor of phytosulfokine-alpha, a sulfated peptide growth factor found in plants. Proc Natl Acad Sci USA, 96(23): 13560-13565.

Yang H P, Matsubayashi Y, Nakamura K, et al. 2001. Diversity of *Arabidopsis* genes encoding precursors for phytosulfokine, a peptide growth factor. Plant Physiol, 127(3): 842-851.

Yang J, Lin R, Sullivan J, et al. 2005. Light regulates COP1-mediated degradation of HFR1, a transcription factor essential for light signaling in *Arabidopsis*. Curr Biol, 17(3): 804-821.

Yang M, Sack F D. 1995. The too many mouths and four lips mutations affect stomatal production in *Arabidopsis*. Plant Cell, 7(12): 2227-2239.

Yang S L, Xiea L F, Mao H Z, et al. 2003. TAPETUM DETERMINANT1 is required for cell specialization in the *Arabidopsis* anther. Plant Cell, 15(12): 2792-2804.

Yang W C, Shi D Q, Chen Y H. 2010. Female gametophyte development in flowering plants. Annu Rev Plant Biol, 61: 89-108.

Yoshida R, Hobo T, Ichimura K, et al. 2002. ABA-activated SnRK2 protein kinase is required for dehydration stress signaling in *Arabidopsis*. Plant Cell Physiol, 43(12): 1473-1483.

Yu F, Qian L, Nibau C, et al. 2012. FERONIA receptor kinase pathway suppresses abscisic acid signaling in *Arabidopsis* by activating ABI2 phosphatase. Proc Natl Acad Sci USA, 109(36): 14693-14698.

Yu L P, Miller A K, Clark S E. 2003. POLTERGEIST encodes a protein phosphatase 2C that regulates CLAVATA pathways controlling stem cellidentity at *Arabidopsis* shoot and flower meristems. Curr Biol, 13(3): 179-188.

Yu L P, Simon E J, Trotochaud A E, et al. 2000. POLTERGEIST functions to regulate meristem development downstream of the CLAVATA loci. Development, 127(8): 1661-1670.

Zegzouti H, Anthony R G, Jahchan N, et al. 2006. Phosphorylation and activation of PINOID by the phospholipid signaling kinase 3-phosphoinositide-dependent protein kinase 1 (PDK1) in *Arabidopsis*. Proc Natl Acad Sci USA, 103(16):6404-6609.

Zegzouti H, Li W, Lorenz T C, et al. 2006. Structural and functional insights into the regulation of *Arabidopsis* AGC VIIIa kinases. J Biol Chem, 281(46): 35520-35530.

Zhang D B, Luo X, Zhu L. 2011. Cytological analysis and genetic control of rice anther development. J Genet Genomics, 38(9): 379-390.

Zhang J, Li W, Xiang T, et al. 2010. Receptor-like cytoplasmic kinases integrate signaling from multiple plant immune receptors and are targeted by a Pseudomonas syringae effector. Cell Host Microbe, 7(4):290-301.

Zhang J, Nodzynski T, Pencík A, et al. 2010. PIN phosphorylation is sufficient to mediate PIN polarity and direct auxin transport. Proc Natl Acad Sci USA, 107(2): 918-922.

Zhang X, Wang J, Huang J, et al. 2012. Rare allele of OsPPKL1 associated with grain length causes extra-large grain and a significant yield increase in rice. Proc Natl Acad Sci USA, 109(52): 21534-21539.

Zhang Y, Primavesi L F, Jhurreea D, et al. 2009. Inhibition of SNF1-related protein kinase1 activity and regulation of metabolic pathways by trehalose-6-phosphate. Plant Physiol, 149(4): 1860-1871.

Zhao C, Nie H, Shen Q, et al. 2014. EDR1 Physically interacts with MKK4/MKK5 and negatively regulates a MAP kinase cascade to modulate plant innate immunity. PLoS Genet 10(5): e1004389.

Zhao D Z, Wang G F, Speal B, et al. 2002. The EXCESS MICROSPOROCYTES1 gene encodes a putative leucine-rich repeat receptor protein kinase that controls somatic and reproductive cell fates in the *Arabidopsis* anther. Genes Dev, 16(15): 2021-2031.

Zhao X A, de Palma J, Oane R, et al. 2008. OsTDL1A binds to the LRR domain of rice receptor kinase MSP1, and is required to limit sporocyte numbers. Plant J, 54(3): 375-387.

Zhao Y D. 2010. Auxin biosynthesis and its role in plant development. Annu Rev Plant Biol, 61: 49-64.

Zhou H W, Nussbaumer C, Chao Y, et al. 2004. Disparate roles for the regulatory A subunit isoforms in *Arabidopsis* protein phosphatase 2A. Plant Cell, 16(3): 709-722.

Zhou H, Lin H, Chen S, et al. 2014. Inhibition of the *Arabidopsis* salt overly sensitive pathway by 14-3-3 proteins. Plant Cell, 26(3): 1166-1182.

Zhou J, Loh Y-T, Bressan R A, et al. 1995. The tomato gene Pti1 encodes a serine/threonine kinase that is phosphorylated by Pto and is involved in the hypersensitive response. Cell, 83(6): 925-935.

Zhu J K, Liu J, Xiong L. 1998. Genetic analysis of salt tolerance in *arabidopsis*. Evidence for a critical role of potassium nutrition. Plant Cell, 10(7): 1181-1191.

Zhu S Y, Yu X C, Wang X J, et al. 2007. Two calcium-dependent protein kinases, CPK4 and CPK11, regulate abscisic acid signal transduction in *Arabidopsis*. Plant Cell, 19(10): 3019-3036.

Zik M, Irish V F. 2003. Flower development: Initiation, differentiation, and diversification. Annu Rev Cell Dev Biol, 19: 119-140.

Zipfel C, Kunze G, Chinchilla D, et al. 2006. Perception of the bacterial PAMP EF-Tu by the receptor EFR restricts Agrobacterium-mediated transformation. Cell, 125(4): 749-760.

Zou J J, Wei F J, Wang C, et al. 2010. *Arabidopsis* calcium-dependent protein kinase CPK10 functions in abscisic acid- and Ca^{2+}-mediated stomatal regulation in response to drought stress. Plant Physiol, 154(3): 1232-1243.

Zourelidou M, Absmanner B, Weller B, et al. 2014. Auxin efflux by PIN-FORMED proteins is activated by two different protein kinases, D6 PROTEIN KINASE and PINOID. eLife, advance online: e02860.

Zourelidou M, Müller I, Willige B C, et al. 2009. The polarly localized D6 PROTEIN KINASE is required for efficient auxin transport in *Arabidopsis thaliana*. Development, 136(4): 627-636.

* 策划编辑：刘小龙　中国科学院上海生命科学研究院生物化学与细胞生物学研究所

抗原受体 V(D)J 基因重排

作　者：邓子木　许　可　刘小龙

中国科学院上海生命科学研究院生物化学与细胞生物学研究所

- 1. 重排激活蛋白 RAG / 79
- 2. BCR 基因重排 / 83
- 3. TCR 基因重排 / 87
- 4. 抗原受体多样性的产生 / 90
- 5. V(D)J 基因重排与疾病 / 91
- 6. V(D)J 基因重排的起源与演化 / 92
- 7. 展望 / 94

摘要

适应性免疫通过抗原受体分子（Ig 与 TCR）来识别多种多样的抗原。早在20世纪初期，已有研究发现 Ig 分子可以分成两个结构域：可变区与恒定区。Dreyer 与 Bennett 在 1965 年提出假设：可变区与恒定区的编码基因可能是由多个分隔存在的基因片断在 DNA 或 RNA 水平重排而成。直到1976年，Hozumi 和 Tonegawa 才证明了 Ig 编码基因的重排现象。

接下来几十年的研究表明：编码抗原受体分子可变区的胚系基因由处于成簇的分隔状态 V、D 与 J 片段构成，这些基因片段包含有重排信号序列（RSS）。重排激活蛋白 RAG 首先识别并剪切 RSS，然后通过非同源末端修复途径将切断的基因片段连接起来重组成新的编码序列，这个过程被称为 V(D)J 基因重排。重排是淋巴细胞早期发育过程中的重大事件，它的成功与否决定了淋巴细胞进一步分化或凋亡，因此 V(D)J 基因重排过程必须受到精确的调控。首先，淋巴细胞通过控制重排激活蛋白的表达与抗原受体基因所处染色质的常染色质化来确保重排发生的细胞特异性与阶段特异性；其次，每一次重排反应都要遵循 12/23 规则，即重排只能发生在两种不同类型的 RSS 之间；最后，为了确保淋巴细胞抗原受体的特异性，重排遵守等位排斥原则，即每一种淋巴细胞克隆只表达一种抗原受体分子。

V(D)J 基因重排研究从根本上阐明了适应性免疫识别特异性、群体多样性的形成机制，揭示了免疫记忆的分子基础，同时也为基因组稳定性的调控提供了新的线索。对 V(D)J 基因重排的研究与认识是分子免疫学发展过程中的重要篇章。

关键词

适应性免疫、抗原受体分子、V(D)J 重排、重排激活蛋白、等位排斥

高等脊椎动物的适应性免疫又称获得性免疫，按照应答方式可将其分成细胞免疫和体液免疫。其中，细胞免疫由 T 淋巴细胞表面的 T 细胞抗原受体分子(TCR)识别抗原开始应答；而体液免疫则由 B 细胞分泌的抗原受体(BCR)——免疫球蛋白(Ig)介导完成。每个个体都具有一个高度多样性的抗原受体库，是特异识别和清除外界病原体的物质基础。

抗原受体库的高度多样性是在 T 和 B 淋巴细胞早期发育过程中经过 V(D)J 基因重排而获得的(Hozumi et al., 1976)。抗原受体分子可变区的胚系编码基因成簇存在，分为 variable(V)、diversity(D) 和 joining(J) 片段。这些基因片段的 5'端、3'端或 5'端和 3'端同时连接有重排信号序列(RSS)，RSS 由保守的七聚体和九聚体及非保守的 12 个或 23 个碱基对(bp)的间隔区构成，因此存在两种 RSS：12-RSS 和 23-RSS。催化 V(D)J 基因重排反应的酶被称为重排激活蛋白(RAG)，包括 RAG1 和 RAG2(Oettinger et al., 1990; Schatz et al., 1989)。RAG 特异结合 RSS 并在 V、D 和 J 基因中选择某些基因片段进行重新组合，这一过程称为 V(D)J 基因重排(图版Ⅵ)。数量众多的 V(D)J 基因片段通过随机组合产生抗原受体可变区的编码序列，重排后的可变区再与恒定区通过转录后剪切等方式拼接成一条完整的抗原受体编码基因。

RAG 介导的重排反应遵循 12/23 规则，即重排只能发生在两种不同类型的 RSS 之间 (Akira et al., 1987)。RAG1 与 RAG2 以复合物的形式识别一对 12-RSS/23-RSS，组成配对复合物(PC)，RAG1 作为内切酶在七聚体的位置剪切 DNA 造成单链缺口，游离的 3'-OH 攻击互补链并通过转酯反应形成发卡结构的编码末端(CE)，以及平头的信号末端(SE)。这个过程是重排的第一阶段，即剪切阶段。第二阶段称为修复阶段(Gellert, 2002)。在完成剪切之后，RAG 并不会立即离开断裂的 DNA 双链，而是继续结合在 DNA 末端，以剪切后复合物(CSC)的形式(Tsai et al., 2002)协助招募非同源重排末端修复途径(NHEJ)相关因子(Rval et al., 2008)。编码末端首先脱离 CSC 并被重新连接产生编码连接产物(CJ)，RAG 与信号末端继续结合组成信号末端复合物(SEC)直到信号末端被重新连接产生信号连接产物(SJ)。

V(D)J 基因重排涉及 DNA 剪切、连接和修复，重排过程受到了严密的调控(Bassing et al., 2002)(图版Ⅶ图 1)。首先，V(D)J 基因重排遵循严格的细胞特异性：尽管在 T 和 B 淋巴细胞前体分化的多个阶段都表达重排激活蛋白 RAG1 与 RAG2，但是在 T 细胞前体中只有 TCR 编码基因发生重排；而在 B 细胞前体中只启动 BCR 的编码基因发生重排。其次，V(D)J 基因重排按特定的顺序进行：重排的各个步骤与 T 和 B 淋巴细胞的分化成熟进程高度协同，在 DN2 胸腺细胞中，重排激活蛋白首先开始 *Tcrb* 的 $D_\beta \rightarrow J_\beta$ 重排，重排产物 DJ_β 接下来在 DN3 胸腺细胞中与上游 V_β 片段再次重排；完成 β 链重排的 DN3 胸腺细胞表达由 TCRβ 与 pre-Tα 共同组成的 pre-TCR，pre-TCR 通过传递信号促进 DN4 细胞继续分化发育到 DP 期并且启动 *Tcrα* 的重排。骨髓中的 B 细胞前体经历类似的分化与重排过程，在 pro-B 细胞中特异地启动免疫球蛋白重链编码基因 *Igh* 的重排。重排成功的 pro-B 细胞表达抗原受体重链与替代轻链的 VpreB 和 λ5 二者组成的 pre-BCR，pre-BCR 促进细胞进入 pre-B 细胞阶段并启动免疫球蛋白轻链编码基因的重排。Pre-B 细胞中首先启动的是 *Igκ* 的重排，如果 *Igκ* 的重排失败，则继续启动 *Igλ* 的重排。因此 *Igκ* 的利用率是 *Igλ* 的 20 倍。最后，V(D)J 基因重排遵守等位排斥原则：抗原受体的一对等位基因只有一条会发生 $V_\beta \rightarrow DJ_\beta$ 的重排，一条链的重排会抑制另外一条等位基因的重排；只有在重排不能获得正确编码的受体时，才启动第二条链的重排，从而避免了一个细胞克隆产生多种抗原受体。等位排斥原则保证了每个淋巴细胞克隆只表达识别单一抗原表位的一种受体，是特异性识别和适应性免疫记忆的分子基础。

1. 重排激活蛋白 RAG

在 20 世纪 70 年代，研究者发现了淋巴细胞抗原受体基因的重排现象，然而介导重排的关键因子却不清楚。Schatz 等(1988)首先将人的基因组 DNA 片段转入 3T3 细胞系，发现某些 DNA 克隆的表达产物具备催化 V(D)J 基因重排的能力。说明了在淋巴细胞中存在特异介导 V(D)J 重排的蛋白酶。通过基因组步移筛选法，他们筛选鉴定到具有催化活性并特异表达在 T 与 B 淋巴细胞中蛋白酶的基因，并命名为 recombination activating gene(RAG1)(Schatz et al., 1989)。然而单独转染 RAG1 的 cDNA 后的重排活性却非常低，这个疑问促使研究者们发现了催化重排反应所需要的另外一个蛋白——RAG2(Oettinger et al., 1990)。重排激活蛋白包括 RAG1 和 RAG2，他们形成异源二聚体发挥功能。在小鼠基因组中，它们的编码基因仅包含一个单独的外显子，并以尾对尾的形式共同定位，它们共同转录并接受协同调控。这对独特的基因最早出现在软骨鱼中，在进化过程中高度保守。

1.1 RAG 核心区结构与功能

以小鼠的重排激活蛋白为例，RAG1 与 RAG2 分别由 1040 个和 527 个氨基酸组成（图版Ⅶ图 2A）。依据体外重排实验结果，将可以催化重排反应的最短部分定义为 RAG 蛋白的核心区(core-RAG1: 384-1008aa; core-RAG2: 1-387aa)(Aidinis et al., 2000)。核心区的 RAG1 又可以进一步分为 3 个部分：位于 N 端的 NBD 区(384-458aa)负责识别并结合 RSS 的九聚体；C 端区负责结合锌离子并介导自身二聚化(761-979aa)(De et al., 2004)；中央区(528-760aa)则主要参与结合 RAG2 及 RSS 的七聚体。通过定点突变筛选与功能鉴定，揭示了 core-RAG1 的 3 个酸性氨基酸(Asp-600、Asp-708 及 Glu-962)是 RAG1 内切核酸酶的活性中心，被称为"DDE 基序"(Landree et al., 1999; Kim et al., 1999)。序列比对分析提示 RAG1 酶的活性区域类似于转座酶及病毒的整合酶(Haren et al., 1999)。小鼠 RAG2 蛋白的核心区位于 N 端(1-387aa)（图版Ⅶ图 2B），包含 6 个 Kelch 重复基序，主要通过结合 RAG1 促进 RAG1 的构象变化，从而剪切 RSS 的七聚体(De et al., 2004; Callebaut et al., 1998)。RAG2 蛋白本身没有 DNA 内切核酸酶活性。

1.2 RAG 与 RSS 的组装过程

由于 core-RAG 的表达量、水溶性及稳定性均优于全长的 RAG，并且能够替代全长 RAG 准确识别 RSS 并催化剪切反应，因此大部分关于重排机制的体外研究都选用 core-RAG 来代替全长 RAG。RAG1 可以形成稳定的同源二聚体结合 RSS；RAG2 则主要以单体存在，对 DNA 基本不表现亲和性。RAG1 与 RAG2 一起能够形成异源二聚体或者四聚体，这种结合不需要借助 DNA 底物，然而由两者组成的复合物对 RSS 的亲和力要远高于单独的 RAG1(Godderz et al., 2003; Swanson et al., 1998, 1999)。

通过 EMSA 方法，研究者发现体外纯化的 RAG1 与 RAG2 能够结合一个拷贝的 RSS 形成单 RSS 复合物(SC)。根据这个复合物中 RAG2 分子的拷贝数不同又可以将 SC 复合物分为 SC1(2×RAG1+1×RAG2)和 SC2(2×RAG1+2×RAG2)(Swanson, 2002a; Mundy et al., 2002)。单独表达并纯化的 RAG1 和 RAG2 与 RSS 共孵育后更倾向形成 SC1；共表达且通过高盐条件共纯化后的 RAG 与 RSS 也倾向于形成 SC1；而共表达且通过温和条件共纯化的 RAG 与 RSS 容易组成 SC2。在 SC1 或 SC2 两种复合物中，RAG 都表现出相同的剪切能力。在体外实验中，SC2 可以继续捕捉一条 RSS 形成 PC 复合物，研究者据此提出了"俘获"模型：RAG 复合物首先结合一个 12-RSS 组成 SC 复合物(SC 或许在遇到合适的 23-RSS 之前就在 12-RSS 的七聚体处剪切并形成单链缺口），通过一轮"选择"和"释放"，SC 最终俘获了合适的 23-RSS，形成 PC 复合物（Curry et al., 2005）。尽管 RAG1 与 RAG2 可以有效地结合 RSS，高速泳动蛋白 HMG1/2 被证实在体外实验中可以通过结合 RAG1 提高 RAG 复合物与 DNA 的亲和力（van Gent et al., 1997），并且协助 RAG 组装成 PC 复合物（Swanson, 2002b），促进 RAG 结合并剪切 23-RSS（Yoshida et al., 2000）。然而，在体内 HMG 是否同样协同 RAG 起始剪切仍有待证实。在组装好的配对复合物中，RSS 是怎样排列的？研究结果表明，无论在体外实验还是体内重排，如果一对 RSS 以倒转方向排列，重排效率随着 RSS 之间距离的缩短而降低，因此推测在配对复合物中 RSS 应该是以平行的方向排列。研究者通过透射电镜也直观地了解到了配对复合物的结构（Grundy et al., 2009）。

1.3 "重排中心"假说

Jones 等（2002）通过体外实验证明，在 HMG 的陪伴下 RAG 结合了一个 12-RSS 后，将更倾向结合 23-RSS 而不是 12-RSS。在结合了 12-RSS 之后，RAG 复合物结构发生变化，处于被"锁定"状态，只有 23-RSS 才是开锁的钥匙；如果 RAG 复合物首先结合的是 23-RSS，则整个复合物对 12-RSS 没有那么大的偏好性。因此研究者推测，为了保证重排遵循 12/23 原则，RAG 复合物在体内应该是首先结合了 12-RSS（或许在七聚体位置上率先引入缺口），接下来挑选 23-RSS 形成 PC 复合物。那么在体内 PC 复合物又是如何组装的？Ji 等（2010）运用染色质免疫共沉淀技术完整地分析了 *Igh*、*Igκ*、*Tcrb* 及 *Tcra* 重排开始之前 RAG1 与 RAG2 的定位情况。他们的结果表明，在重排开始之前，RAG1 能够特异地识别 RSS 所在区域，主要定位在 J 区，以及靠近 J 区的 D 区附近；RAG2 的定位则稍微广泛，与 H3K4Me3 的分布重合。这些发现证实 RAG 通过优先定位于 12-RSS 来维持 12/23 原则，暗示接下来的组装过程中，RAG 复合物（SC）通过筛选 23-RSS 来完成配对复合物的装配。

针对 V(D)J 基因重排过程，研究者很早就提出了重排中心假说（Jung et al., 2006）：V(D)J 基因重排起始之前，一部分抗原受体基因片段的 RSS 首先经历了组蛋白活化修饰、胚系转录，以及核小体移位等变化变得高度"易接近"，这个"易接近"的区域被称为重排中心，RAG 及 NHEJ 因子在重排中心完成对 RSS 的剪切与修复工作。通过分析 RAG1 与 RAG2 在各细胞中的定位，Ji 等（2010）的工作还为"重排中心假说"提供了直接的证据：RAG1 与 RAG2 在细胞核内不是随机分布的，而是非常集中定位于抗原受体基因的 J 片段附近。RAG1 特异地识别这些 J 片段"易接近"的 12-RSS 进入重排中心，RAG2 通过结合组蛋白 H3K4Me3 也定位到了重排中心。二者结合了 12-RSS 之后可能首先形成 SC 复合物，此后，D 或者 V 片段在一系列活化修饰、反式因子及染色质三维结构改变的帮助下进入重排中心，SC 复合物选择并捕获了与 12-RSS 相匹配的另一个基因片段的 23-RSS，启动重排（Schatz et al., 2011）。

1.4 RAG1 非核心区结构与功能

RAG1 非核心区的 1-265aa 区域包含了 4 个保守的半胱氨酸：C1、C2、C3 和 CH，这些保守氨基酸可能介导 RAG1 与其他蛋白质的结合(Roman et al., 1997)。在 RAG1 非核心区的 265-383aa 区域包含一个 C3HC4 指环及一个锌指结构。已有研究表明锌指结构与 RAG1 自身二聚化相关。指环结构具有 E3 泛素连接酶活性，可以催化自身 233 位赖氨酸发生泛素化，调控自身蛋白质水平(Yurchenko et al., 2003)。在 2003 年，Dudly 等(2003)报道了 core-RAG1 敲入小鼠的 T 和 B 淋巴细胞分化虽然受到影响，但相当一部分的 T 和 B 淋巴细胞仍然能够正常分化和成熟，提示 RAG1 蛋白的非核心区对于体内重排并不是必需的。接下来的研究发现 core-RAG1 小鼠的 V(D)J 基因重排的效率和准确率均有所下降，因此研究者普遍认为 RAG1 非核心区的主要作用是调控重排的效率及准确性(Talukder et al., 2004)。

人 RAG1 的 N 端非核心区内的指环结构突变(*hC328Y*)导致新生儿的严重免疫缺陷：患者体内仅剩一部份 T 淋巴细胞，B 淋巴细胞几乎检测不到(Villa et al., 2001)，这与上述 core-RAG1 敲入小鼠观察到的现象不同。这种看似"矛盾"的现象引起了不少科学家的兴趣。体外实验表明，当指环结构发生突变导致 E3 泛素连接酶失去活性时，RAG1 催化 V(D)J 基因重排的能力也随之降低(Simkus et al., 2009a; Simkus et al., 2007)。这表明指环结构的 E3 泛素连接酶活性对 V(D)J 基因重排是不可或缺的。那么 RAG1 的指环结构到底通过泛素化哪些底物进而调控重排呢？已鉴定到的 RAG1 的泛素化底物有 KPNA1(Simkus et al., 2009b)、RAG1 自身及组蛋白 H3(包括 H3.3)(Jones et al., 2011; Grazini et al., 2010)，然而这些底物通过哪种机制调控基因重排依旧未知。

1.5 RAG2 非核心区结构与功能

RAG2 蛋白非核心区(388-527aa)包含一个 PHD 结构域(414-487aa)，以及一个连接核心区和 PHD 结构域的铰链区(402-407aa)。

RAG2 主要表达于 G_0/G_1 细胞期。在 G_1/S 过渡期，cyclineA/Cdk2 催化 RAG2 非核心区的 490 位色氨酸发生磷酸化修饰，进一步招募 Skp2-SCF 复合物，介导自身多泛素化并被降解(Li et al., 1996; Lee et al., 1999)。当细胞开始进入 S 期，RAG2 迅速降解，因而重排反应只发生在 G_0/G_1 期细胞中(Schlissel et al., 1993; Lee et al., 2004)。在 V(D)J 基因重排过程中，这种伴随细胞周期进行调控的意义在于确保断裂的双链 DNA 由 NHEJ 修复途径来完成修复，因为在 G_1 期细胞中，同源重组修复途径(HR)被关闭，从而减少了染色体移位的发生几率。在 RAG2T490A 敲入小鼠体内，RAG2 蛋白的积累使重排不再仅限于 G_1 期细胞，因此可以检测到大量通过 HR 途径修复的错误重排产物(Zhang et al., 2011)。最近的研究发现，小鼠的全长 RAG2 被 core-RAG2 替代后，重排引发的染色体丢失及易位几率显著上升，在 p53 缺失的背景下，小鼠胸腺发生肿瘤的比例明显增加(Deriano et al., 2011)。造成这种现象的原因可能主要是因为重排发生在非 G_1 期，HR 修复途径的参与导致重排的修复精确性下降；另外也有可能是 RAG2 的非核心区也参与断裂 DNA 的修复。

三维结构分析显示，RAG2 的 PHD 结构域包含由芳香族氨基酸侧链形成的框架（aromatic cage），负责结合甲基化的赖氨酸。RAG2 的 PHD 结构域对二甲基化的 H3K4 的亲和力远小于三甲基化修饰的 H3K4(Matthews et al., 2007)。染色质免疫共沉淀结果表明，RAG2 在染色体上定位与 H3K4me3 的定位高度重合，提示抗原受体基因正是通过组蛋白 H3K4 的三甲基化修饰来招募 RAG2(Liu et al., 2007)。另外，当 RAG2 的 PHD 结构域结合了甲基化修饰的 H3K4 之后，RAG 复合物的重排能力被激活(Shimazaki et al., 2009)，提示 PHD 结构域与 H3K4Me3 的结合也许伴随着构象的改变。当 PHD 结构域突变导致丧失了螯合锌离子或者降低对 H3K4 的结合能力后，基因重排基本不发生，患者的 T 和 B 淋巴细胞发育停滞，最后导致免疫缺陷(Couedel et al., 2010)。

2. BCR 基因重排

B 淋巴细胞在骨髓中分化成熟，其前体细胞依次经过 ELP 和 CLP 分化到 pro-B 细胞期，启动 Ig 重链(Igh)基因重排，重排成功的细胞进入 pre-B 细胞期，开始 Ig 轻链($Ig\kappa$ 与 $Ig\lambda$)基因重排。Igh 基因位于小鼠第 12 号染色体，全长约 2.5Mb，包括可变区的 152 个 V_H 片段(全长 2Mb)，17~20 个 D_H 片段(长约 60Kb)，4 个 J_H 片段(长约 2Kb)以及恒定区的 8~9 个 C_H 片段(2000Kb)。V_H 基因根据序列保守分析可分为 16 个家族(Johnston et al., 2006; Ye, 2004)(图版ⅧA)。每个 V_H 与 D_H 都拥有各自的启动子，转录方向相同。小鼠 $Ig\kappa$ 位于第 6 号染色体，全长约 3.2Mb，包含了 174 个 V_κ、5 个 J_κ、1 个 C_κ(图版ⅧB)。$Ig\lambda$ 位于小鼠的第 16 号染色体，全长 240Kb，包括了 3~8 个 V_λ，3~5 个 J_λ，2~4 个 C_λ 片段(图版ⅧC)(Lefranc et al., 2005)。

2.1 RSS 序列与 VH 的利用率

RAG1 通过结合 RSS 的九聚体来接近抗原编码基因片段，RAG2 促进 RAG1 在 RSS 的七聚体与编码序列的边界切断 DNA，不同物种间 RSS 的七聚体与九聚体是高度保守的。通过构建 RSS 突变体并检测其重排效率，研究者发现七聚体的前 3 个碱基序列 CAC 对于重排反应是至关重要的，七聚体剩余的 4 个碱基序列与九聚体的序列保守性相对较低(Hesse et al., 1989)。此外，RSS 间隔区序列并不是完全随机的，也存在一定保守性(Ramsden et al., 1994)。一部分 RSS 间隔区序列可能影响编码基因使用效率，通过重排竞争实验，研究者发现某些 VH 家族的重排效率受到 23-RSS 间隔区序列控制(Connor et al., 1995)。同样的调控作用也出现在一部分 $Ig\kappa$ 与 $Tcrb$ 的 V 基因片段中(Posnett et al., 1994; Nadel et al., 1998)。

2.2 转录水平调控V(D)J基因重排效率

Igh 基因簇中分布着大量调控基因片段转录的顺式元件，比如启动子和增强子。在这些启动子或增强子的作用下，部分基因片段在重排发生之前转录产生非编码RNA(Yancopoulos et al., 1985)。当 $D_H \rightarrow J_H$ 重排开始时，*Igh* 的胚系转录率先起始于两个区域：处在内含子内的增强子 Eμ(Iμ 转录本)(Lennon et al., 1985)和距离 J_H 最近的DQ52 的启动子 PDQ52(μ0 转录本)(Thompson et al., 1995)。在 $D_H \rightarrow J_H$ 重排结束之后，将产生新的非编码 Dμ 转录本(Reth et al., 1984)。尽管这些非编码 RNA 的生物学功能还不清楚，但胚系转录水平却和重排效率密切相关。根据抗原受体基因片段的转录水平与重排效率的相关性，研究者们提出了编码基因片段的"易近性"假说：胚系基因的转录促进原本紧密包装的染色质结构变为松散的单个核小体结构，从而使重排激活蛋白质 RAG 容易接近 RSS 并介导重排(Krangel, 2003)。Pro-B 细胞阶段只有 *Igh* 发生转录，*Igh* 基因座的染色质结构开放，而 *Igκ* 与 *Tcrb* 基因座仍处于异染色质状态，所以RAG 更容易接近活化的 *Igh*，从而保证了重排的细胞特异性和阶段特异性。同样的现象也出现在胸腺细胞的 *Tcr* 基因重排过程中，在 DN 胸腺细胞中，*Tcrb* 的 $D_\beta J_\beta$ 基因簇通过胚系转录首先"打开"染色质结构，利于重排激活蛋白的接近，而 *Tcra* 与 *Igh* 基因座的染色质结构则处与非活化的关闭状态(Stanhope-Baker et al., 1996)。体外实验为这个假说提供了直接的证据：将外源纯化的 RAG 复合物与不同阶段的淋巴细胞的细胞核物质共孵育，RAG 复合物优先催化该细胞阶段特异的抗原受体链发生重排(StanhopeBaker et al., 1996)。

基因片段的转录水平升高促进了染色质解聚，然而也有研究表明，在有些情况下转录水平升高是染色质解聚后的结果。基因片段区间的转录(inergenic transcription)也是打开染色质结构的重要手段之一。*Igh* 的 V_H 区包含了大量长约 500bp 的 V_H 片段，这些 V_H 片段之间却隔着 10~20kb 的 intergenic 序列，即使每个 V_H 都在各自启动子的驱动下起始胚系转录，理论上也很难维持在 $V_H \rightarrow DJ_H$ 重排过程中整个 V 区域染色质结构的打开。因此 V_H 片段的胚系转录就很可能是重排时整个 V_H 区染色质结构打开的结果。RT-PCR 和RNA-FISH 的结果显示，在整个 *Igh* 存在反向的基因区间转录(Bolland et al., 2004)，这种反向转录出现于重排之前，随着重排结束而消失(Bolland et al., 2007)。与以往认知的反向转录不同，这种大规模的反向转录并不是为了拮抗正义链的表达，而是为了重塑 V区染色质结构，为重排创造条件(Giallourakis et al., 2010)。在 *Tcra* 基因重排过程中也发现了类似的基因片段区间转录，却是正义链的转录(Abarrategui et al., 2006)。综上所述，无论是正义链转录还是反义链转录，基因片段区间的转录是基因重排时重塑染色质的有效手段。

2.3 表观遗传修饰与V(D)J基因重排

染色体组蛋白的表观遗传修饰被称为组蛋白密码。修饰后的组蛋白作为识别信号招募反式因子，调控染色质的结构与转录活性。组蛋白 H3 与 H4 的乙酰化是最早被发现的与重排相关的组蛋白表观修饰（Chowdhury et al., 2001; Maes et al., 2001; Johnson et al., 2003）。受体基因座的 H3 与 H4 的乙酰化水平升高伴随着转录激活，是有效重排的必要条件之一。相反，当 H3 与 H4 的乙酰化水平下降，则有利于关闭未重排的基因座，以及维持等位排斥。在 pro-B 细胞完成一条重链的重排后，V_H 区的乙酰化水平伴随 IL-7 信号的减弱而被下调，*Igh* 重排被抑制。在 DN 胸腺细胞向 DP 期过渡过程中，*Tcrb* 的乙酰化水平不断降低，*Tcrb* 的重排也被关闭（Tripathi et al., 2002）。

组蛋白的某些甲基化修饰也与染色质结构改变有关。组蛋白 H3K9 的二甲基化是典型的异染色质标记。在 pro-B 细胞中，PAX5 通过降低 V_H 区 H3K9 的甲基化水平而促进 $V_H \rightarrow DJ_H$ 的重排（Morshead et al., 2003）。组蛋白甲基转移酶 G9a 促进抗原受体基因片段的 H3K9 甲基化，从而抑制了胚系转录及重排（Thomas et al., 2008）。另外，组蛋白甲基转移酶 Ezh2 通过促进组蛋白 H3K27 甲基化修饰，同样抑制 V(D)J 基因重排（Su et al., 2003; Mandal et al., 2011）。并非所有的组蛋白甲基化修饰都是促进关闭染色质，比如组蛋白 H3K4 的甲基化促进染色质的活化。在 pro-B 细胞中，组蛋白 H3K4 的二甲基化与三甲基化对 $D_H \rightarrow J_H$ 重排的作用不同（Chakraborty et al., 2009）。H3K4 的二甲基化主要调节染色质结构打开过程（Morshead et al., 2003），而 H3K4 的三甲基化直接参与招募 RAG2 定位到抗原受体基因座（Ji et al., 2010; Morshead et al., 2003）。

DNA 甲基化修饰也是基因活化程度的重要标志。CpG 甲基化水平上升通常抑制基因转录，与 V(D)J 基因重排效率降低相关（Hsieh et al., 1992）。在 pre-B 细胞中，*Igκ* 的 $J_κ$-$C_κ$ 最先发生 DNA 去甲基化并且移位至常染色质，开始重排过程（Fitzsimmons et al., 2007）。在敲除 *Tcrb* 的 Dβ 上游的启动子或者 Cβ 的 3'端增强子之后，*Tcrb* 基因座的 CpG 甲基化水平升高，$D_β1$-$J_β1$ 区的重排受到抑制（Bories et al., 1996; Bouvier et al., 1996）。DNA 的去甲基化则常伴随着重排效率提升（Whitehurst et al., 2000）。

即使染色质由高度聚集的结构变为松散的单核小体结构，被随机缠绕在核小体上的 RSS 也并不是轻易就变成可以被 RAG 复合物剪切的线性状态（Baumann et al., 2003）。核小体的包装给 RSS 的"易近性"设置了最直接的一道空间位阻（Golding et al., 1999）。一些促进核小体重构的因素能为 RAG 与 RSS 的相互结合创造有利条件。如启动子和增强子通过招募 SWI/SNF 重塑复合物来帮助 RAG 启动重排（Osipovich et al., 2009），该复合物通过水解 ATP 来促进核小体的滑动与移位（Patenge et al., 2004），从而逐渐解开缠绕在核小体上的 RSS（Morshead et al., 2003）。组蛋白末端修饰可以直接解开 DNA 与核小体的结合，也可以影响核小体与核小体之间的结合，还可以改变核小体与其他蛋白质的结合。这一系列的修饰导致包装完整的核小体结构不断转变，最终释放出线性 RSS（Kwon et al., 2000; Nightingale et al., 2007），促进重排。

2.4 核定位调控 V(D)J 基因重排

显微镜和电镜技术的发展使得直观了解 V(D)J 基因重排过程成为可能。通过原位荧光杂交技术，研究者们清楚地看到抗原受体基因所处的染色质定位随着淋巴细胞前体的分化进程发生变化。在前体 B 淋巴细胞中，Igh 与 $Ig\kappa$ 所处的染色质定位于细胞核边缘的异染色质区，随着前体细胞分化至 pro-B 细胞阶段，Igh 的一对等位基因迁移至细胞核中心常染色质区，并启动 $D_H \to J_H$ 的转录和重排(Kosak et al., 2002)。当一条等位基因完成重排之后，pro-B 细胞中表达的重链与替代轻链组成 pre-BCR，pre-BCR 信号促进另一条没有发生 $V_H \to DJ_H$ 重排的等位基因移向异染色质区(Roldan et al., 2005)。上述现象再次表明染色质空间定位的变化决定重排的细胞特异性和发育阶段特异性。近年来的研究表明，RAG1 与 ATM 调控重排起始阶段染色质定位的动态变化。在 RAG1 的协助下，抗原受体基因所在的同源染色体首先经历配对的过程。当完成了一条链的重排之后，ATM 会协助另一条链移动至异染色质区，保证基因重排遵循等位排斥原则(Hewitt et al., 2009)。

免疫荧光与荧光原位杂交技术还促进发掘了很多参与重排的关键分子。早在 2000 年，研究人员发现 NBS1 与组蛋白 H2AX 共定位于重排剪切之后形成的 DNA 损伤处(Chen et al., 2000)。后续的研究揭示这一类 DNA 损伤标志分子主要维持重排中间产物 DNA 末端的稳定性，以便于 NHEJ 修复过程顺利进行，保护细胞免受因 DNA 末端移位造成的癌变或者凋亡(Helmink et al., 2011)。

2.5 反式因子调控 V(D)J 基因重排

近年来的 DNA 序列分析逐渐发现 Igh 基因座中分布着各种反式因子(比如 PAX5、E2A、CTCF 和 Rad52 等)的保守结合序列(Ebert et al., 2011)。功能研究证明这些反式因子通过结合 Igh 基因座中的顺式元件调控着重排。E-蛋白家族转录因子通过特异地结合增强子而得名。E2A 是 E-蛋白的典型代表，在 T 和 B 淋巴细胞重排过程中通过羧基端 helix-loop-helix(bHLH) 结构域结合 E-box(CANNTG) 并调控基因表达(Langerak et al., 2001; Zhuang et al., 1994)。已知 Igh 基因组的 Eμ 增强子、$Ig\kappa$ 增强子、Tcrb 增强子、Tcra 增强子，以及 $Ig\kappa$ 的 V 区都含有 E-box 保守序列。E-家族蛋白主要通过调控转录水平来提高受体基因的"易近性"从而促进重排。IL-7 受体在 pro-B 细胞表面高表达，与基质细胞分泌的 IL-7 作用后转导信号。E2A 与 EBF 共同响应 IL-7 信号(Kee et al., 1998)，通过结合 Igh 的增强子促进 D_H 与 J_H 区转录，提高 D_H-J_H 重排效率(Romanow et al., 2000)。

在 $V_H \to DJ_H$ 重排阶段,持续的 IL-7 信号驱动 E2A 和 EBF 诱导 PAX5 表达(Singh et al., 2007)。PAX5 是调控 Igh 的 V 区重排的关键因子(Cobaleda et al., 2007)。作为一个功能多样的转录因子，PAX5 的羧基端同时包含有转录激活结构域和转录抑制结构域。在 PAX5 缺失的 pro-B 细胞中，远端 V_H 片段附近的组蛋白是高度乙酰化的，转录水平也和野生型相当，然而 V 区的重排受到阻碍(Hesslein et al., 2003)，说明除了调控转录，PAX5 还通过其他方式促进 V_H 的重排。据统计，94%的 Igh 的 V_H 编码区包含 PAX5 的保守结合序列(Guo et al., 2011)。在 $V_H \to DJ_H$ 重排时，PAX5 介导远端 V_H 片段通过成环和聚集等拓扑结构的改变促进其在三维空间上接近 DJ_H(Sayegh et al., 2005)。一旦重排成功，染色质重新解聚，避免二次重排发生(Roldan et al., 2005)。另外一个功能类似于 PAX5 的反式因子 YY1 结合 V_H 片段基因区间的增强子，通过介导增强子与启动子的结合来促进远端 V_H 的逆向转录，并且改变染色质的拓扑结构(Verma-Gaur et al., 2012)。尽管 YY1 与 PAX5 在功能上重叠，二者的结合位点和作用方式却是互相独立的。除了重塑 Igh 的 V 区染色质结构，PAX5 还能够结合 RAG1 从而帮助 RAG1 定位到 V 区基因片段上(Zhang et al., 2006)。

STAT5 作为 IL-7 信号的重要功能执行者，通过结合 Oct1 被招募至 V_H 区的启动子上，提高组蛋白 H3 乙酰化水平，作为共激活因子促进远端 V_H 胚系转录，从而提高远端 V_H 重排效率(Bertolino et al., 2005)。STAT5 不仅在 pro-B 阶段促进 $V_H \rightarrow DJ_H$ 的重排，还招募 Ezh2 到 Igκ 的增强子 $iE_κ$ 上，通过提高组蛋白 H3K27 的甲基化水平来关闭 $J_κ$，确保轻链编码基因在 pro-B 细胞期的沉默状态(Mandal et al., 2011; Malin et al., 2010)。

当 Igh 重排完成之后，IL-7 信号继续促进细胞进入增殖期，此后 RAG 蛋白被降解 (Ochiai et al., 2012)。Pre-BCR 在细胞表面表达后通过传导信号继续调控 pre-B 细胞的分化与轻链的重排。Pre-BCR 信号上调 IRF4 和 IRF8 的表达，IRF4/IRF8 能够抑制细胞增殖，并重新上调 RAG 表达。IRF4 与 IRF8 一方面通过结合 Igk 的 3'增强子，激活 Igk 的胚系转录；另一方面 IRF4 通过上调趋化因子受体 CXCR4 的表达促使 pre-B 细胞迁移远离表达 IL-7 的基质细胞，从而下调 IL-7 信号，解除了 STAT5 对 Igκ 的抑制(Johnson et al., 2008)。

3. TCR 基因重排

T 淋巴细胞抗原受体(TCR)分为两种，即 TCRαβ 受体和 TCRγδ 受体。小鼠的 Tcrb 位于第 6 号染色体，全长约 700kb，5'包含了 $34V_β$，长约 425kb。V 区的 3'下游存在两簇 $D_β$-$J_β$-$C_β$，长约 25kb。在整个基因的 3'端有一个与 5'端 $V_β$ 反向的 $V_β14$。通常 $V_β \rightarrow DJ_β$ 的重排产生的是环出的信号连接产物，而 $V_β14 \rightarrow DJ_β$ 重排产生的是倒转的信号连接产物(图版ⅧD)。Tcrd 散布在第 14 号染色体的 Tcra 基因簇内，全长 270kb，包括 6 个 $V_δ$，10 个与 Tcra 共用的 $V_{α/δ}$，2 个 $D_δ$，2 个 $J_δ$，1 个 $C_δ$(图版ⅧF)。Tcrg 位于第 13 号染色体，全长 200kb，包括 7 个 $V_γ$，4 个 $J_γ$，4 个 $C_γ$(图版ⅧG)。(Glusman et al., 2001; Bosc et al., 2003)。90%~95%外周 T 淋巴细胞表达 TCRαβ 受体，以此识别由 MHC 递呈的抗原并介导免疫应答(Davis et al., 1988)，其余的 T 淋巴细胞表达 TCRγδ 受体。本文将着重介绍 TCRαβ 链的重排。

3.1 顺式元件与 Tcrb 基因重排

T 和 B 淋巴细胞均由前体细胞 CLP 分化而来，CLP 从骨髓迁移至胸腺中经历 DN 期的 β 链基因重排和 DP 期的 α 链基因重排才能够进一步分化成熟，最后迁移至外周发挥免疫功能。与 Igh 重排相似，Tcrb 的重排效率也与胚系转录水平相关。$C_β2$ 与 $V_β14$ 之间

的增强子 E_β 控制着 D_β-J_β 区的转录水平(Krimpenfort et al., 1988)。在小鼠中敲除 E_β 之后，$D_\beta \rightarrow J_\beta$ 重排效率下降，胸腺细胞发育被阻滞(Bories et al., 1996; Bouvier et al., 1996)。此外，在 E_β 敲除小鼠的重排产物中发现了一些异常重排的发生，比如 $D_\beta 1 \rightarrow D_\beta 2$ 的重排，DP 期胸腺细胞中 $V_\beta 14$ 与 D_β 的剪切。因此 E_β 除了维持重排效率，还可能参与了 RSS 的配对过程或者中间产物的修复过程(Hempel et al., 1998; Ryu et al., 2003)。在 $E_\beta 1$-$E_\beta 7$ 中包含多个转录因子的保守结合序列，比如 GATA、ATF、CREB、bHLH、ETS 及 RUNX 家族(Takeda et al., 1990)，这些转录因子通过结合 E_β 参与调控 *Tcrb* 的重排。位于 $D_\beta 1$ 上游的启动子 $PD_\beta 1$ 通过启动转录来促进 $D_\beta 1$ 的重排，敲除 $PD_\beta 1$ 则阻止 $D_\beta 1 \rightarrow J_\beta$ 的重排，然而 $D_\beta 2$ 的重排仍能顺利进行，因为位于 $D_\beta 2$ 下游的序列调控了 $D_\beta 2$ 转录(Sikes et al., 1998)。

3.2 Beyond 12/23 原则

Tcrb 重排通过两个有序的步骤完成：首先发生的是 $D_\beta \rightarrow J_\beta$ 的重排，然后才发生 $V_\beta \rightarrow DJ_\beta$ 的重排，获得完整的可变区编码序列 VDJ_β。这种重排顺序使得 V_β 的 3' 23-RSS 与 J_β 的 5' 12-RSS 基本不发生直接重排，确保重排不会跳过 D_β，从而获得重排产物高度多样性。这种顺序性重排看似超出了重排的 12/23 原则，因此被称为 beyond 12/23 原则(B12/13)。但是为什么不能首先进行 $V_\beta \rightarrow D_\beta$ 重排，然后再完成 $VD \rightarrow J_\beta$ 重排来获得 VDJ_β 呢？经过长期的观察总结，研究者发现顺序性重排总是与等位排斥现象相伴，如 *Tcrb* 和 *Igh* 的 V(D)J 基因重排具有顺序性，同样遵循等位排斥规律；而 *Tcrd* 的重排没有顺序性，同样也没有等位排斥现象。因此推测重排的顺序性可能与等位排斥有关(Jackson et al., 2006; Khor et al., 2005)。早期研究结果表明，B12/23 原则的建立主要依靠 D_β 两侧 RSS 序列。一系列定点突变小鼠模型为阐释 B12/23 原则奠定了基础。在小鼠的 $D_\beta 2$-$J_\beta 2$-$C_\beta 2$ 被敲除的基础上，首先将 $D_\beta 1$ 的 3'端 23-RSS 敲除，结果 $D_\beta 1 \rightarrow J_\beta 1$ 的重排被阻断，但 $V_\beta \rightarrow D_\beta 1$ 的重排却依旧能够进行，说明 B12/23 原则不依赖于 $D_\beta \rightarrow J_\beta$ 的重排(Sleckman et al., 2000)。当整个 $D_\beta 1$ 连带着两侧的 RSS 序列都被敲除后，V_β 依旧无法启动直接到 $J_\beta 1$ 的重排，提示 V_β 只能特异地匹配 D_β 的 12-RSS(Bassing et al., 2000)。当 $V_\beta 14$ 的 23-RSS 被 D_β 的 23-RSS 代替，且 $D_\beta 1$ 片段也被敲除的情况下，$V_\beta 14 \rightarrow J_\beta 1$ 的重排显著升高，B12/23 原则被打破(Wu et al., 2003)。这些说明 V_β 在 *Tcrb* 重排起始时已经在空间上接近 J_β，然而 D_β 的 23-RSS 却保持着足够的竞争力抢先完成与 J_β 的重排，维持了 B12/23 原则。上述现象已经充分说明了 D_β 及两侧的 RSS 在维持 B12/23 原则中的关键地位，那么这种顺序性重排是如何实现的？Wang 等(2008)的工作揭示 c-Fos/AP-1 调控了 *Tcrb* 的重排顺序。c-Fos 是一个多功能的转录因子，可以与 Jun 家族蛋白组成 AP-1 异源二聚体(Chen et al., 1998)。序列分析及 EMSA 实验证明，在 D_β 基因片段 3'的 23-RSS 中有一个高度保守的 AP-1 结合位点。c-Fos/AP-1 可以与重排激活蛋白 RAG 相互作用，促进招募 RAG 到含有 AP-1 位点的 D_β 23-RSS，从而促进了 $D_\beta \rightarrow J_\beta$ 重排，保证了 $D_\beta \rightarrow J_\beta$ 重排的优先性和 *Tcrb* 基因重排的顺序性。在 *c-Fos* 基因敲除的小鼠中，*Tcrb* 的重排效率严重下降，重排的顺序也严重错乱。

3.3 Pre-TCR 信号调控

TCR β 链重排成功后，表达产物与 pre-Tα 形成异源二聚体 Pre-TCR。Pre-TCR 信号将促进 DN 胸腺细胞分化至 DP 期；如果重排失败或者 Pre-TCR 信号异常，DN 胸腺细胞将走向凋亡，这一过程被称为 β 选择。

Notch 信号和 pre-TCR 信号是 DN 胸腺细胞完成 β 选择的必需信号。Notch 信号主要通过 PI3K-AKT 及 C-Myc 调控细胞的增殖、存活及代谢（Juntilla et al., 2008）。除此之外，Notch 信号促使转录因子 CSL 结合到 pre-Tα 编码基因的上游启动子上，促进 pre-Tα 的表达（Yashiro-Ohtani et al., 2010）。pre-Tα 结构类似于 TCRα 链，在 Tcra 的重排开始之前，pre-Tα 代替 TCRα 链与 TCRβ 组成 pre-TCR。Pre-TCR 同 CD3 组成复合物，通过 Lck、ZAP70 及 SLP76 等分子传递信号（von Boehmer, 2005）。

当 DN 胸腺细胞接受到 pre-TCR 信号后，促使 DN3 胸腺细胞进入细胞周期，RAG2 响应周期变化从 S 期开始迅速降解，重排反应关闭。没有重排的 $V_β$ 基因座在这一时期从细胞核中央的常染色区移位到边缘异染色质区（Michie et al., 2002）；同时，Tcrb 等位基因不再重排，实现等位排斥。在 pre-Tα 敲除的小鼠体内，胸腺细胞数量只有野生型对照鼠的 1/10，绝大多数的胸腺细胞被阻滞在 DN3 细胞期。有少量的细胞可以在缺少 pre-TCR 的条件下分化至 DP 期（Fehling et al., 1995），一个原因可能是 TCRγδ 具备类似于 pre-TCR 的功能，能够促使 DN 细胞分化至 DP；另外一个原因可能是 DP 细胞得益于提前重排的 Tcra，TCRαβ 可以部分代替 pre-TCR 行使功能（Bueret al., 1997）。有意思的是，在这部分残存的 DP 细胞中，Tcrb 两条等位基因都发生 $V_β→DJ_β$ 重排的细胞比例显著上升（60%），而且同时表达两条不同 β 链的细胞比例大幅增加（30%）（Krotkova et al., 1997; Aifantis et al., 1997）。依据上述现象，Pre-TCR 信号的生物学功能可以分为 3 个部分：①促进 β 链重排成功的细胞进入细胞周期；②促进细胞分化成熟；③抑制 Tcrb 等位基因的重排，从而实现等位排斥。

3.4 *Tcra* 基因重排

Tcrb 基因重排在 DN 阶段完成，Tcra 基因重排则在 DP 胸腺细胞中进行。小鼠的 Tcra 基因位于第 14 号染色体，全长约 1700kb，包含了 98 个 $V_α$ 片段（约 1200kb），分属 23 个家族；60 个 $J_α$ 片段（64kb）；一个 $C_α$ 片段（Warr et al., 2003）。Tcra 基因簇并不包含 D 片段，按照 $V_α$ 与 $J_α$ 彼此之间的距离，可以粗略的将 $V_α$ 与 $J_α$ 分别划分为远端、中间与近端三部分区域（图版ⅧE）。Tcra 基因重排具有明显的区域偏好性：近端的 $V_α$ 更倾向于选择近端的 $J_α$ 进行重排；而远端的 $V_α$ 则更容易同远端 $J_α$ 重排（Roth et al., 1991; Pasqual et al., 2002）。这种选择偏好性可能与染色质结构的开放先后顺序有关，位于 $J_α$ 上游的启动子 TEA（T early α）最初响应位于 $J_α$ 下游的增强子 $E_α$（Hernandez-Munain et al., 1999），通过启动转录非编码 RNA 从近端 $J_α$ 片段打开染色质结构，促进近端 $V_α$ 与近端 $J_α$ 的重排（Villey et al., 1996）。同时 TEA 的转录也抑制了远端 $J_α$ 活化（Abarrategui et al., 2007）。在 Tcra 重排过程中，V 区染色质拓扑结构发生改变，使得远端 $V_α$ 通过染色质成环或者聚集在三维结构上更靠近远端 $J_α$ 区（Skok et al., 2007）。

4. 抗原受体多样性的产生

4.1 组合多样性

胚系基因簇中众多的 V、D、J 片段通过重新排列组合构成了抗原受体多样性的来源之一：组合多样性。值得注意的是，在重排过程中并非每个片段的使用几率都是相同的，而且两条抗原受体链不一定能够配对成功，所以在实际情况中组合多样性要略低于计算数。

4.2 连接多样性

基因重排的中间产物编码末端与信号末端都通过 NHEJ 修复途径被重新修复连接，然而由于二者的末端结构不同，连接的准确性也不同(图版Ⅸ图 1)。通常情况下，平头的信号末端可以被连接酶准确地修复(Roth et al., 1993)，连接点很少发生碱基的掺入或者丢失。而发卡结构的编码末端却无法作为连接酶的底物(Roth et al., 1992)。发卡结构首先需要内切酶 Artemis 随机剪切产生带有回文结构(palindrome)的黏性 DNA 末端(Ma et al., 2002)。这些回文序列如果经过修复依然保留在连接点处，则被称为 P-核苷酸。黏性的 DNA 末端接下来可能接受 DNA 外切酶、末端转移酶 TdT 的进一步处理，发生碱基切除或者增加之后再被修复。通过 TdT 引入的非模板编码核苷酸序列被称为 N-核苷酸(Komori et al., 1993)。碱基的随机切除与引入的 P/N-核苷酸是抗原受体多样性产生的另外一个重要原因(Cabaniols et al., 2001)。

4.3 高频突变

在成熟 B 淋巴细胞接受抗原刺激之后的二次免疫应答阶段，胞嘧啶核苷脱氨酶(AID)选择性地促进免疫球蛋白可变区编码基因(尤其是互补决定区 CDR)中胞嘧啶的去氨基化，突变形成的尿嘧啶通过错配和碱基剪切被修复(Liu et al., 2008)。这种突变与修复导致免疫球蛋白与抗原结合的特异性发生改变，那些与抗原亲和力更高的 B 淋巴细胞克隆被选择存活下来，并且进一步向记忆性 B 淋巴细胞和浆细胞分化。B 淋巴细胞中突变频率大约是正常体细胞的 10^6，该过程主要发生在外周淋巴器官的生发中心(Odegard et al., 2006)。

4.4 受体修正

一直以来研究者普遍认为，V(D)J 基因重排伴随 T 和 B 淋巴细胞的早期分化，是只发生在胸腺或骨髓中的事件。然而这个观点已发生改变，成熟的 T 和 B 淋巴细胞的受体基因还可能被第二次编辑，被称为受体修正（editing）。当 T 淋巴细胞在外周淋巴系统中与抗原递呈细胞作用后，如果 TCR 与自身抗原结合过强，一部分细胞会下调抗原受体的表达，通过上调 CXCR5 迁移到外周淋巴器官的生发中心并进入 G_0/G_1 期。此后 CD40 信号可能诱导细胞重新表达 RAG1、RAG2 及 TdT 等一系列与重排相关的分子。重排结束后，T 淋巴细胞会高表达新的抗原受体并继续参与淋巴循环。受体修正发生的几率与生物体的遗传背景等有密切的关系，受体修正是完善初始免疫系统有效而常用的方法（Hale et al., 2010）。

另外，最新的研究显示，在小鼠的肠固有层驻有处于 pro-B 及 pre-B 阶段的未成熟 B 淋巴细胞，这些细胞发生与在骨髓中 pro-B、pre-B 细胞相对应的重排过程。这个重排过程产生的免疫球蛋白谱与肠道菌群密切相关，轻链种类也受到肠道环境影响（Wesemann et al., 2013）。

5. V(D)J 基因重排与疾病

5.1 重症联合免疫缺陷

重症联合免疫缺陷（SCID）是一种细胞免疫和体液免疫同时缺陷的严重免疫疾病。患者体内缺少成熟的 T 和 B 淋巴细胞，因此无法抵御外界病原的入侵，主要表现为呼吸系统和肠道的反复感染。目前可以通过造血干细胞移植缓解和治疗该疾病，但也存在由于组织不相容造成的抗宿主病的风险。

据统计大约有 30%的原发性重症联合免疫缺陷是由 V(D)J 重排缺陷所导致的（Fischer, 2004）。其中一部分缺陷是由于 *Rag1* 或 *Rag2* 基因突变造成酶活性缺失所引起的（Villa et al., 2001）；另外一部分缺陷则是由 NHEJ 修复因子突变导致重排连接过程受阻而造成的。NHEJ 修复途径缺陷导致的另外一种症状是患者的造血系统与皮肤细胞对电离辐射十分敏感，因此也被称为 RS-SCID。通过对原发性 SCID 患者的基因缺失分析，使我们更加全面地了解 NHEJ 途径。比如 Artemis 与 Cernonnus 的发现都源自于对 SCID 患者基因缺陷的分析（Moshous et al., 2001; Buck et al., 2006）。

5.2 Omenn 综合症（OS）

Omenn 综合症属于一种原发性免疫缺陷，Omenn 于 1965 年首先报道了此免疫缺陷，并因此而得名（Omenn, 1965）。OS 患者出生时生命体征基本正常，然而婴儿通常在出生 3 周之后开始发病，治愈率低下。数十年的观察及研究表明 OS 患者的主要表现为红皮病、嗜酸性粒细胞增多、肝脾变大、淋巴结肿大、免疫球蛋白 IgE 过高、单克隆 T 淋巴细胞异常扩增及淋巴细胞无能（de Villartay, 2009）。V(D)J 基因重排失败也是导致 Omenn

综合症的主要原因之一(Simkus et al., 2007)。在 Omenn 综合症患者中，淋巴细胞早期发育受阻，只有极少淋巴细胞分化成熟，但是成熟的淋巴细胞常常过度活化，其调控机制一直不清楚。

5.3 RAG 引发染色体移位

V(D)J 基因重排过程中产生的 DNA 双链断裂是一种潜在的危险，RAG 引发的染色体移位是淋巴系统癌变的原因之一(Tycko et al., 1990)。基因组中存在一些类似于 RSS 的碱基序列，被称为"cryptic RSS"(Lewis et al., 1997a)。在某些 T 或 B 淋巴细胞性白血病中，发生染色体移位的基因附近都存在"cryptic RSS"，而且移位的染色体交界还往往留有 TdT 作用后产生的 N-核苷酸，说明 RAG 在错误的位置催化了重排反应(Zhang et al., 2008)。然而这种由于 RSS 识别错误导致细胞恶化发生的几率比较低。大多数移位导致的癌变病例中，染色体移位是由 NHEJ 修复途径中某些因子的突变而造成的，RAG 剪切形成的 DNA 双链断裂可能采取了出错几率较高的修复途径从而造成染色体移位 (Jhappan et al., 1997; Li et al., 1998; Zhu et al., 2002)。除了以上两种机制造成的染色体移位，RAG 与信号末端组成的 SEC 也有可能通过 RAG1 的"转座活性"将信号末端重新整合进入基因组，造成插入突变或者染色体移位(Hiom et al., 1998)。尽管 RAG1 在体外可以有效催化转座反应(Agrawal et al., 1998)，但是在体内，这种转座方式却非常罕见。据估计，在 pre-B 细胞中，大约 50 000 次 VJ 重排反应才可能出现一次转座反应(Reddy et al., 2006)。细胞采取多层次调控机制防止发生这种转座(Tsai et al., 2003)，比如 RAG2 通过非核心区稳定 SEC 复合物从而抑制转座发生(Elkin et al., 2003)。

RAG 通过重排反应引起染色体移位导致淋巴细胞癌变的原因包括以下两个方面：某些转录活性极高的 *Ig* 或者 *Tcr* 的启动子转而调控原癌基因的表达，结果导致细胞的异常分化、增值或存活；染色体移位可能融合产生了嵌合型原癌基因，激活了原癌基因表达产物的下游信号通路。

6. V(D)J 基因重排的起源与演化

通过 V(D)J 基因重排建立的适应性免疫系统最早出现在软骨鱼体内。无颌生物体内既没有 RAG 也不没有抗原受体编码基因，提示 RAG 介导的 V(D)J 基因重排形成于从无颌鱼过渡到软骨鱼的阶段(Rast et al., 1998)(图版Ⅸ图 2)。

6.1 RAG 的起源

自从发现 RAG 以来，RAG 的起源就一直受到免疫学家的广泛关注。早在 20 世纪 70 年代末，通过分析免疫球蛋白轻链编码基因的重排方式，Sakano 等(1979)最先提出重排可能起源于转座子的假说。其后有研究者提出 RAG1 起源于 DNA 转座酶, 在不断演变过程中依然保留着转座活性(Lewis et al., 1997b)。与此相对应, RSS 序列类似于转座子两侧的反向重复序列; 而 V(D)J 基因重排的剪切与连接过程十分类似于转座过程(Kapitonov et al., 2005; Lu et al., 2006; Grundy et al., 2007)。在海胆基因组内发现了最古老的 RAG1 与 RAG2 类似编码基因序列, 并且该序列两侧也存在类似转座酶识别的重复序列。更为重要的是, 体外生化实验表明 RAG1 具转座酶活性(Agrawal et al., 1998)。然而上述假设却存在两点缺陷: 首先, 多数转座子家族的转座酶并没有类似 RAG2 的成员存在; 其次, 海胆属于低等的棘皮生物, 在进化树上更接近软骨鱼的原索动物文昌鱼及无颌鱼类七鳃鳗、八目鳗的基因组内却没有发现这种类似 RAG 的转座酶基因。是因为进化过程中物种灭绝导致进化上的空白还是因为 RAG 另有起源仍然不得而知。

于是科学家(Fugmann et al., 2006)提出, RAG1 与 RAG2 也许并非共同起源, RAG1 可能意外插入到 RAG2 编码基因附近。RAG1 的前体可能来自于 DNA 病毒内切酶, 病毒编码一个类似于 RAG1 的内切酶, 通过侵入生物体后将自身 DNA 编码序列留在了脊椎生物祖先的基因组内(Dreyfus, 2009)。那么功能性的 RAG1 最早出现在哪种生物体内? 研究者将目光转向头索生物文昌鱼。文昌鱼是由无脊椎向脊椎动物过渡的低等动物, 是脊椎动物最简单模型的代表。Zhang 等首先鉴定到文昌鱼基因组中与脊椎动物 RAG1 基因同源的 DNA 片段 bfRAG1L。虽然其表达的产物 bfRAG1L 只有脊椎动物 RAG1 的 1/3 大小, 但包含有 RAG1 的核心结构域和一个病毒相关的 II 型核酸酶功能域 DXN(D/E)XK, 能够降解 DNA 和 RNA。更为重要的是, bfRAG1L 具备 RAG1 的基本属性, 包括与 RAG2 相互作用以及定位于核中。将 bfRAG1L 改造成小鼠核心 RAG1 类似物 Ch-bfRAG1L 后, Ch-bfRAG1L 能够识别抗原受体基因并能在 RAG1 缺失的小鼠中介导基因重排, 获得 T 和 B 淋巴细胞。证明文昌鱼的 bfRAG1L 在功能上类似于脊椎动物 RAG1 的核心结构域, 为 RAG1 的起源提供了关键的分子证据, 并为 RAG1 基因的起源与进化提供了新的观点。

6.2 抗原受体分子进化

从发现 V(D)J 基因重排到揭示其调控机制, 大部分研究工作聚焦在人类和小鼠的免疫系统中。如前所说, 目前已知适应性免疫最早出现于软骨鱼, 软骨鱼的受体基因重排模式与结果让我们更好地了解早期适应性免疫系统(Greenberg et al., 1995)。和在高等哺乳动物中的发现类似, 早期的适应性免疫系统也是通过 T 淋巴细胞表面抗原受体(αβ 和 γδ)和 B 淋巴细胞分泌的免疫球蛋白(IgM 与 IgW)行使功能。然而除此之外, 低等有颌脊椎生物体内还存在一些特有的类似于抗原受体的分子, 比如鲨鱼的 Ig-NAR 和 NAR-TCR, 以及有袋类哺乳动物的 TCRu。Ig-NAR 由两条免疫球蛋白重链组成, 没有轻链, 每条重链包含 5 个恒定区形成的一个较长的 CDR3 区(Criscitiello et al., 2006; Parra et al., 2007)。比起免疫球蛋白, 这种分子结构更接近于进化上比软骨鱼稍低等的原索生物七鳃鳗与八目鳗体内的抗原受体分子 VLR。因此 Ig-NAR 被认为是从 VLR 到 BCR 进化的中间态。在骆驼科生物体内也存在 Ig-NAR, 然而此时的分子结构已经更接近免疫球蛋白，说明抗原受体分子存在趋同进化过程。

6.3 V(D)J基因重排模式的进化

尽管RAG剪切RSS的过程在各物种中高度相似，但软骨鱼的抗原受体基因片段的排列方式及重排模式与高等哺乳动物相比差别很大。软骨鱼免疫球蛋白重链可变区的四个编码基因片段以 V_H-D_H1-D_H2-J_H 的方式排列，全长大约2kb，重排过程依旧遵循12/23原则，形成VDDJ重排产物(Flajnik, 2002)。在软骨鱼中大约有15~200个相似的基因簇散布在基因组内，检测发现重排产物均来自同簇基因片段，并不发生基因簇之间的重排，因此染色质不需要经历重塑过程(Lee et al., 2008; Anderson et al., 1994)。同时，V(D)J基因重排没有先后顺序限制，V→D→J的重排是同时发生的(Eason et al., 2004)。每个B淋巴细胞依然只表达一个抗原受体分子，也就是说等位排斥原则在进化早期已经建立(Malecek et al., 2008)。

7. 展　　望

V(D)J基因重排已经历了5亿年的演化，重排过程的多层次精密调控已逐步建立并不断完善。重排等位排斥原则确保"一个淋巴细胞克隆只表达一种抗原受体分子"，为特异性免疫识别和免疫记忆提供了分子基础，使得适应性免疫成为可能。然而等位排斥是如何建立和维持的仍然不清楚。数十年的研究表明，淋巴细胞通过非同步复制和不同核定位来促进一条抗原受体等位基因率先启动重排，重排成功的抗原受体通过转导反馈信号关闭另一条链的重排。两条等位基因如何能够分别维持常、异染色质的状态？淋巴细胞是随机选择了等位基因中的一条来起始重排吗？即使重排成功，细胞又如何保证同一条基因之内不会发生第二次重排？这些科学问题依旧有待于我们进一步的探索和研究(Brady et al., 2010)。

由于受到技术方法的限制，我们对RAG-RSS装配过程的认知大都来自体外生化实验。目前还没有直接的证据证明在体内V(D)J基因重排过程中，RAG与RSS确实形成SC、PC及CSC这些复合物。RAG在不同的复合物中是否存在构像变化？这些构像变化如何影响其复杂的生物学功能？在体内这些复合物的组装是否需要依赖于其他因素？这些问题需要更多分子层面的技术创新来提供答案。

不少研究表明RAG参与了一部分淋巴瘤的染色体移位过程。如果"剪切"与"连接"两步反应中间没有顺利衔接，染色体移位的几率将大幅升高。RAG不仅仅是普通的限制性内切酶，在重排反应两个阶段的转换过程中，RAG蛋白也许参与维持DNA双链断裂的稳定性，确保DNA末端被控制在固定的范围内被迅速修复。目前已知RAG2的非核心区在维持基因组稳定性方面发挥重要作用，但RAG1在这个过程扮演怎样的角色？相比较RAG2的非核心区功能的研究，RAG1的非核心区的功能仍然有待深入探索。

参 考 文 献

Abarrategui I, Krangel M S 2006. Regulation of T cell receptor-alpha gene recombination by transcription. Nat Immunol, 7: 1109-1115.

Abarrategui I, Krangel M S. 2007. Noncoding transcription controls downstream promoters to regulate T-cell receptor alpha recombination. Embo J, 26: 4380-4390.

Agrawal A, Eastman Q M, Schatz D G. 1998. Transposition mediated by RAG1 and RAG2 and its implications for the evolution of the immune system. Nature, 394: 744-751.

Aidinis V, Dias D C, Gomez C A, et al. 2000. Definition of minimal domains of interaction within the recombination-activating genes 1 and 2 recombinase complex. J Immunol, 164: 5826-5832.

Aifantis I, Buer J, von Boehmer H, et al. 1997. Essential role of the pre-T cell receptor in allelic exclusion of the T cell receptor beta locus (vol 7, pg 601, 1997). Immunity, 7: U9-U9.

Akira S, Okazaki K, Sakano H. 1987. Two pairs of recombination signals are sufficient to cause immunoglobulin V-(D)-J joining. Science, 238: 1134-1138.

Anderson M, Amemiya C, Luer C, et al. 1994. Complete genomic sequence and patterns of transcription of a member of an unusual family of closely related, chromosomally dispersed Ig gene clusters in Raja. Int Immunol, 6: 1661-1670.

Bassing C H, Alt F W, Hughes M M, et al. 2000. Recombination signal sequences restrict chromosomal V(D)J recombination beyond the 12/23 rule. Nature, 405: 583-586.

Bassing C H, Swat W, Alt F W. 2002. The mechanism and regulation of chromosomal V(D)J recombination. Cell, 109: S45-S55.

Baumann M, Mamais A, McBlane F, et al. 2003. Regulation of V(D)J recombination by nucleosome positioning at recombination signal sequences. Embo J, 22: 5197-5207.

Bertolino E, Reddy K, Medina K L, et al. 2005. Regulation of interleukin 7-dependent immunoglobulin heavy-chain variable gene rearrangements by transcription factor STAT5. Nat Immunol, 6: 836-843.

Blair J E, Hedge S B. 2005. Molecular phylogeny and divergence times of deuterostome animals. Molecular biology and evolution, 22: 2275-2284.

Bolland D J, Wood A L, Afshar R, et al. 2007. Antisense intergenic transcription precedes Igh D-to-J recombination and is controlled by the intronic enhancer Emu. Mol Cell Biol, 27: 5523-5533.

Bolland D J, Wood A L, Johnston C M, et al. 2004. Antisense intergenic transcription in V(D)J recombination. Nature Immunology, 5: 630-637.

Bories J C, Demengeot J, Davidson L, et al. 1996. Gene-targeted deletion and replacement mutations of the T-cell receptor beta-chain enhancer: The role of enhancer elements in controlling V(D)J recombination accessibility. Proceedings of the National Academy of Sciences of the United States of America, 93: 7871-7876.

Bosc N, Lefranc M P. 2003. The mouse (Mus musculus) T cell receptor alpha (TRA) and delta (TRD) variable genes. Dev Comp Immunol, 27: 465-497.

Bouvier G, Watrin F, Naspetti M, et al. 1996. Deletion of the mouse T-cell receptor beta gene enhancer blocks alpha beta T-cell development. Proceedings of the National Academy of Sciences of the United States of America, 93: 7877-7881.

Brady B L, Steinel N C, Bassing C H. 2010. Antigen Receptor Allelic Exclusion: An Update and Reappraisal. J Immunol, 185: 3801-3808.

Buck D, Malivert L, de Chasseval P, et al. 2006. Cernunnos, a novel nonhomologous end-joining factor, is mutated in human immunodeficiency with microcephaly. Cell, 124: 287-299.

Buer J, Aifantis I, DiSanto J P, et al. 1997. Role of different T cell receptors in the development of pre-T cells. J Exp Med, 185: 1541-1547.

Cabaniols J P, Fazilleau N, Casrouge A, et al. 2001. Most alpha/beta T cell receptor diversity is due to terminal deoxynucleotidyl transferase. J Exp Med, 194: 1385-1390.

Callebaut I, Mornon J P. 1998. The V(D)J recombination activating protein RAG2 consists of a six-bladed propeller and a PHD fingerlike domain, as revealed by sequence analysis. Cell Mol Life Sci, 54: 880-891.

Chakraborty T, Perlot T, Subrahmanyam R, et al. 2009. A 220-nucleotide deletion of the intronic enhancer reveals an epigenetic hierarchy in immunoglobulin heavy chain locus activation. J Exp Med, 206: 1019-1027.

Chen H T, Bhandoola A, Difilippantonio M J, et al. 2000. Response to RAG-mediated VDJ cleavage by NBS1 and gamma-H2AX. Science, 290: 1962-1965.

Chen L, Glover J N M, Hogan P G, et al. 1998. Structure of the DNA binding domains from NFAT, Fos and Jun bound specifically to DNA. Nature, 392: 42-48.

Chowdhury D, Sen R. 2001. Stepwise activation of the immunoglobulin mu heavy chain gene locus. Embo J, 20: 6394-6403.

Cobaleda C, Schebesta A, Delogu A, et al. 2007. Pax5: the guardian of B cell identity and function. Nature Immunology, 8: 463-470.

Connor A M, Fanning L J, Celler J W, et al. 1995. Mouse VH7183 recombination signal sequences mediate recombination more frequently than those of VHJ558. J Immunol, 155: 5268-5272.

Couedel C, Roman C, Jones A, et al. 2010. Analysis of mutations from SCID and Omenn syndrome patients reveals the central role of the Rag2 PHD domain in regulating V(D)J recombination. J Clin Invest, 120: 1337-1344.

Criscitiello M F, Saltis M, Flajnik M F. 2006. An evolutionarily mobile antigen receptor variable region gene: doubly rearranging NAR-TcR genes in sharks. P Natl Acad Sci USA, 103: 5036-5041.

Curry J D, Geier J K, Schlissel M S. 2005. Single-strand recombination signal sequence nicks in vivo: Evidence for a capture model of synapsis. Nature Immunology, 6: 1272-1279.

Davis M M, Bjorkman P J. 1988. T-Cell Antigen Receptor Genes and T-Cell Recognition. Nature, 334: 395-402.

De P, Rodgers K K. 2004. Putting the pieces together: identification and characterization of structural domains in the V(D)J recombination protein RAG1. Immunol Rev, 200: 70-82.

de Villartay J P. 2009. V(D)J Recombination Deficiencies. V(D)J Recombination, 650: 46-58.

Deriano L, Chaumeil J, Coussens M, et al. 2011. The RAG2 C terminus suppresses genomic instability and lymphomagenesis. Nature, 471: 119-123.

Dreyfus D H. 2009. Immune System: Success Owed to a Virus? Science, 325: 392-393.

Dudley D D, Sekiguchi J, Zhu C M, et al. 2003. Impaired V(D)J recombination and lymphocyte development in core RAG1-expressing mice. J Exp Med, 198: 1439-1450.

Eason D D, Litman R T, Luer CA, et al. 2004. Expression of individual immunoglobulin genes occurs in an unusual system consisting of multiple independent loci. Eur J Immunol, 34: 2551-2558.

Ebert A, McManus S, Tagoh H, et al. 2011. The Distal V-H Gene Cluster of the Igh Locus Contains Distinct Regulatory Elements with Pax5 Transcription Factor-Dependent Activity in Pro-B Cells. Immunity, 34: 175-187.

Elkin S K, Matthews A G, Oettinger M A. 2003. The C-terminal portion of RAG2 protects against transposition in vitro. Embo J, 22: 1931-1938.

Fehling H J, Krotkova A, Saint-Ruf C, et al. 1995. Crucial role of the pre-T-cell receptor alpha gene in development of alpha beta but not gamma delta T cells. Nature, 375: 795-798.

Fischer A. 2004. Human primary immunodeficiency diseases: a perspective. Nature Immunology, 5: 23-30.

Fitzsimmons S P, Bernstein R M, Max E E, et al. 2007. Dynamic changes in accessibility, nuclear positioning, recombination, and transcription at the Ig kappa locus. J Immunol, 179: 5264-5273.

Flajnik M F. 2002. Comparative analyses of immunoglobulin genes: surprises and portents. Nat Rev Immunol, 2, 688-698.

Fugmann S D, Messier C, Novack L A, et al. 2006. An ancient evolutionary origin of the Rag1/2 gene locus. Proceedings of the National Academy of Sciences of the United States of America, 103: 3728-3733.

Gellert M. 2002. V(D)J recombination: RAG proteins, repair factors, and regulation. Annual Review of Biochemistry, 71: 101-132.

Giallourakis C C, Franklin A, Guo C G, et al. 2010. Elements between the IgH variable (V) and diversity (D) clusters influence antisense transcription and lineage-specific V(D)J recombination. Proceedings of the National Academy of Sciences of the United States of America, 107: 22207-22212.

Glusman G, Rowen L, Lee I, et al. 2001. Comparative genomics of the human and mouse T cell receptor loci. Immunity, 15: 337-349.

Godderz L J, Rahman N S, Risinger G M, et al. 2003. Self-association and conformational properties of the VDJ recombination activating protein RAG1. Faseb J, 17: C191-C191.

Golding A, Chandler S, Ballestar E, et al. 1999. Nucleosome structure completely inhibits in vitro cleavage by the V(D)J recombinase. Embo J, 18: 3712-3723.

Grazini U, Zanardi F, Citterio E, et al. 2010. The RING Domain of RAG1 Ubiquitylates Histone H3: A Novel Activity in Chromatin-Mediated Regulation of V(D)J Joining. Mol Cell, 37: 282-293.

Greenberg A S, Avila D, Hughes M, et al. 1995. A new antigen receptor gene family that undergoes rearrangement and extensive somatic diversification in sharks. Nature, 374: 168-173.

Grundy G J, Hesse J E, Gellert M. 2007. Requirements for DNA hairpin formation by RAG1/2. Proceedings of the National Academy of Sciences of the United States of America, 104: 3078-3083.

Grundy G J, Ramon-Maiques S, Dimitriadis E K, et al. 2009. Initial Stages of V(D)J Recombination: The Organization of RAG1/2 and RSS DNA in the Postcleavage Complex. Mol Cell, 35: 217-227.

Guo C G, Alt F W, Giallourakis C. 2011. PAIRing for Distal Igh Locus V(D)J Recombination. Immunity, 34: 139-141.

Hale J S, Fink P J. 2010. T-cell receptor revision: friend or foe? Immunology, 129: 467-473.

Haren L, Ton-Hoang B, Chandler M. 1999. Integrating DNA: Transposases and retroviral integrases. Annu Rev Microbiol, 53: 245-281.

Helmink BA, Tubbs A T, Dorsett Y, et al. 2011. H2AX prevents CtIP-mediated DNA end resection and aberrant repair in G1-phase lymphocytes. Nature, 469: 245-249.

Hempel W M, Stanhope-Baker P, Mathieu N, et al. 1998. Enhancer control of V(D)J recombination at the TCR beta locus: differential effects on DNA cleavage and joining. Gene Dev, 12: 2305-2317.

Hernandez-Munain C, Sleckman B P, Krangel M S. 1999. A developmental switch from TCR delta enhancer to TCR alpha enhancer function during thymocyte maturation. Immunity, 10: 723-733.

Hesse J E, Lieber M R, Mizuuchi K, et al. 1989. V(D)J recombination: a functional definition of the joining signals. Genes Dev, 3: 1053-1061.

Hesslein D G T., Pflugh D L, Chowdhury D, et al. 2003. Pax5 is required for recombination of transcribed, acetylated, 5' IgH V gene segments. Gene Dev, 17: 37-42.

Hewitt S L, Yin B, Ji Y, et al. 2009. RAG-1 and ATM coordinate monoallelic recombination and nuclear positioning of immunoglobulin loci. Nat Immunol, 10: 655-664.

Hiom K, Melek M, Gellert M. 1998. DNA transposition by the RAG1 and RAG2 proteins: A possible source of oncogenic translocations. Cell, 94: 463-470.

Hozumi N, Tonegawa S. 1976. Evidence for somatic rearrangement of immunoglobulin genes coding for variable and constant regions. Proceedings of the National Academy of Sciences of the United States of America, 73: 3628-3632.

Hsieh C L, Lieber M R. 1992. CpG methylated minichromosomes become inaccessible for V(D)J recombination after undergoing replication. Embo J, 11: 315-325.

Jackson A M, Krangel M S. 2006. Turning T-cell receptor beta recombination on and off: more questions than answers. Immunol Rev, 209: 129-141.

Jhappan C, Morse H C 3rd, Fleischmann R D, et al. 1997. DNA-PKcs: a T-cell tumour suppressor encoded at the mouse scid locus. Nature genetics, 17: 483-486.

Ji Y H, Resch W, Corbett E, et al. 2010. The In Vivo Pattern of Binding of RAG1 and RAG2 to Antigen Receptor Loci (vol 141, pg 419, 2010). Cell, 143: 170-170.

Johnson K, Angelin-Duclos C, Park S, et al. 2003. Changes in histone acetylation are associated with differences in accessibility of V-H gene segments to V-DJ recombination during B-cell ontogeny and development. Mol Cell Biol, 23: 2438-2450.

Johnson K, Hashimshony T, Sawai C M, et al. 2008. Regulation of immunoglobulin light-chain recombination by the transcription factor IRF-4 and the attenuation of interleukin-7 signaling. Immunity, 28: 335-345.

Johnston C M, Wood A L, Bolland D J, et al. 2006. Complete sequence assembly and characterization of the C57BL/6 mouse Ig heavy chain V region. J Immunol, 176: 4221-4234.

Jones J M, Bhattacharyya A, Simkus C, et al. 2011. The RAG1 V(D)J recombinase/ubiquitin ligase promotes ubiquitylation of acetylated, phosphorylated histone 3.3. Immunol Lett, 136: 156-162.

Jones J M, Gellert M. 2002. Ordered assembly of the V(D)J synaptic complex ensures accurate recombination. Embo J, 21: 4162-4171.

Jung D, Giallourakis C, Mostoslavsky R, et al. 2006. Mechanism and control of V(D)J recombination at the immunoglobulin heavy chain locus. Annu Rev Immunol, 24: 541-570.

Juntilla M M, Koretzky G A. 2008. Critical roles of the PI3K/Akt signaling pathway in T cell development. Immunol Lett, 116: 104-110.

Kapitonov V V, Jurka J. 2005. RAG1 core and V(D)J recombination signal sequences were derived from Transib transposons. Plos Biol, 3: 998-1011.

Kee B L, Murre C. 1998. Induction of early B cell factor (EBF) and multiple B lineage genes by the basic helix-loop-helix transcription factor E12. J Exp Med, 188: 699-713.

Khor B, Sleckman B P. 2005. Intra- and inter-allelic ordering of T cell receptor beta chain gene assembly. Eur J Immunol, 35: 964-970.

Kim D R, Dai Y, Mundy C L, et al. 1999. Mutations of acidic residues in RAG1 define the active site of the V(D)J recombinase. Gene Dev, 13: 3070-3080.

Komori T, Okada A, Stewart V, et al. 1993. Lack of N-Regions in Antigen Receptor Variable Region Genes of Tdt-Deficient Lymphocytes. Science, 261: 1171-1175.

Kosak S T, Skok J A, Medina K L, et al. 2002. Subnuclear compartmentalization of immunoglobulin loci during lymphocyte development. Science 296,: 158-162.

Krangel M S. 2003. Gene segment selection in V(D)J recombination: accessibility and beyond. Nat Immunol, 4: 624-630.

Krimpenfort P, Dejong R, Uematsu Y, et al. 1988. Transcription of T-Cell Receptor Beta-Chain Genes Is Controlled by a Downstream Regulatory Element. Embo J, 7: 745-750.

Krotkova A, vonBoehmer H, Fehling H J. 1997. Allelic exclusion in pT alpha-deficient mice: No evidence for cell surface expression of two T cell receptor (TCR)-beta chains, but less efficient inhibition of endogenous V beta->(D)J beta rearrangements in the presence of a functional TCR-beta transgene. J Exp Med, 186: 767-775.

Kwon J, Morshead K B, Guyon J R, et al. 2000. Histone acetylation and hSWI/SNF remodeling act in concert to stimulate V(D)J cleavage of nucleosomal DNA. Mol Cell, 6: 1037-1048.

Landree M A, Wibbenmeyer J A, Roth D B. 1999. Mutational analysis of RAG1 and RAG2 identifies three catalytic amino acids in RAG1 critical for both cleavage steps of V(D)J recombination. Gene Dev, 13: 3059-3069.

Langerak A W, Wolvers-Tettero I L M, van Gastel-Mol E J, et al. 2001. Basic helix-loop-helix proteins E2A and HEB induce immature T-cell receptor rearrangements in nonlymphoid cells. Blood, 98: 2456-2465.

Lee G S, Neiditch M B, Salus S S, et al. 2004. RAG proteins shepherd double-strand breaks to a specific pathway, suppressing error-prone repair, but RAG nicking initiates homologous recombination. Cell, 117: 171-184.

Lee J, Desiderio S. 1999. Cyclin A/CDK2 regulates V(D)J recombination by coordinating RAG-2 accumulation and DNA repair. Immunity, 11: 771-781.

Lee V, Huang J L, Lui M F, et al. 2008. The evolution of multiple isotypic IgM heavy chain genes in the shark. J Immunol, 180: 7461-7470.

Lefranc M P, Giudicelli V, Kaas Q, et al. 2005. IMGT, the international ImMunoGeneTics information system(R). Nucleic Acids Research, 33: D593-D597.

Lennon G G, Perry R P. 1985. C-Mu-Containing Transcripts Initiate Heterogeneously within the Igh Enhancer Region and Contain a Novel 5'-Nontranslatable Exon. Nature, 318: 475-478.

Lewis S M, Agard E, Suh S, et al. 1997a. Cryptic signals and the fidelity of V(D)J joining. Mol Cell Biol, 17: 3125-3136.

Lewis S M, Wu G E. 1997b. The origins of V(D)J recombination. Cell, 88: 159-162.

Li G C, Ouyang H, Li X, et al. 1998. Ku70: a candidate tumor suppressor gene for murine T cell lymphoma. Mol Cell, 2: 1-8.

Li Z, Dordai D I, Lee J, et al. 1996. A conserved degradation signal regulates RAG-2 accumulation during cell division and links V(D)J recombination to the cell cycle. Immunity, 5: 575-589.

Liu M, Duke J L, Richter D J, et al. 2008. Two levels of protection for the B cell genome during somatic hypermutation. Nature, 451: 841-U811.

Liu Y, Subrahmanyam R, Chakraborty T, et al. 2007. A plant homeodomain in Rag-2 that binds hypermethylated lysine 4 of histone H3 is necessary for efficient antigen-receptor-gene rearrangement. Immunity, 27: 561-571.

Lu C P, Sandoval H, Brandt V L, et al. 2006. Amino acid residues in Rag1 crucial for DNA hairpin formation. Nature structural & molecular biology, 13: 1010-1015.

Ma Y M, Pannicke U, Schwarz K, et al. 2002. Hairpin opening and overhang processing by an Artemis/DNA-dependent protein kinase complex in nonhomologous end joining and V(D)J recombination. Cell, 108: 781-794.

Maes J, O'Neill L P, Cavelier P, et al. 2001. Chromatin remodeling at the Ig loci prior to V(D)J recombination. J Immunol, 167: 866-874.

Malecek K, Lee V, Feng W, et al. 2008. Immunoglobulin heavy chain exclusion in the shark. Plos Biol, 6: e157.

Malin S, McManus S, Cobaleda C, et al. 2010. Role of STAT5 in controlling cell survival and immunoglobulin gene recombination during pro-B cell development. Nature Immunology, 11: 171-U197.

Mandal M, Powers S E, Maienschein-Cline M, et al. 2011. Epigenetic repression of the Igk locus by STAT5-mediated recruitment of the histone methyltransferase Ezh2. Nature Immunology, 12: 1212-U1107.

Matthews A G, Kuo A J, Ramon-Maiques S, et al. 2007. RAG2 PHD finger couples histone H3 lysine 4 trimethylation with V(D)J recombination. Nature, 450: 1106-1110.

Michie A M, Zuniga-Pflucker J C. 2002. Regulation of thymocyte differentiation: pre-TCR signals and beta-selection. Semin Immunol, 14: 311-323.

Morshead K B, Ciccone D N, Taverna S D, et al. 2003. Antigen receptor loci poised for V(D)J rearrangement are broadly associated with BRG1 and flanked by peaks of histone H3 dimethylated at lysine 4. P Natl Acad Sci USA, 100: 11577-11582.

Moshous D, Callebaut I, de Chasseval R, et al. 2001. Artemis, a novel DNA double-strand break repair/V(D)J recombination protein, is mutated in human severe combined immune deficiency. Cell, 105: 177-186.

Mundy C L, Patenge N, Matthews A G W, et al. 2002. Assembly of the RAG1/RAG2 synaptic complex. Mol Cell Biol, 22: 69-77.

Nadel B, Tang A, Escuro G, et al. 1998. Sequence of the spacer in the recombination signal sequence affects V(D)J rearrangement frequency and correlates with nonrandom Vkappa usage in vivo. J Exp Med, 187: 1495-1503.

Nightingale K P, Baumann M, Eberharter A, et al. 2007. Acetylation increases access of remodelling complexes to their nucleosome targets to enhance initiation of V(D)J recombination. Nucleic Acids Res, 35: 6311-6321.

Ochiai K, Maienschein-Cline M, Mandal M, et al. 2012. A self-reinforcing regulatory network triggered by limiting IL-7 activates pre-BCR signaling and differentiation. Nature Immunology, 13: 300-U124.

Odegard V H, Schatz D G. 2006. Targeting of somatic hypermutation. Nat Rev Immunol, 6: 573-583.

Oettinger M A, Schatz D G, Gorka C, et al. 1990. Rag-1 and Rag-2, adjacent genes that synergistically activate V(D)J recombination. Science, 248: 1517-1523.

Omenn G S. 1965. Familial Reticuloendotheliosis with Eosinophilia. New Engl J Med, 273: 427-&.

Osipovich O A, Subrahmanyam R, Pierce S, et al. 2009. Cutting edge: SWI/SNF mediates antisense Igh transcription and locus-wide accessibility in B cell precursors. J Immunol, 183: 1509-1513.

Parra Z E, Baker M L, Schwarz R S, et al. 2007. A unique T cell receptor discovered in marsupials. P Natl Acad Sci USA, 104: 9776-9781.

Pasqual N, Gallagher M, Aude-Garcia C, et al. 2002. Quantitative and qualitative changes in V-J alpha rearrangements during mouse thymocytes differentiation: implication for a limited T cell receptor alpha chain repertoire. J Exp Med, 196: 1163-1173.

Patenge N, Elkin S K, Oettinger M A. 2004. ATP-dependent remodeling by SWI/SNF and ISWI proteins stimulates V(D)J cleavage of 5 S arrays. J Biol Chem, 279: 35360-35367.

Posnett D N, Vissinga C S, Pambuccian C, et al. 1994. Level of human TCRBV3S1 (V beta 3) expression correlates with allelic polymorphism in the spacer region of the recombination signal sequence. J Exp Med, 179: 1707-1711.

Ramsden D A, Baetz K, Wu G E. 1994. Conservation of sequence in recombination signal sequence spacers. Nucleic Acids Res, 22: 1785-1796.

Rast J P, Litman G W. 1998. Towards understanding the evolutionary origins and early diversification of rearranging antigen receptors. Immunol Rev, 166: 79-86.

Reddy Y V R., Perkins E J, Ramsden D A. 2006. Genomic instability due to V(D)J recombination-associated transposition. Gene Dev, 20: 1575-1582.

Reth M G, Alt F W. 1984. Novel immunoglobulin heavy chains are produced from DJH gene segment rearrangements in lymphoid cells. Nature, 312: 418-423.

Roldan E, Fuxa M, Chong W, et al. 2005. Locus 'decontraction' and centromeric recruitment contribute to allelic exclusion of the immunoglobulin heavy-chain gene. Nat Immunol, 6: 31-41.

Roman C A J, Cherry, S R., Baltimore, D. 1997. Complementation of V(D)J recombination deficiency in RAG-1(-/-) B cells reveals a requirement for novel elements in the N-terminus of RAG-1. Immunity, 7: 13-24.

Romanow WJ, Langerak A W, Goebel P, et al. 2000. E2A and EBF act in synergy with the V(D)J recombinase to generate a diverse immunoglobulin repertoire in nonlymphoid cells. Mol Cell, 5: 343-353.

Roth D B, Menetski J P, Nakajima P B, et al. 1992. V(D)J Recombination - Broken DNA-Molecules with Covalently Sealed (Hairpin) Coding Ends in Scid Mouse Thymocytes. Cell, 70: 983-991.

Roth D B., Zhu C M, Gellert M. 1993. Characterization of Broken DNA-Molecules Associated with V(D)J Recombination. Proceedings of the National Academy of Sciences of the United States of America, 90: 10788-10792.

Roth M E, Holman P O, Kranz D M. 1991. Nonrandom Use of J-Alpha Gene Segments - Influence of V-Alpha and J-Alpha Gene Location. J Immunol, 147: 1075-1081.

Rval P, Kriatchko A N, Kumar S, et al. 2008. Evidence for Ku70/Ku80 association with full-length RAG1. Nucleic Acids Research, 36: 2060-2072.

Ryu C J, Haines B B, Draganov D D, et al. 2003. The T cell receptor beta enhancer promotes access and pairing of D beta and J beta gene segments during V(D)J recombination. Proceedings of the National Academy of Sciences of the United States of America, 100: 13465-13470.

Sakano H, Huppi K, Heinrich G, et al. 1979. Sequences at the Somatic Recombination Sites of Immunoglobulin Light-Chain Genes. Nature, 280: 288-294.

Sayegh C E, Jhunjhunwala S, Riblet R, et al. 2005. Visualization of looping involving the immunoglobulin heavy-chain locus in developing B cells. Genes Dev, 19: 322-327.

Schatz D G, Baltimore D. 1988. Stable Expression of Immunoglobulin Gene V(D)J Recombinase Activity by Gene-Transfer into 3t3 Fibroblasts. Cell, 53: 107-115.

Schatz D G, Ji Y H. 2011. Recombination centres and the orchestration of V(D)J recombination. Nat Rev Immunol, 11: 251-263.

Schatz D G, Oettinger M A, Baltimore D. 1989. The V(D)J recombination activating gene, Rag-1. Cell, 59: 1035-1048.

Schlissel M, Constantinescu A, Morrow T, et al. 1993. Double-Strand Signal Sequence Breaks in V(D)J Recombination Are Blunt, 5'-Phosphorylated, Rag-Dependent, and Cell-Cycle-Regulated. Gene Dev, 7: 2520-2532.

Shimazaki N, Tsai A G, Lieber M R. 2009. H3K4me3 Stimulates the V(D)J RAG Complex for Both Nicking and Hairpinning in trans in Addition to Tethering in cis: Implications for Translocations. Mol Cell 34: 535-544.

Sikes M L, Gomez R J, Song J, et al. 1998. A developmental stage-specific promoter directs germline transcription of D beta J beta gene segments in precursor T lymphocytes. J Immunol, 161: 1399-1405.

Simkus C, Anand P, Bhattacharyya A, et al. 2007. Biochemical and folding defects in a RAG1 variant associated with Omenn syndrome. J Immunol, 179: 8332-8340.

Simkus C, Bhattacharyya A, Zhou M, et al. 2009a. Correlation between recombinase activating gene 1 ubiquitin ligase activity and V(D)J recombination. Immunology, 128: 206-217.

Simkus C, Makiya M, Jones J M. 2009b. Karyopherin alpha 1 is a putative substrate of the RAG1 ubiquitin ligase. Mol Immunol, 46: 1319-1325.

Singh H, Pongubala J M R, Medina K L. 2007. Gene Regulatory Networks that Orchestrate the Development of B Lymphocyte Precursors. Mechanisms of Lymphocyte Activation and Immune Regulation Xi: B Cell Biology, 596: 57-62.

Skok J A, Gisler R, Novatchkova M, et al. 2007. Reversible contraction by looping of the Tcra and Tcrb loci in rearranging thymocytes. Nat Immunol, 8: 378-387.

Sleckman B P, Bassing C H, Hughes M M, et al. 2000. Mechanisms that direct ordered assembly of T cell receptor beta locus V, D, and J gene segments. Proceedings of the National Academy of Sciences of the United States of America, 97: 7975-7980.

Stanhope-Baker P, Hudson K M, Shaffer A L, et al. 1996. Cell type-specific chromatin structure determines the targeting of V(D)J recombinase activity in vitro. Cell, 85: 887-897.

Su IH, Basavaraj A, Krutchinsky A N, et al. 2003. Ezh2 controls B cell development through histone H3 methylation and Igh rearrangement. Nature Immunology, 4: 124-131.

Swanson P C, Desiderio S. 1998. V(D)J recombination signal recognition: Distinct, overlapping DNA-protein contacts in complexes containing RAG1 with and without RAG2. Immunity, 9: 115-125.

Swanson P C, Desiderio S. 1999. RAG-2 promotes heptamer occupancy by RAG-1 in the assembly of a V(D)J initiation complex. Mol Cell Biol, 19: 3674-3683.

Swanson P C. 2002a. A RAG-1/RAG-2 tetramer supports 12/23-regulated synapsis, cleavage, and transposition of V(D)J recombination signals. Mol Cell Biol, 22: 7790-7801.

Swanson P C. 2002b. Fine structure and activity of discrete RAG-HMG complexes on V(D)J recombination signals. Mol Cell Biol, 22: 1340-1351.

Takeda J, Cheng A, Mauxion F, et al. 1990. Functional-Analysis of the Murine T-Cell Receptor-Beta Enhancer and Characteristics of Its DNA-Binding Proteins. Mol Cell Biol, 10: 5027-5035.

Talukder S R, Dudley D D, Alt F W, et al. 2004. Increased frequency of aberrant V(D)J recombination products in core RAG-expressing mice. Nucleic Acids Research, 32: 4539-4549.

Thomas L R, Miyashita H, Cobb R M, et al. 2008. Functional analysis of histone methyltransferase g9a in B and T lymphocytes. J Immunol, 181: 485-493.

Thompson A, Timmers E, Schuurman R K, et al. 1995. Immunoglobulin heavy chain germ-line JH-C mu transcription in human precursor B lymphocytes initiates in a unique region upstream of DQ52. Eur J Immunol, 25: 257-261.

Tripathi R, Jackson A, Krangel M S. 2002. A change in the structure of Vbeta chromatin associated with TCR beta allelic exclusion. J Immunol, 168: 2316-2324.

Tsai C L, Drejer A H, Schatz D G. 2002. Evidence of a critical architectural function for the RAG proteins in end processing, protection, and joining in V(D)J recombination. Genes Dev, 16: 1934-1949.

Tsai C L, Schatz D G. 2003. Regulation of RAG1/RAG2-mediated transposition by GTP and the C-terminal region of RAG2. Embo J, 22: 1922-1930.

Tycko B, Sklar J. 1990. Chromosomal translocations in lymphoid neoplasia: a reappraisal of the recombinase model. Cancer Cells, 2: 1-8.

van Gent D C, Hiom K, Paull T T, et al. 1997. Stimulation of V(D)J cleavage by high mobility group proteins. Embo J, 16: 2665-2670.

Verma-Gaur J, Torkamani A, Schaffer L, et al. 2012. Noncoding transcription within the Igh distal V-H region at PAIR elements affects the 3D structure of the Igh locus in pro-B cells. Proceedings of the National Academy of Sciences of the United States of America, 109: 17004-17009.

Villa A, Sobacchi C, Notarangelo L D, et al. 2001. V(D)J recombination defects in lymphocytes due to RAG mutations: severe immunodeficiency with a spectrum of clinical presentations. Blood, 97: 81-88.

Villey I, Caillol D, Selz F, et al. 1996. Defect in rearrangement of the most 5' TCR-J alpha following targeted deletion of T early alpha (TEA): Implications for TCR alpha locus accessibility. Immunity, 5: 331-342.

von Boehmer H. 2005. Opinion - Unique features of the pre-T-cell receptor alpha-chain: not just a surrogate. Nat Rev Immunol, 5: 571-577.

Wang X M, Xiao G, Zhang Y F, et al. 2008. Regulation of Tcrb recombination ordering by c-Fos-dependent RAG deposition. Nature Immunology, 9: 794-801.

Warr G W, Clem L W, Soderhall K. 2003. The international ImMunoGeneTics Database IMGT. Dev Comp Immunol, 27: 1.

Wesemann D R, Portuguese A J, Meyers R M, et al. 2013. Microbial colonization influences early B-lineage development in the gut lamina propria. Nature, 501: 112-115.

Whitehurst C E, Schlissel M S, Chen J. 2000. Deletion of germline promoter PD beta 1 from the TCR beta locus causes hypermethylation that impairs D beta 1 recombination by multiple mechanisms. Immunity, 13: 703-714.

Wu C, Bassing C H, Jung D, et al. 2003. Dramatically increased rearrangement and peripheral representation of V beta 14 driven by the 3 ' D beta 1 recombination signal sequence. Immunity, 18: 75-85.

Yancopoulos G D, Alt F W. 1985. Developmentally Controlled and Tissue-Specific Expression of Unrearranged Vh Gene Segments. Cell, 40: 271-281.

Yashiro-Ohtani Y, Ohtani T, Pear W S. 2010. Notch regulation of early thymocyte development. Semin Immunol, 22: 261-269.

Ye J. 2004. The immunoglobulin IGHD gene locus in C57BL/6 mice. Immunogenetics, 56: 399-404.

Yoshida T, Tsuboi A, Ishiguro K, et al. 2000. The DNA-bending protein, HMG1, is required for correct cleavage of 23 bp recombination signal sequences by recombination activating gene proteins in vitro. Int Immunol, 12: 721-729.

Yurchenko V, Xue Z, Sadofsky M. 2003. The RAG1 N-terminal domain is an E3 ubiquitin ligase. Gene Dev 17: 581-585.

Zhang L, Reynolds T L, Shan X, et al. 2011. Coupling of V(D)J recombination to the cell cycle suppresses genomic instability and lymphoid tumorigenesis. Immunity, 34: 163-174.

Zhang M, Swanson P C. 2008. V(D)J recombinase binding and cleavage of cryptic recombination signal sequences identified from lymphoid malignancies. J Biol Chem, 283: 6717-6727.

Zhang Z, Espinoza C R, Yu Z, et al. 2006. Transcription factor Pax5 (BSAP) transactivates the RAG-mediated V(H)-to-DJ(H) rearrangement of immunoglobulin genes. Nat Immunol, 7: 616-624.

Zhu C, Mills K D, Ferguson D O, et al. 2002. Unrepaired DNA breaks in p53-deficient cells lead to oncogenic gene amplification subsequent to translocations. Cell, 109: 811-821.

Zhuang Y, Soriano P, Weintraub H. 1994. The Helix-Loop-Helix Gene E2a Is Required for B-Cell Formation. Cell, 79: 875-884.

* 策划编辑：汤其群　复旦大学上海医学院生物化学与分子生物学系

脂肪细胞的发育分化与代谢

作　　者：黄海艳　汤其群

复旦大学上海医学院生物化学与分子生物学系
代谢分子医学教育部重点实验室

▶ 1. 脂肪组织的生理功能与疾病 / 105
▶ 2. 脂肪组织的分类 / 107
▶ 3. 脂肪细胞发育分化的调节 / 108
▶ 4. 脂肪组织与代谢 / 123
▶ 5. 结语 / 126

摘要

脂肪组织具有储能、供能、产热、泌乳、免疫反应等多种重要的作用。作为脂肪组织主要细胞成分的脂肪细胞发育分化异常会引起肥胖、血脂异常、胰岛素抵抗、Ⅱ型糖尿病、高血压等心血管疾病，非酒精性脂肪肝，以及某些癌症等代谢综合征的表现。哺乳动物体内存在白色脂肪、棕色脂肪和米色脂肪3种类型脂肪组织。不同类型脂肪组织的脂肪细胞来源、细胞形态特征、分子标记、发育调节、诱导因素及其对代谢的作用各不相同，该文将对此进行介绍，并就这些脂肪组织对肥胖及相关疾病的发生、预防和潜在临床意义进行讨论。

关键词

脂肪细胞、发育分化、代谢、调节

1. 脂肪组织的生理功能与疾病

1.1 脂肪组织的生理功能

脂肪组织是具有不同的解剖结构、特殊的血管和神经支配、复杂的细胞组分和高度生理可塑性的大器官(Pankov, 1999; Cinti, 2011)。脂肪细胞是脂肪组织的主要细胞成分，被巨噬细胞、成纤维细胞、前脂肪细胞及其他血管基质组分环绕(Hausman et al., 2001; Nishimura et al., 2007; Cinti, 2012)。脂肪组织是多部位的器官，主要分布于皮下(皮下脂肪组织)和躯干(内脏脂肪组织)。脂肪组织具有多种重要的生理功能，例如产热、泌乳、免疫反应、作为代谢的能量来源等(Cinti, 2001; Cinti, 2005)。

1.2 脂肪组织异常与疾病

脂肪组织发育异常与许多疾病密切相关，脂肪组织过多会引起肥胖。过多的脂质沉积又是胰岛素抵抗、Ⅱ型糖尿病、高血压等心血管疾病、非酒精性脂肪肝，以及某些癌症的主要危险因素(Hotamisligil and Erbay, 2008)。

1.2.1 脂肪组织与肥胖

脂肪的生成贯穿生物体的一生，在机体出生以前及出生以后均存在，即使在成年时期仍可产生新的脂肪细胞。性腺周围的脂肪组织在出生后发育，而皮下的脂肪在胚胎14~18天就已经发育(Wang et al., 2013)。当机体能量摄入长期超过能量消耗时，多余的能量会以甘油三酯的形式储存在脂肪细胞内，进而产生肥胖。脂肪细胞体积增大和数目增多均可引起肥胖(Hirsch and Batchelor, 1976; Shepherd et al., 1993)。高脂饮食时，皮下的脂肪以肥大为主(Wang et al., 2013)。脂肪细胞肥大与脂肪组织功能异常有关，是引起肥胖个体代谢综合征的主要因素(Sun et al., 2011)。而脂肪细胞数量的增加则是由于脂肪组织血管基质中存在的干细胞定向为前脂肪细胞，后者在一定环境因素的作用下，分化为成熟的脂肪细胞(Yu et al., 1997)。

1.2.2 脂肪组织与糖尿病

研究证明，适量的脂肪组织为人体健康所必需，过多或过少的脂肪组织都会引起代谢综合征类似症状，如胰岛素抵抗。大量的研究发现，脂肪组织的脂代谢紊乱损害胰岛素敏感性(Vessby, 2003)。进食后，脂肪组织以甘油三酯的形式储存能量，当机体其他组织需要能量时，甘油三酯则分解为游离脂肪酸供能。当脂肪组织能量储存和释放的调节机制受到损害时，血浆游离脂肪酸(FFA)水平增加，引发脂肪酸代谢紊乱和脂质的异位沉积。升高的 FFA 不仅可以诱导胰岛 β 细胞凋亡(Cnop, 2008)，还可以导致肌肉组织胰岛素受体底物 1(insulin receptor substrate 1，IRS-1)丝氨酸磷酸化水平升高，酪氨酸磷酸化水平降低，PI3K 激活减弱(Ragheb et al., 2009)。而脂质在肝脏和肌肉组织的异位沉积也是导致胰岛素抵抗的原因(Boden and Shulman, 2002)。肥胖相关的慢性系统性炎症(也称为代谢性炎症)是人类和啮齿类动物胰岛素抵抗和 II 型糖尿病的重要病理生理基础 (Hotamisligil, 2006; Shoelson et al., 2006; Gregor and Hotamisligil, 2011; Ouchi et al., 2011)。尽管在肝脏、肌肉也会呈现肥胖诱导的轻度炎症，但是白色脂肪组织是介导系统性代谢炎症的主要部位(Odegaard and Chawla, 2013)。脂肪组织分为躯干和外周脂肪组织。躯干脂肪组织包括胸部和腹部的皮下脂肪组织及胸腔和腹腔的脂肪组织。而外周的脂肪组织包括上肢和下肢的皮下脂肪组织(Garg, 2004)。不同部位的脂肪组织对胰岛素抵抗的发生作用不同。研究认为，内脏的脂肪组织是胰岛素抵抗的主要因素。躯干皮下脂肪组织对胰岛素抵抗的发生也有作用(Patel and Abate, 2013)。

1.2.3 肥胖与心血管疾病

动脉粥样硬化等血管疾病也是脂肪组织功能异常的重要临床结果(Apovian and Gokce, 2012; Nakamura et al., 2013; Yoo and Choi, 2014)。心血管并发症是饮食诱导的肥胖、胰岛素抵抗及 II 型糖尿病个体发病和死亡的主要原因。这些并发症包括大血管的特异性病变(如动脉粥样硬化、心肌肥大等)和小血管的特异性病变(如视网膜病变、肾病变、神经病变等)(Poirier et al., 2006; Marinou et al., 2010)。脂肪组织异常引起心血管病变的机制包括：高血糖、高胰岛素血症、胰岛素抵抗、炎症及循环中细胞因子浓度的改变等，而胰岛素抵抗和炎症是肥胖和心血管疾病之间的重要枢纽(Calabro et al., 2009; Wang and Nakayama, 2010; Reaven, 2011; DeBoer, 2013)。

1.2.4 肥胖与肿瘤

肥胖能增加某些类型癌症的发病风险，包括乳腺癌、肾癌、卵巢癌、食管癌、胰腺癌、前列腺癌、肝癌、大肠癌、黑色素瘤等(Sanchez et al., 2014)。脂肪组织作为内分泌器官产生和分泌多肽、激素、脂肪细胞因子等，其中瘦素和脂联素含量最丰富并参与肿瘤的进程(Dalamaga et al., 2012; Drew, 2012)。

2. 脂肪组织的分类

根据脂肪组织的结构和功能不同，哺乳动物体内脂肪组织可分为3种，分别是白色脂肪组织(white adipose tissue, WAT)、棕色脂肪组织(brown adipose tissue, BAT)和米色脂肪组织(beige adipose tissue)，其中的脂肪细胞分别称为白色脂肪细胞、棕色脂肪细胞和米色脂肪细胞(图版Ⅹ图1)。

2.1 白色脂肪

白色脂肪组织在哺乳动物体内分布较为广泛，白色脂肪组织通常占正常成人体重约20%。白色脂肪组织的发育在子宫内开始，但主要在出生后发育。白色脂肪组织中的脂肪细胞称为白色脂肪细胞。白色脂肪细胞内为单一的大脂滴，占据细胞90%的体积，细胞核位于细胞边缘，线粒体较少、较小、较长。白色脂肪细胞表达特征性基因 *leptin*、*adiponectin*、*resistin*(Beltowski, 2003)、*LPL*、*G3PDH*(Dani et al., 1990)等。其主要功能是以甘油三酯的形式储存机体摄取的多余能量及饥饿时提供能源，与其他重要代谢活性器官一起维持机体能量稳态；同时成熟的脂肪细胞能以内分泌、旁分泌、自分泌的方式分泌大量的脂肪因子(细胞因子、激素和其他炎症标志分子)，参与调节机体稳态(Ahima, 2006; Lago et al., 2007; Lago et al., 2009; Lancha et al., 2012)。

2.2 经典棕色脂肪

多种哺乳动物(如小鼠等)在成年后体内仍保留一定量棕色脂肪组织，以更好地适应低温胁迫(Cannon and Nedergaard, 2004)。人在婴儿期，棕色脂肪主要分布在肾周和背部，成年人活跃的棕色脂肪存在于颈部、锁骨上部和脊柱旁侧(Nedergaard et al., 2007; Cypess et al., 2009; Saito et al., 2009; van Marken Lichtenbelt et al., 2009; Virtanen et al., 2009)。棕色脂肪组织中的脂肪细胞称为棕色脂肪细胞。棕色脂肪细胞为多边形，细胞直径为15~60μm，细胞内含有多个小脂滴，富含线粒体。棕色脂肪细胞中的线粒体数目、线粒体大小、线粒体嵴的密度均远大于白色脂肪细胞(Cinti, 2009)。线粒体使棕色脂肪组织表现为棕色(大鼠、小鼠等，在人中则不明显)和具有高度氧化能力。棕色脂肪组织的主要功能是消耗能量和产热。棕色脂肪细胞表面密布交感神经纤维。棕色脂肪的产热通过交感神经系统作用于棕色脂肪细胞上 β3 肾上腺素能受体进行调节，并最终由解偶联蛋白1(uncoupling protein, UCP1)完成的。UCP1是成熟棕色脂肪细胞的特异性标志，也是白色脂肪棕色化或米色脂肪激活的主要表现。除了UCP1外，*Eva1*、*Pdk4*、*Ebf3*、*Hspb7*也是棕色脂肪细胞特异性基因(Lean, 1989; Enerback, 2010)。棕色脂肪组织具有很高的葡萄糖摄取能力，因此，可用脱氧葡萄糖-正电子发射断层显像(18F-fluorodeoxyglucose positron emission tomography, FDG-PET)方法进行检测(Frontini and Cinti, 2010; Tseng et al., 2010; Bartelt et al., 2011)。

2.3 米色脂肪

在白色脂肪和肌肉中存在诱导性棕色脂肪(inducible, BAT)，也被称为米色脂肪(beige)或brite(brown-in-white)(Guerra et al., 1998; Almind et al., 2007; Petrovic et al., 2010; Schulz et al., 2011; Wu et al., 2012)。米色脂肪和经典棕色脂肪关键区别在于：米色脂肪细胞通常低表达UCP1，在某些因素诱导下，表达高水平的UCP1(Frontini and Cinti, 2010; Cinti, 2011; Cinti, 2012)，进而消耗能量、产热，表现为多脂滴，其产热效率与经典棕色脂肪细胞几乎一样(Petrovic et al., 2010; Cinti, 2012; Wu et al., 2012)。成人棕色脂肪主要是指"米色脂肪"，散落在成人体内脊柱两侧，以及锁骨附近的皮肤下，而成人棕色脂肪的激活实际指的是白色脂肪的棕色化(Wu et al., 2012)。

小鼠白色脂肪中的米色脂肪细胞具有不均一性。例如，在诱导因素作用下，同一只小鼠的腹股沟白色脂肪和腹膜后白色脂肪会出现许多高表达UCP1的多脂滴细胞，但在性腺周围的白色脂肪中，仅有少量UCP1阳性的米色脂肪细胞(Cinti, 2011; Cinti, 2012)。Wu等(Gburcik et al., 2012; Wu et al., 2012)报道米色脂肪特异性表达与发育相关的转录因子Hoxc9、脂类代谢通路重要成员Slc27a1，以及细胞表面蛋白TMEM26和CD137。这种基因表达的差异提示米色脂肪与经典的棕色脂肪的来源不同。

3. 脂肪细胞发育分化的调节

3.1 白色脂肪细胞的发育分化

3.1.1 白色脂肪细胞的来源

每个脂肪垫都很复杂，不仅由成熟的脂肪细胞组成，也含有脂肪细胞的前体细胞、成纤维细胞、神经、血管细胞、巨噬细胞及其他细胞，这些细胞统称为基质血管组分(stromal vascular fraction, SVF)。一直以来，未分化的脂肪细胞的前体细胞被认为存在于基质血管组分中。近几年，根据细胞表面的标志蛋白，利用细胞流式FACS(fluorescence activated cell sorting)方法将脂肪细胞的前体细胞与其他基质血管组分区开来。例如，利用FACS发现白色脂肪组织中存在$CD45^-CD31^-Ter119^-$(Lin$^-$)$CD29^+Sca1^+CD34^+CD24^+$($CD24^+$)和(Lin$^-$)$CD29^+Sca1^+CD34^+CD24^-$($CD24^-$)两组细胞群，而只有$CD24^+$的细胞移植到脂肪发育不良的小鼠体内能够形成功能性脂肪组织(Rodeheffer et al., 2008)。$CD24^+$细胞群是脂肪细胞的前体细胞，可定向为 CD24⁻ 的前脂肪细胞。CD24⁺ 细胞群和 CD24⁻ 细胞群都表达 PdgfRα，而成熟脂肪细胞不再表达 PdgfRα，所以白色脂肪细胞来源于 PdgfRα⁺ 细胞(Berry and Rodeheffer, 2013)。谱系追踪可最终判断脂肪细胞的发育起源。谱系追踪提示脂肪细胞早期前体细胞表达 PPARγ，定位于血管的周细胞(Tang et al., 2008)，高表达周细胞的标志蛋白 PDGFRβ(platelet-derived growth factor receptor β, PDGFRβ)。谱系追踪实验显示白色和棕色脂肪细胞均来自内皮细胞，表达有前脂肪细胞的标志蛋白 Zfp423(Tran et al., 2012)。脂肪细胞除了来源于中胚层外，也有证据表明某些白色脂肪细胞来源于神经外胚层。来源于小鼠胚胎干细胞的神经上皮细胞能分化成脂肪细胞(Billon et al., 2006; Takashima et al., 2007)，而体内实验也证明，头部位于唾液腺和耳之间的脂肪组织来源于 Sox10⁺ 的细胞(在神经脊高表达，而中胚层不表达)(Billon et al., 2007)。

3.1.2 白色脂肪细胞的发育调节

白色脂肪组织在出生后不久开始形成。白色脂肪细胞经历间充质干细胞定向为前脂肪细胞，及进一步分化为脂肪细胞的发育过程(图版X图2)。

体内研究前脂肪细胞发育分化很困难。在动物的脂肪组织中，约 1/3 为成熟脂肪细胞，而人的脂肪组织中成熟脂肪细胞约占一半，其余部分为血管、神经、成纤维细胞和处于不同发育阶段的前脂肪细胞。将前脂肪细胞与成纤维细胞分离开非常困难，体内试验也很难使前脂肪细胞处于同一发育阶段，这都限制了脂肪细胞发育的体内研究。因此，不同前脂肪细胞系被广泛应用于研究脂肪细胞分化的分子事件。利用细胞模型的优势在于细胞同一性好，细胞处于分化的同一阶段，因此对外界刺激可看到确定的结果，又因为可传代，因此提供了稳定的前脂肪细胞来源。3T3-L1 和 3T3-F442A 细胞系是常见的体外前脂肪细胞模型，被广泛用于研究和阐明脂肪细胞分化的转录级联反应(Rosen et al., 2000; Gregoire, 2001)。下面简单介绍该级联反应中的重要分子(图版X图3)。

3.1.2.1 C/EBPβ

C/EBP 家族转录因子含有保守的羧基端碱性亮氨酸拉链区域(basic zipper domain, bZIP)和氨基端的激活区域。羧基端区域由参与 DNA 识别的碱性区域和邻近的介导单体间形成二聚体的亮氨酸拉链结构组成。单体蛋白之间通过亮氨酸拉链形成二聚体对于序列特异的 DNA 结合是非常重要的。在脂肪细胞分化级联反应中，CCAAT 增强子结合蛋白 β(CCAAT enhancer binding protein, C/EBPβ)在脂肪细胞分化过程中起重要作用。C/EBPβ 是转录因子 C/EBP 家族的重要成员，其 C 端具有高度保守的 DNA 结合域和二聚化功能域。利用细胞模型研究发现，在脂肪细胞分化程序的早期 C/EBPβ 即有表达，C/EBPβ 首先始动有丝分裂克隆扩增(mitotic clonal expansion, MCE)(Tang and Lane, 1999; Zhang et al., 2004)，生长抑制的前脂肪细胞重新进入细胞周期，经历两轮左右的有丝分裂后，退出细胞周期，进入细胞分化的终末阶段，随后 C/EBPβ 激活许多脂肪细胞表型基因(如 *422/aP2*、*SCD1*、*Glut4*、*ob* 基因)的转录激活因子 C/EBPα 和 PPARγ 的表达(Tang and Lane, 2012)。C/EBPβ 通过作用于 C/EBPα 和 PPARγ 启动子上的 C/EBP 调节元件发挥对C/EBPα和PPARγ的转录激活作用。因此，C/EBPβ通过促进有丝分裂克隆扩增和激活C/EBPα及PPARγ的转录，促进脂肪细胞的分化。

在脂肪细胞分化的早期，C/EBPβ获得DNA结合活性要经过一段时间的延迟。首先因为抑制因子CHOP-10与C/EBPβ相互作用，抑制C/EBPβ的DNA结合活性，其次处于低磷酸化状态的C/EBPβ具有很低的DNA结合活性。汤其群课题组研究发现，在3T3-L1前脂肪细胞中，C/EBPβ转录激活C/EBPα和PPARγ的级联反应中，C/EBPβ发生程序性磷酸化才能具有转录活性，首先Thr-188位点被磷酸化（先被MAPK磷酸化，然后由cdk2/cyclinA维持该位点磷酸化），接着其Ser-184或Thr-179位点再被GSK3β磷酸化，之后C/EBPβ才具有DNA结合能力和转录活性(Tang et al., 2005; Li et al., 2007)。

3.1.2.2　CHOP-10

CHOP-10属于C/EBP家族转录因子，能延迟C/EBPβ DNA结合活性的获得。CHOP-10碱性DNA结合区域含有一个甘氨酸和两个脯氨酸，使得CHOP-10缺少DNA结合能力，并成为C/EBP家族成员的抑制调节因子。当CHOP-10与C/EBP形成异二聚体时，CHOP-10-C/EBP异二聚体则不能与经典的C/EBP位点5'-(A/G)TTGCG(C/T)AA(C/T)-3'结合，从而抑制后者转录激活的功能(Guyton et al., 1996)。CHOP-10在生长抑制的前脂肪细胞表达很高，前脂肪细胞经诱导分化后，胎牛血清通过YY1(Yin Yang 1)下调CHOP-10的表达(Huang et al., 2005; Huang et al., 2008)，C/EBPβ获得DNA结合活性，并结合到着丝粒上，发挥转录激活功能(Tang and Lane, 1999)。

3.1.2.3　C/EBPα和PPARγ

在C/EBPα和PPARγ基因启动子近端都含有C/EBP调控元件，C/EBPβ结合在该元件上激活转录(Christy et al., 1991; Clarke et al., 1997; Tang et al., 1999)。C/EBPα和PPARγ可分别通过C/EBP和PPAR调控元件直接调控许多脂肪细胞表型和功能基因的表达(MacDougald and Lane, 1995; Hwang et al., 1997; Rosen and Spiegelman, 2000)。

3.1.2.4　固醇调节元件结合蛋白

固醇调节元件结合蛋白(Sterol regulatory element-binding protein, SREBP)包含3个成员：SREBP-1a、SREBP-1c和SREBP-2。SREBP主要在肝脏和脂肪细胞中表达，通过调节脂肪代谢相关酶的表达来调控物体内的脂肪合成。

SREBP-1c主要的靶基因涉及脂肪酸代谢，例如脂肪酸合成酶(fatty acid synthase, FASN)的基因。SREBP-2与胆固醇代谢相关的基因有关，例如HMG-CoA还原酶和LDLR的基因。SREBP-1a的靶基因包括上述两类基因(Tontonoz et al., 1993; Horton, 2002)。在脂肪细胞分化程序中，SREBP-1c mRNA的表达是在C/EBPα和PPARγ表达之后激活的(Kim and Spiegelman, 1996)，与脂肪细胞终末分化有关。SREBP前体嵌于内质网膜，与SREBP裂解激活蛋白(SREBP cleavage-activating protein, SCAP)紧密结合成SCAP/SREBP复合物。当细胞内胆固醇水平高时，胰岛素诱导基因蛋白(insulin-induced gene, Insig)与 SCAP 作用，使 SCAP/SREBP 复合物滞留在内质网。当胆固醇浓度低时，SCAP 与胆固醇分离，使 Insig 与 SCAP 的作用中断，SCAP 伴随 SREBP 从内质网转到高尔基体；再经位于高尔基体的两种蛋白酶(S1P、S2P)的水解，释放出 N 端 bHLH-Zip 区域，成为核型 SREBP(nuclear SREBP, nSREBP)。nSREBP 快速转入核内，与靶基因的 SRE 序列相结合，启动靶基因转录。目前传统研究认为，nSREBP-1 的结合位点应包含 SRE(5′-TCACNCCAC-3′)和(或)E 盒(5′-CANNTG-3′)两种。最近证据(Yellaturu et al., 2009; Yellaturu et al., 2009)表明，胰岛素通过增加 mRNA 的更新抑制 Insig-2a 水平，从而调节 SREBP-1c/ADD1 从内质网的释放。Insig-2a 水平的降低使得 SCAP·SREBP-1c/ADD1 从内质网输出到高尔基体进行蛋白质水解，成为 nSREBP 被转运入核，转录激活脂肪生成的相关基因。

3.1.2.5　WNT 蛋白

WNT 能促进成骨、成肌、成软骨分化，却唯独抑制间充质发育过程中脂肪的分化(Zhou et al., 2004; Akimoto et al., 2005; Kennell and MacDougald, 2005; Kirton et al., 2007; Zhou et al., 2008)。Wnt 通过经典的和非经典的信号通路来抑制成脂。

(1) WNT10B

WNT10B 通过抑制 C/EBPα 和 PPARγ 来维持前脂肪细胞的未分化状态(Bennett et al., 2003; Kawai et al., 2007)。WNT10B 通过经典 Wnt 信号通路影响脂肪发育(Rosen and Spiegelman, 2000)。FABP4-WNT10B 转基因小鼠在正常饮食情况下，脂肪组织明显减小，对高脂饮食诱导的肥胖也是耐受的(Wright et al., 2007)。这些小鼠即使有 ob/ob 背景，体重也没有明显增加，内脏和皮下脂肪的量与 ob/ob 小鼠相比明显减少。通过对 WNT10B 基因测序发现，在超过 200 个肥胖患者中，*WNT10B* 基因含有 *C256Y* 突变，使 WNT10B 不能激活经典的 Wnt 信号通路，因此不能抑制成脂，发生严重肥胖。而在正常的对照组中则没有发现此突变(Christodoulides et al., 2006)。

(2) WNT5A

WNT5A 能激活 CAMKII 依赖的信号通路，诱导 NLK(nemo-like kinase)，后者使组蛋白甲基转移酶磷酸化，促进共抑制复合体形成，通过组蛋白 H3K9 甲基化使 PPARγ 失活(Takada et al., 2007)，抑制脂肪细胞分化。在动物模型上 WNT5A 也与肥胖密切相关，与肥胖小鼠相比，WNT5A 在正常体重小鼠的骨骼肌、肝脏和脂肪的表达明显增加(Almind and Kahn, 2004)。WNT5A 在诱导多能的间充质干细胞向成骨方向分化的同时，抑制其向脂肪分化(Arnsdorf et al., 2009; Bilkovski et al., 2010; Santos et al., 2010)。而巨噬细胞不仅通过分泌促炎因子影响间充质干细胞向脂肪细胞的分化，也能通过分泌 WNT5A 来影响脂肪分化(Bilkovski et al., 2011)。在细胞外基质中，Sfrp(secreted frizzled-related protein)蛋白能阻止 Wnt 蛋白与受体的结合，其中，SFRP5 能与 Wnt5a 和 Wnt11 结合，并阻断它们的作用(Li et al., 2008)。在 ob/ob 肥胖小鼠和 Zucker 糖尿病大鼠中，Sfrp5 的表达减少，而 Wnt5a 表达增加。

3.2 棕色脂肪细胞发育分化的调节

3.2.1 棕色脂肪细胞的来源

利用遗传图方法证实肩胛间区棕色脂肪组织来源于轴旁中胚层(Atit et al., 2006)。谱系追踪实验显示,棕色脂肪组织中的脂肪细胞源自 $Myf5^+$ 成肌细胞(Seale et al., 2008; Kajimura et al., 2009),因此棕色脂肪组织中的脂肪细胞与骨骼肌细胞有共同的分化来源。

3.2.2 棕色脂肪细胞的发育调节

棕色脂肪组织在出生前就已经发育分化,以应对出生时的寒冷环境。棕色脂肪的发育在胎羊中被广泛研究,在孕中期可见多腔脂滴的脂肪细胞出现(Pope et al., 2014),但不表达 UCP1(Clarke et al., 1997)。进入孕末期成熟阶段,出现单一脂滴和多脂滴两种细胞的混合体,前者充满脂滴,后者富含线粒体和高表达 UCP1(Gemmell et al., 1972)。孕末期,棕色脂肪细胞环绕在大的、单脂滴的白色脂肪细胞周围(Gemmell and Alexander, 1978)。出生时,白色脂肪细胞明显减少,而 UCP1 表达水平达最高(Clarke et al., 1997)。出生后,棕色脂肪细胞逐渐消失,产后一个月可见其散在分布在白色脂肪细胞间(Pope et al., 2014)。

棕色脂肪组织消耗能量和产热是通过 UCP1 完成的。电子传递链产生 H^+ 跨线粒体内膜的势能,在偶联状态下,势能驱动 H^+ 通过 ATP 合成酶而重新回到线粒体基质中,同时势能转化为 ATP 的化学能。UCP1 是棕色脂肪细胞线粒体内膜的一种异化的质子通道,可以使质子通过该通道回流,从而破坏呼吸链电子传递过程中所建立的跨内膜的质子电化学梯度,物质氧化与 ATP 生成解偶联,不能驱动 ATP 合酶合成 ATP,而电化学梯度储存的能量以热能形式释放,用于维持体温。

UCP1 的表达主要在转录水平进行调节。了解 UCP1 转录激活对于理解棕色脂肪的发育调节非常重要。在 *UCP1* 基因 5'侧翼序列含有近端调节区域和远端增强子区域。近端调节区域含有 C/EBP 调节位点和 cAMP 调节元件(cAMP response element, CRE)(Yubero et al., 1994; Yubero et al., 1998)。远端增强子区域含有另外2个cAMP(cyclic adenosine monophosphate)调节元件和复杂的核受体结合位点组合,如 PPAR(peroxisome proliferator-activated receptor)激活剂、维甲酸类及甲状腺激素等的结合位点(Cassard-Doulcier et al., 1993; Kozak et al., 1994; Sears et al., 1996)。下面将介绍几个与棕色脂肪的发育分化密切相关的转录因子。

3.2.2.1 PRDM16

PRDM16(PRD1-BF1-RIZ1 homologous domain containing 16)能够调控棕色脂肪的定向分化。PRDM16 又名 MEL1(MDS1/EVI1-LIKE GENE 1),最早是在研究骨髓异常增生综合征和急性髓性白血病的过程中被发现(Mochizuki et al., 2000)。PRDM16 除包含 PR 结构域外,还拥有 2 个锌指 DNA 结合结构域、富含脯氨酸结构域、阻遏物结构域及 C

端的酸性结构域，这些结构特征说明 PRDM16 是一种重要的转录调节因子。

PRDM16 在棕色脂肪组织中的表达水平明显比白色脂肪组织高(Seale et al., 2007)。Seale 等(2007)发现 PRDM16 可有效促进棕色脂肪细胞的分化。PRDM16 缺乏可造成棕色脂肪细胞分化明显受阻，而过表达则使棕色脂肪细胞数量增多。PRDM16 能诱导线粒体生物合成、增加细胞呼吸和棕色脂肪细胞特异性基因的表达，从而诱导完整的棕色脂肪细胞分化程序。PRDM16 既能诱导 cAMP 依赖的产热基因(UCP1、PGC-1 和 Dio2)的表达，也能诱导 cAMP 不敏感的棕色脂肪特异性基因(cidea 和 elovl3)的表达。此外，PRDM16 还能诱导 PGC-1α 基因表达，或直接与 PGC-1α 和 PGC-1β 结合，提高他们的转录活性(Seale et al., 2007)。另外，PRDM16 可通过锌指结构与许多经典转录因子(如 PPARα、PPARγ、p53 及一些 C/EBP 家族成员)结合，提高这些转录因子的转录活性(Seale et al., 2008; Kajimura et al., 2009)。BMP7 作为重要的棕色脂肪诱导因子与其诱导关键棕色脂肪调节因子 PRDM16 和 PGC-1α 有关(Tseng et al., 2008)。

PRDM16 在促进棕色脂肪相关基因表达的同时，还会抑制白色脂肪相关基因的表达。PRDM16 与 CtBP1/2 形成复合物，并结合到白色脂肪特异表达的基因(如 resistin)的启动子上，抑制其表达；PRDM16 如与 PGC-1 结合，将有效地激活棕色脂肪基因，因此对控制棕色脂肪细胞和白色脂肪细胞的转换起重要作用(Kajimura et al., 2008)。

PRDM16 也控制骨骼肌细胞和棕色脂肪细胞的转换。棕色脂肪前体细胞如果缺失 PRDM16 则向肌肉分化，而在成肌细胞过表达 PRDM16，会使其分化为棕色脂肪细胞。PRDM16-C/EBPβ 形成的复合体能促进从前成肌细胞向棕色脂肪细胞的转化(Kajimura et al., 2009)。C/EBPβ(LAP)在棕色脂肪细胞和成肌细胞中都有丰富表达，C/EBPβ 缺失能明显抑制 PRDM16 诱导的棕色脂肪细胞分化，C/EBPβ 缺失小鼠的棕色脂肪与 PRDM16 缺失小鼠的棕色脂肪具有同样的分子特性：棕色脂肪特异性基因表达减少，而骨骼肌特异性基因表达增加。C/EBPβ 的表达受胎盘特异性蛋白 8(placenta-specific 8，Plac8)的调节。Plac8 能与 C/EBPβ 相互作用并结合到 C/EBPβ 启动子上诱导其转录。Plac8 缺失小鼠对寒冷不耐受，棕色脂肪组织发育缺陷，产热功能受损，最终产生肥胖（图版XI图1）(Jimenez-Preitner et al., 2011)。

3.2.2.2 PGC-1α

过氧化物酶体增生物激活受体γ共激活因子1(PPARγ-co-activator-1, PGC-1)家族有3个成员，PGC-1α、PGC-1β、PGC 相关共激活因子。其中 PGC-1α、PGC-1β 在肝脏、脂肪组织、骨骼肌、心肌、脑等组织表达丰富，与代谢密切相关。

PGC-1α 是一种核转录共激活因子，通过辅助激活核受体和非核受体发挥作用。PGC-1α 的 N 端为蛋白激活区域及介导核激素受体相互作用的 LXXLL 基序，C 端为 RNA 结合序列和富含丝氨酸-精氨酸的 RS 区。PGC-1α 是大多数细胞线粒体生物合成和氧化代谢的主要调节因子，能诱导许多线粒体基因和产热基因 UCP1 的表达。

与白色脂肪组织相比，PGC-1α 在棕色脂肪组织中的表达更为特异。PGC-1α 能与 PPARγ 和 TRβ(thyroid hormone receptor β)结合并作用于 UCP1 的启动子上，激活 UCP1 表达。冷暴露(Puigserver et al., 1998)或棕色脂肪细胞分化过程中 PGC-1α 表达增加(Uldry et al., 2006)。配体依赖性的核受体共抑制因子 RIP140，通过与 PGC-1α 相互作用，抑制 PGC-1α 转录活性。RIP140 缺失小鼠，白色脂肪组织会出现棕色化(Leonardsson et al., 2004)。甾体激素受体共激活因子(steroid receptor co-activator, SRC)家族中的 SRC2/TIF2/GRIP1 也能抑制 PGC-1α 转录活性。SRC2 功能缺失会导致适应性产热和能量消耗增加(Picard et al., 2002)。Rb 蛋白和 p107 也能负调节 PGC-1α 表达，pRb 缺失的成纤维细胞或胚胎干细胞会出现棕色样脂肪细胞表型，线粒体含量增加，UCP1、PGC-1α、线粒体基因表达增加(Hansen et al., 2004)。p107 缺失小鼠的白色脂肪组织也会出现棕色样脂肪细胞(Scime et al., 2005)。twist-1 也是 PGC-1α 在棕色脂肪组织中的负调节因子，缺失 twist-1 能诱导棕色脂肪特异性基因的表达；过表达 twist-1 会以 PGC-1α 依赖的方式抑制棕色脂肪特异性基因的表达(Pan et al., 2009)。

3.2.2.3 核受体在棕色脂肪细胞发育中的作用

核受体 PPAR、甲状腺激素受体、RXR/RAR 等对棕色脂肪细胞发育分化具有重要的调节作用(图版Ⅺ图 2)。

(1) 核受体的结构

核受体(NR)包括 5 个结构域：从 N 端到 C 端依次为 A/B、C、D、E 和 F 结构域。①A/B 结构域：为调节结构域，该结构域的序列在不同的 NR 之间高度可变，包含激活功能域 1(activation function 1, AF-1)。AF-1 是配体非依赖性转录激活功能域，该区特定位置的酪氨酸和丝/苏氨酸残基能被不同信号通路相关激酶磷酸化，从而影响受体与配体的亲和力和转录活性。②C 结构域：为高度保守的区域，含 2 个锌指结构，介导 NR 与靶基因结合，故又称 DBD。③D 结构域：为铰链区(hinge region, HR)，与 NR 的细胞内转运和亚细胞定位有关。④E 结构域：在空间结构上高度保守，包含 NR 二聚体化区、配体结合域和配体依赖性转录激活功能域(ligand-dependent transcriptional activation function domain，又称 AF-2)。该结构域除了与配体结合外，还可结合辅活化子和辅抑制子。⑤F 结构域：位于 C 端，该结构域的序列在不同的 NR 之间是可变的。

(2) PPAR 与棕色脂肪细胞发育

PPAR(peroxisome proliferator-activated receptor)是一类以脂肪酸为配体的核受体，包括 PPARα、PPARγ 和 PPARβ/δ。PPARα 主要在肝脏中表达，心脏、肾脏和棕色脂肪组织中也存在，它的功能是调控脂肪酸的合成；PPARγ 在脂肪组织中大量表达，是调控脂肪细胞形成、胰岛素敏感性的关键分子；PPARβ/δ 表达的特异性并不强，在代谢旺盛的组织中，如肌肉、肝脏和脂肪组织含量比较高(Sears et al., 1996)，主要调控能量代谢。

PPAR 与激动剂结合后入核，与 RXR 形成异二聚体，并与靶基因特异 DNA 区域 PPRE(peroxisome proliferator hormone response element)结合(Berger and Moller, 2002)。PPARα 是棕色脂肪细胞分化末期(Valmaseda et al., 1999)和白色脂肪细胞棕色化时特异性被诱导的基因(Petrovic et al., 2010)。PPARα 在棕色脂肪细胞中高表达，其靶基因主要涉及脂肪酸摄取和脂肪酸氧化，因此被认为是棕色脂肪组织区分于白色脂肪组织的标志蛋白。PPARα 能直接调节 UCP1(Barbera et al., 2001)和 PGC-1α 的表达(Hondares et al., 2011)。

(3) 甲状腺激素受体

甲状腺激素对生长、分化、发育和保持代谢平衡具有深远影响。甲状腺激素有 T3 和 T4 两种主要形式，其中 T3 的生物活性最强，而血液中主要以 T4 形式存在，它是由甲状腺合成并分泌到血液中。在外周组织中 T4 被 I 型或 II 型脱碘酶转换为 T3。甲状腺激素的生理作用主要是通过甲状腺激素受体(thyroid hormone receptor, TR)进行的。TR 主要位于细胞核中，与维甲酸 X 受体(RXR)及其他核受体形成异二聚体结合到靶基因启动子区域的甲状腺激素反应元件(thyroid hormone response element, TRE)上，从而调控基因转录。TR 由 TRα 和 TRβ 两个基因编码，分别会产生 TRβ1、TRβ2、TRβ3 和 TRα1、TRα2 几种异构体，其中 TRβ 的 3 种异构体和 TRα1 能够和 T3 结合(Cheng et al., 2010)，而 T3 对 UCP1 的诱导依赖于 TRβ(Lee et al., 2012)。在 *UCP1* 基因的远端增强子区域含有结合甲状腺激素受体的反应元件。甲状腺激素与交感神经系统可协同参与棕色脂肪的产热(Silva and Rabelo, 1997; Lanni et al., 2003)。一方面，DIO2 被肾上腺素能系统激活，产生足够量的 T3，以用于最大限度的产热。此外，肾上腺素能系统也被 T3 调节，放大肾上腺素能信号传导(Rubio et al., 1995a, 1995b)。

(4) RXR/RAR

维甲酸(retinoid acid, RA)是一类重要的维生素 A 代谢中间物。包括 9-顺式维甲酸(9-*cis* retinoid acid)和全反式维甲酸(all-*trans* retinoid acid, ATRA)两种异构体。维甲酸通过 RAR(retinoic acid receptor)和 RXR(retinoid X receptor)行使功能。维甲酸也是 *UCP1* 基因的转录激活因子(Alvarez et al., 1995; Puigserver et al., 1996; Kumar and Scarpace, 1998)，为 β 肾上腺素能激动剂和 PPARγ 激动剂诱导人 *UCP1* 基因所必需(del Mar Gonzalez-Barroso et al., 2000)。在 *UCP1* 启动子 5'端的增强子上存在非经典的 RA 反应元件(RA response element)(Chambon, 1996; Larose et al., 1996; Rabelo et al., 1996)和 PPAR 反应元件(PPAR response element)(Sears et al., 1996)。前者可以和 RAR/RXR 异二聚体结合，激活相关基因的表达。后者可以和 PPAR/RXR 异二聚体结合，在 RA 与 PPAR 配体同时存在的条件下达到最强的激活作用(Mukherjee et al., 1997)。维甲酸会以 p38 依赖的方式诱导小鼠胚胎成纤维细胞(mouse embryonic fibroblasts, MEF)来源脂肪细胞的 UCP1 表达(Mercader et al., 2010)。ATRA 通过 RAR 诱导小鼠白色和棕色脂肪细胞 UCP1 的表达，而与 PPARδ 无关，也不依赖于 PGC-1α。

3.2.2.4 microRNA 调节棕色脂肪发育

microRNA 的表达通常具有发育和组织特异性。最近一些研究显示，miRNA 调节包括脂肪细胞在内的许多细胞分化和代谢过程(Esau et al., 2004; Xie et al., 2009; McGregor and Choi, 2011; Hilton et al., 2013)。脂肪组织特异性敲除 Dicer 的小鼠，白色脂肪组织明显减少，棕色脂肪调节产热的基因(*UCP1*、*PGC-1α*、*COX1B*、*CIDEA*)的转录明显下降，PPARγ2 和脂肪酸合成酶(fatty acid synthase, FAS)也下降(Mudhasani et al., 2011)，肩胛间区的棕色脂肪组织增大，出现白色化，提示 miRNA 在脂肪细胞发育过程中发挥重要作用(Mori et al., 2014)。

(1) miR-193b-365

miR-193b 和 miR-365 是由单个 pri-miRNA 共同转录而来，在 BAT 中高表达。在培养的前棕色脂肪敲低 dicer，会出现白色脂肪细胞的表型，这种作用可部分被 miR-365 逆转(Mori et al., 2014)。利用锁核酸技术抑制 miR-193a/b 和（或）miR-365 在棕色脂肪组织的 SVF 或纯化的前棕色脂肪细胞的表达，棕色脂肪细胞分化受到明显抑制。尽管在白色脂肪组织中低表达，miR-193b 和 miR-365 对白色脂肪细胞分化也是必需的。miR-193a/b 可能通过调节 RUNX1t1(Runt-related transcription factor 1; translocated to 1, RUNX1t1)发挥抑制脂肪分化作用(Sun et al., 2011)。

(2) miR-133

miR-133 除了在肌肉中高表达外，在棕色脂肪细胞也有较高水平表达(Walden et al., 2009)，冷暴露后 miR-133 水平明显下降。miR-133 的 5'区域有高度保守的种子序列与 PRMD16 mRNA 的 3' UTR 保守序列互补。因此，miR-133 能抑制 PRDM16 的表达，进而抑制棕色脂肪细胞分化(Trajkovski et al., 2012)。

(3) miR-155

深度测序显示，miR-155 在棕色脂肪细胞分化过程中下降。miR-155 能直接下调棕色脂肪细胞和皮下脂肪前脂肪细胞的 C/EBPβ 表达，分别抑制棕色和米色脂肪细胞分化(Chen et al., 2013)。棕色脂肪特异性 miR-155 转基因小鼠的棕色脂肪细胞量减少，UCP1 表达下降。miR-155 敲除小鼠对冷暴露引起的体温下降耐受，甘油释放增加，细胞呼吸加强，棕色脂肪组织中脂滴大小和数目下降，棕色脂肪组织和米色脂肪组织中 UCP1 和 PGC-1α 表达增加，进一步从体内证实 miR-155 对棕色和米色脂肪细胞分化的影响。

3.2.3 棕色脂肪细胞发育的诱导因素

一些因素能促进棕色脂肪细胞的发育分化，这些因素包括冷暴露、肾上腺素受体激动剂、FGF21、BMP7 等。

3.2.3.1 肾上腺素能受体激动剂

棕色脂肪细胞上有 β3 肾上腺素能受体(β3-AR)，在啮齿类动物，β3-AR 受体激活能刺激棕色脂肪非颤抖性产热(Yoshida et al., 1994)，因此，具有抗肥胖和抗糖尿病的作用(Arch et al., 1984; Yoshida et al., 1994)。

β-AR 提高棕色脂肪分解代谢和使白色脂肪棕色化的具体机制非常复杂，存在许多信号间的协同作用(Collins et al., 2010; Kozak, 2011)。β-AR 可通过 cAMP/PKA 信号的下游分子 p38(Cao et al., 2004)和 cAMP 反应结合蛋白(cAMP response element-binding protein, CREB)(Rim and Kozak, 2002)调节 PGC-1α 的表达，后者对线粒体生物合成和 UCP1 的最大诱导是必需的(Puigserver et al., 1998; Uldry et al., 2006; Chang et al., 2012)，该信号通路不仅激活棕色脂肪组织发育分化，也促进白色脂肪棕色化。除了直接激活转录因子外，PKA 也能激发脂解，为核受体 PPARα 和 PPARδ 提供脂肪酸作为配体，使脂肪酸的供给与氧化能力相匹配(Mottillo et al., 2012)。HSL(hormone-sensitive lipase)或 PPARα 敲除，β3-AR 对线粒体生物合成和脂肪酸氧化基因的诱导明显延迟(Li et al., 2005; Roth Flach et al., 2013)。同样，过表达 ATGL (adipose triglyceride lipase，ATGL) 能增加白色脂肪细胞分解代谢，而脂肪组织特异性缺失 ATGL 则促进 BAT 出现 WAT 表型，PPARα 靶基因表达减少。尽管 β3-AR 激动剂与棕色脂肪的激活有关，但是完整的产热反应不但需要激活棕色脂肪组织，也需要激活白色脂肪组织 (Grujic et al., 1997)。β3-AR 激动剂的抗糖尿病作用需要长期治疗 (Yoshida et al., 1994)，并且与显著的 WAT 重建有关 (Granneman et al., 2005; Li et al., 2005)。

寒冷暴露也可以激活啮齿类的棕色脂肪组织。寒冷可被皮肤感知，并将信号通过外周神经元传递到脊髓，进一步上传至下丘脑突触前区域。下丘脑的一部分信号到达皮层，用于温度的感知和定位。另外一部分信号到达脑干的延髓中缝苍白核，投射到外周的交感神经系统，最后，神经节后的 SNS (sympathetic nervous system) 释放去甲肾上腺素，通过作用于脂肪细胞上的 G 蛋白偶联的 β-AR，激活 cAMP/PKA/CREB 信号通路，进而调控产热基因的转录，激活 BAT (Morrison, 2011)。人也有棕色脂肪组织，人棕色脂肪组织 (human BAT, hBAT) 也有丰富的血管和神经支配 (Zingaretti et al., 2009)，这提示人棕色脂肪组织可能成为肥胖和糖尿病治疗的靶点。冷暴露激活棕色脂肪组织的内源途径很复杂，因此特异性激活棕色脂肪组织药物的研发成为一种新的选择。去甲肾上腺素因具有很多心血管系统副作用，所以不适合用于激活人棕色脂肪组织，而脂肪组织特异 β3-AR 激动剂则成为激活棕色脂肪组织的最佳选择，但非常遗憾的是：这些激动剂都没能成功激活人棕色脂肪组织 (Weyer et al., 1998; Larsen et al., 2002; Redman et al., 2007; Arch, 2008)。麻黄碱和异丙肾上腺素等拟交感神经药在提高代谢率的剂量也不能激活人棕色脂肪组织 (Cypess et al., 2012; Vosselman et al., 2012; Carey et al., 2013)。异丙肾上腺素拟交感神经药能增加心血管和代谢活性，能量消耗与冷暴露相似，但不能激活棕色脂肪组织，提示其他组织对 β-肾上腺素能激活引起的产热也有贡献 (Vosselman et al., 2012)。

3.2.3.2 FGF21

FGF (fibroblast growth factor) 和 FGFR (fibroblast growth factor receptor) 与发育、转化和血管生成有关。研究显示 FGF/FGFR 调节内分泌相关组织和器官的功能。人 FGF 家族有 22 个成员，FGF21 主要由肝脏产生，在肥胖和糖尿病动物模型中具有抗糖尿病和减少脂肪的作用。

FGF21 通过 FGFR1c 和 β-Klotho 发挥对代谢的调节作用 (Adams et al., 2012; Ding et al., 2012; Foltz et al., 2012)。β-Klotho 主要在脂肪组织、肝脏和胰腺表达并发挥作用。β-Klotho 能特异与 FGFR 结合，是 FGF21 发挥代谢活性的共因子。细胞缺少 β-Klotho，不能对 FGF21 作出反应，引入 β-Klotho 后，能完全恢复 FGF21 的作用。

尽管肝脏是 FGF21 旁分泌和自分泌作用的首选器官，脂肪组织的 FGF21 和 FGFR 表达比肝脏都低，但 FGF21 在脂肪组织也发挥重要作用。棕色脂肪组织也表达 FGFR1c 和 β-Klotho，外源给予 FGF21 能激活棕色脂肪组织 (Fisher et al., 2011)。冷暴露和肾上腺素能激活产热，棕色脂肪组织 FGF21 增加，并伴有血浆中 FGF21 增加 (Chartoumpekis et al., 2011)。棕色脂肪组织被产热激活后，棕色脂肪组织成为循环中 FGF21 的来源。去甲肾上腺素通过 β-肾上腺素能受体，cAMP 介导的 PKA 和 p38 MAPK 激活，使 ATF2 结合

到 *FGF21* 启动子上，诱导 *FGF21* 基因的转录(Hondares et al., 2011)。小鼠在出生后，血浆中 FGF21 和肝 FGF21 显著增加，能驱动出生后棕色脂肪组织依赖的产热(Hondares et al., 2010)。白色脂肪组织也表达 FGF21、FGFR1c 和 β-Klotho(Fisher et al., 2011)。FGF21 敲除小鼠对冷暴露的反应受损，白色脂肪棕色化消失。白色脂肪组织产生的 FGF21 通过增加翻译后修饰提高 PGC-1α 活性，进而增加 UCP1 和其他产热基因的表达(Fisher et al., 2012)。AMPK 的激活及随后的 SIRT1 介导的 PGC-1α 去乙酰化，对 FGF21 提高 PGC-1α 活性和增加线粒体功能是必需的(Chau et al., 2010)。

3.2.3.3 BMP 信号通路

BMP(bone morphogenetic protein)为 TGF-β(transforming growth factor-β)超家族中成员。BMP 主要通过 BMP-Ⅰ型(BMPR-Ⅰ)和 BMP-Ⅱ型(BMPR-Ⅱ)两种跨膜的丝氨酸/苏氨酸受体传递信号。Ⅰ型受体包括 BMPR-ⅠA、BMPR-ⅠB 和活化素Ⅰ型受体；Ⅱ型受体包括 BMPR-Ⅱ、活化素Ⅱ型受体及活化素ⅡB 型受体(Attisano et al., 1992; ten Dijke et al., 1994; Kawabata et al., 1995)。当受体与 BMP 结合后，形成由两对Ⅰ、Ⅱ型复合体组成的异四聚体激活受体复合物(Gilboa et al., 2000; Nohe et al., 2002)。BMPR-Ⅱ可使 BMPR-Ⅰ的 GS 区域(含特征性 SGSGS 标志的区域)磷酸化，引起 BMPR-Ⅰ激酶的活化(Wrana et al., 1994; Massague, 1998)。BMP 受体下游最重要的信号分子是 Smads(Heldin et al., 1997; Massaous and Hata, 1997)和 p38 MAPK(Iwasaki et al., 1999; Nohe et al., 2002; Guicheux et al., 2003; Hata et al., 2003)。

Tseng 等(2008)通过比较多种 BMP 发现，仅 BMP7 能促进棕色脂肪细胞分化过程中 UCP1 的表达。BMP7 能够强烈激活 p38MAPK 和它下游的 ATF-2(activating transcription factor-2)，并通过 PGC1 来激活 UCP1 的表达及线粒体生成。此外，将 BMP7 处理的间充质干细胞移植到裸鼠体内，可以形成棕色脂肪组织。BMP7 敲除小鼠胚胎棕色脂肪缺乏，完全不表达 UCP1；而 BMP7 过表达能显著增加小鼠体内棕色脂肪，增加能量消耗，减轻体重。Myf5 驱动的条件性敲除棕色脂肪组织中Ⅰ型 BMP 受体 ACVR1 和 BMPR1A 会抑制前棕色脂肪细胞的分化，并且 BMP7 诱导的下游信号分子 Smad 和 p38 MAPK 的磷酸化也明显下降(Schulz et al., 2013)。

BMP7 自身能促进间充质干细胞向棕色脂肪细胞定向分化，也能与其他分化试剂，如罗格列酮等一起促进原代肩胛间棕色脂肪的 SVF 和胚胎成纤维细胞出现棕色脂肪细胞表型(Tseng et al., 2008)。BMP-7 虽能促进前棕色脂肪细胞的终末分化，但对前白色脂肪细胞的进一步分化则没有作用(Tseng et al., 2008)。

BMP 蛋白也与能量稳态密切相关。在成熟棕色脂肪细胞和中枢神经系统表达的 BMP8B 对正常棕色脂肪细胞的发育是不需要的，在机体需热增加时(冷环境或饱食)，BMP8B 由棕色脂肪细胞产生，通过自分泌的方式激活 P38MAPK，进而加强 β 肾上腺素能刺激的脂解和产热反应。除了对外周棕色脂肪的作用外，BMP8B 在下丘脑腹内侧核有表达，增加棕色脂肪交感神经的支配并促进产热(Whittle et al., 2012)。BMP7 也能通过促进棕色脂肪相关的能量消耗来正调节能量稳态。在脑的一些区域，如 VMH，也有 BMP7 表达，提示 BMP7 可能像 BMP8B 一样通过中枢来调节棕色脂肪组织功能(Ohyama et al., 2008)。最近的研究表明，BMP7 也能调节食欲(Townsend et al., 2012)。全身给予 BMP7 的高脂饮食诱导肥胖小鼠，能量消耗增加的同时，小鼠的摄食亦降低；给予 BMP7 的 ob/ob 小鼠，体重降低、食欲下降，提示 BMP-7 以非瘦素依赖的机制发挥作用(Townsend et al., 2012)。

3.3 米色脂肪细胞发育分化的调节

啮齿类动物的白色脂肪组织在某些诱导因素作用下出现棕色样脂肪细胞称为 brite (brown-in-white)或者米色脂肪细胞(beige adipocyte)(Barbatelli et al., 2010; Petrovic et al., 2010; Wu et al., 2012)，也称为诱导性棕色脂肪。在肥胖耐受小鼠的骨骼肌肌束间也散在分布棕色样脂肪细胞(Almind et al., 2007)。米色脂肪细胞在接受一定的刺激后与棕色脂肪一样，可以燃烧脂肪并产热，而成人以米色脂肪为主要产热脂肪，因此如何激活米色脂肪细胞从而治疗肥胖症及相关代谢疾病已经受到越来越多的重视。

3.3.1 米色脂肪细胞的来源

用 $CD45^-CD31^-Ter119^-(Lin^-)CD29^+Sca1^+$ 细胞表面蛋白从皮下白色脂肪和骨骼肌中分离出的前体细胞(称为 $Sca-1^+$ 前体细胞，ScaPC)在 BMP7 作用下可形成诱导性棕色脂肪细胞(Schulz et al., 2011)。

不同部位的米色脂肪细胞来源不同，Seale 等(2008)发现，冷刺激会引起白色脂肪组织中出现散在的棕色样脂肪细胞，这些细胞表达 UCP1，而这些细胞并非来源于 $Myf5^+$ 的细胞，说明米色脂肪细胞与棕色脂肪细胞的来源不同，但这只局限于腹股沟皮下和附睾周围脂肪组织。肩胛间区和腋窝的白色脂肪细胞则来源于 $Myf5^+$ 的细胞 (Sanchez-Gurmaches et al., 2012)。

白色脂肪组织中的棕色样脂肪细胞的前体为 $PDGFR\alpha^+$ 的细胞(Lee et al., 2012)，β 肾上腺素能受体激动剂(CL316, 243)处理后会出现增殖(Granneman et al., 2005)，但由此产生的米色脂肪细胞在附睾周围脂肪中仅占多腔脂肪细胞的 25% 左右，而在腹股沟脂肪中所占的比例更少(Granneman et al., 2005; Lee et al., 2012)。最近也有文献报道，米色脂肪具有平滑肌样特点，谱系追踪显示，至少部分米色脂肪具有平滑肌样来源($Myh11^+$)(Long et al., 2014)。

转分化是米色脂肪另外一种可能来源。Cinti 等(Himms-Hagen et al., 2000; Cinti, 2009; Barbatelli et al., 2010)用 CL316, 243 处理大鼠一周后，腹膜后的脂肪细胞介于白色脂肪细胞和棕色脂肪细胞之间。冷暴露后，白色脂肪组织中，减少的白色脂肪细胞数目与增加的棕色样脂肪细胞的数目相当，因此米色脂肪更可能由白色脂肪转分化而来的。仍需要更多的研究来阐明米色脂肪的来源。

3.3.2 米色脂肪细胞的发育调节

UCP1 也是米色脂肪激活的主要表现。许多诱导棕色脂肪细胞发育的因素也能诱导米色脂肪细胞的形成，例如 β3 肾上腺素能受体激动剂、PPAR 激动剂、维甲酸、FGF21

等，另外 BMP 家族的某些成员、饮食及运动等都能诱导米色脂肪的形成(图版Ⅺ图3)。

3.3.2.1 β3肾上腺素能受体激动剂对米色脂肪的发育调节

β3 肾上腺素能受体除了结合去甲肾上腺素，介导交感神经系统在棕色脂肪细胞的作用外，还能诱导冷暴露小鼠白色脂肪棕色化(Barbatelli et al., 2010)。持续冷暴露或 β 肾上腺素能受体激动剂可以使白色脂肪内出现棕色样脂肪细胞，散在分布在白色脂肪组织内部(Barbatelli et al., 2010; Petrovic et al., 2010)。慢性给与β3 肾上腺素能受体激动剂(BRL26830A、CGP-12177、316243)可以诱导UCP1在白色脂肪组织内的异位表达(Cousin et al., 1992; Guerra et al., 1998; Pico et al., 1998; Himms-Hagen et al., 2000; Kullmann et al., 2009)，即出现米色脂肪细胞。β3 肾上腺素能受体缺失会明显抑制冷暴露引起的白色脂肪组织内 UCP1 的表达，但对棕色脂肪影响不大，提示 β3 肾上腺素能受体对白色脂肪棕色化的作用要比其对棕色脂肪的激活作用更重要，这种差异可能是棕色脂肪组织内其他 β 肾上腺素能受体代偿的结果。脂肪组织特异性敲除 PRDM16 会明显抑制冷暴露或 β3 肾上腺素能受体激动剂诱导的皮下米色脂肪细胞的出现(Cohen et al., 2014)。最近有报道称，低温能直接激活脂肪细胞产热基因表达，而这种作用特异性的见于白色和米色脂肪细胞，棕色脂肪细胞则没有这种作用，白色和米色脂肪细胞对低温的直接感知不依赖于 β 肾上腺素能受体和其下游的 cAMP/PKA/CREB 信号通路(Ye et al., 2013)。

心脏利尿钠肽和 β 肾上腺素能受体激动剂都能刺激人脂肪细胞脂解，提示心脏利尿钠肽对白色脂肪棕色化也有作用。心房利尿钠肽(ANP)和脑利尿钠肽(BNP)能激活人脂肪细胞的 PGC-1α 和 UCP1 表达，诱导线粒体合成及解偶联呼吸。低浓度 ANP 与 β 肾上腺素能受体激动剂具有叠加作用。冷暴露小鼠，血循环中心脏利尿钠肽水平增加，棕色脂肪组织和白色脂肪组织中心脏利尿钠肽受体增加，而心脏利尿钠肽清除受体(NP clearance receptor, Nprc)表达降低。NPR-C 敲除小鼠，白色脂肪组织和棕色脂肪组织体积变小，但 UCP1 表达增加(Bordicchia et al., 2012)。

在白色脂肪组织中，β 肾上腺素能受体激动剂能增加 C/EBPβ 的表达，这种作用与 C/EBPβ 的转录抑制因子 Hoxc8 的下调表达有关。缺失 C/EBPβ 与产热缺陷有关。增加白色脂肪细胞 C/EBPβ 的量能触发棕色脂肪相关基因转录(Carmona et al., 2005; Karamanlidis et al., 2007; Kajimura et al., 2009)。

3.3.2.2 PPAR 激动剂对米色脂肪的发育调节

除了对棕色脂肪的作用，PPARγ 激动剂对啮齿类动物和人的白色脂肪棕色化也发挥作用(Fukui et al., 2000; Wilson-Fritch et al., 2003; Sell et al., 2004; Bogacka et al., 2005)。PPARγ 激动剂能诱导白色脂肪组织线粒体的生物合成、形成丰富的嵴、诱导 UCP1 及其他线粒体基因表达、促进白色脂肪组织中形成多腔脂肪细胞等(Wilson-Fritch et al., 2003; Wilson-Fritch et al., 2004; Bogacka et al., 2005; Koh et al., 2009)。PPARγ 激动剂通过稳定 PRDM16(Ohno et al., 2012)，抑制白色脂肪细胞特异性基因的表达(Vernochet et al., 2009)，以及 SirT1 依赖的 PPARγ 去乙酰化(Qiang et al., 2012)促进棕色样脂肪细胞的出现。TLE3(transducin-like enhancer of split 3)作为白色脂肪选择的辅助因子，能破坏

PRDM16 和 PPARγ 相互作用。在棕色脂肪中过表达 TLE3，能拮抗 PRDM16 的作用，抑制棕色脂肪特异基因的表达，诱导白色脂肪选择性基因的表达，脂肪酸氧化和产热受损。相反，皮下白色脂肪缺失 TLE3 的小鼠，表现为皮下白色脂肪产热增加，并能保护年龄引起的棕色脂肪功能受损(Villanueva et al., 2013)。PPARγ 可被抗糖尿病药物 TZD 类选择性激活(Spiegelman, 1998)。罗格列酮可作用于 PGC-1α 的启动子远端与 PPARγ/RXR 结合的 PPRE，提高 PGC-1α 表达(Teruel et al., 2003; Hondares et al., 2006)，进而增加棕色脂肪特异性基因的表达。PPARγ 与棕色或白色脂肪细胞决定基因的结合与脂肪部位相关，提示 TZD 促进白色或棕色脂肪细胞表型是表观预编程的(Siersbaek et al., 2012)。TZD 能否促进人棕色脂肪细胞的募集和激活还不清楚，但 PPARγ 激动剂能促进人前脂肪细胞系出现棕色脂肪细胞表型(Elabd et al., 2009)。

一般认为 PPARγ 控制与脂肪细胞分化相关的 *UCP1* 基因的表达，而 PPARα 控制成熟脂肪细胞 *UCP1* 基因的表达(Villarroya et al., 2007)。PPARα 通过诱导 PGC-1α 和 PRDM16 促进白色脂肪棕色化(Li et al., 2005; Xue et al., 2005; Hondares et al., 2011)。PPARα 激动剂(bezafibrate、GW7647、WY-14,643)可诱导人和小鼠白色脂肪细胞线粒体生物合成和脂肪酸氧化(Cabrero et al., 2001; Ribet et al., 2010; Ohno et al., 2012)，最终促进白色脂肪棕色化。

3.3.2.3 BMP 家族对米色脂肪的发育调节

BMP 家族中的 BMP2 和 BMP4 能促进间充质干细胞向前细胞的定向(Ahrens et al., 1993; Sottile and Seuwen, 2000; Tang et al. 2004)，或促进前脂肪细胞向成熟白色脂肪细胞的分化(Wang et al., 1993)。Tang 等(2004)研究发现，BMP4 诱导多能 C3H10T1/2 细胞产生的前脂肪样细胞经皮下注入裸鼠体内，也可分化为脂肪组织。Bowers 等(2006)用 5-氮杂胞苷处理 C3H10T1/2 细胞筛选到一株高表达 BMP4 的前脂肪细胞系，这种内源的 BMP4 对于多能干细胞获得前脂肪细胞特性是必需的。BMP4 促进多能干细胞定向为前脂肪细胞主要是通过激活 Smad 信号通路，P38/MAPK 信号通路也参与部分作用(Huang et al., 2009)，赖氨酰氧化酶(lysyl oxidase, LOX)对该过程非常重要(Huang et al., 2009)。那么，BMP4 究竟诱导多能干细胞定向为前白色脂肪细胞、前棕色脂肪细胞还是米色脂肪细胞呢？BMP4 的作用与 BMP7 的作用相似，能诱导 C3H10T1/2 细胞和小鼠前白色脂肪细胞形成棕色样脂肪细胞，表达米色脂肪的标志蛋白 TMEM26(Xue et al., 2014)。BMP4 和 BMP7 也能诱导来源于人皮下白色脂肪的原代脂肪组织干细胞表达 UCP1，抑制白色脂肪特异性标志蛋白 TCF21 表达，然而线粒体含量，氧耗均不增加。这也说明小鼠的实验结果在人身上不是总适用的(Elsen et al., 2014; Obregon, 2014)。而汤其群课题组的研究显示，脂肪组织内特异性高表达 BMP4 后，皮下白色脂肪组织出现棕色化(Qian et al., 2013)，脂滴和细胞变小，线粒体生成增加，脂肪酸氧化基因表达增加。BMP4 所引起的脂肪组织发育变化是通过 PGC-1α 这一关键基因起作用。在前脂肪细胞 3T3-L1 分化过程中加入 BMP4，也可使细胞获得棕色样脂肪细胞的表型。BMP4 这种新功能的阐明可为临床干预肥胖和改善胰岛素敏感性提供新思路。

3.3.2.4 维甲酸对米色脂肪的发育调节

全反式维甲酸可以引起小鼠白色脂肪的棕色化，表现为 UCP1 表达增加，多脂滴细胞增加，线粒体基因、产热基因、脂肪酸氧化相关基因增加(Mercader et al., 2006)，但对于人的脂肪细胞系或原代白色脂肪细胞的 UCP1 表达则没有作用，甚至抑制 UCP1 表达。视黄醛也有类似维甲酸促进白色脂肪棕色化的作用，视黄醛通过选择性激活 RAR、招募 PGC-1α 并诱导 *UCP1* 启动子活性来诱导白色脂肪细胞中 UCP1 的 mRNA 和蛋白表达。视黄醛脱氢酶 1(retinaldehyde dehydrogenase1, Aldh1a1)是将视黄醛转化为维甲酸的限速酶。抑制 Aldh1a1 在白色脂肪细胞的表达，会出现棕色样脂肪细胞改变。缺乏 Aldh1a1 或被抑制了 Aldh1a1 活性的小鼠对低温更能耐受。

3.3.2.5 免疫细胞对米色脂肪发育的调节

最新研究结果显示，脂肪中的免疫细胞对米色脂肪的发育尤为重要。运动或慢性冷暴露时，脂肪组织的嗜酸性粒细胞能释放 IL-4，活化 M2 型巨噬细胞并释放儿茶酚胺，激活 β 肾上腺素能信号通路，促进白色脂肪棕色化及产热过程(Qiu et al., 2014)。冷暴露或运动亦能诱导 Metrnl(meteorin-like)，过表达 Metrnl 会促进米色脂肪的形成，该作用也依赖于嗜酸性粒细胞产生的 IL-4/IL-13 对 M2 型巨噬细胞的活化。这些结果都说明免疫系统能调节米色脂肪的形成(Rao et al., 2014)。

3.3.2.6 体育锻炼对米色脂肪发育的调节

运动能改善糖尿病、心血管疾病及其他代谢性疾病。运动不仅能增加骨骼肌葡萄糖消耗，改善胰岛素敏感性；运动也能增加白色脂肪的棕色化，进一步改善代谢。

(1) Irisin 的作用

运动增加腹股沟皮下脂肪 UCP1 的表达是通过 PGC-1α 完成的。运动对腹股沟皮下脂肪的 UCP1 诱导作用比性腺周围脂肪(eWAT)更强(Ringholm et al., 2013)。PGC-1α 通过刺激肌肉组织中 *FNDC5* 基因的表达，后者经水解后成为 Irisin 释放入血。Irisin 的增加对于运动引起的白色脂肪棕色化是必需的(Xu et al., 2011; Bostrom et al., 2012)，但并不影响 BAT 中的棕色脂肪特异性基因的表达。肌肉敲除 *Mstn*(myostatin) 的小鼠脂肪体积减少，而白色脂肪组织中棕色脂肪的特异性基因(*UCP1* 和 *PGC-1α*)明显增加，米色脂肪的基因(*TMEM26* 和 *CD137*)表达也增加，出现明显的白色脂肪棕色化。在肌肉内敲除 *Mstn*，AMPK 的表达和磷酸化增加，激活 PGC-1α 和 FNDC5(Shan et al., 2013)，最终由 FNDC5 水解产物 Irisin 驱动白色脂肪棕色化。冷暴露也能增加人血循环中 Irisin 水平，这与冷暴露引起的肌肉颤抖有关(Lee et al., 2014)。Irisin 激活米色脂肪细胞的产生的具体机制还不清，Irisin 的受体的鉴定将会是今后其分子机制和临床应用的研究热点。

(2) IL-6 的作用

IL-6 对运动引起的白色脂肪棕色化也发挥作用。每天重复给予 IL-6 能增加 iWAT(inquinal adipose tissue)的 UCP1 含量，IL-6 对于运动所引起的 iWAT 的 UCP1 含量增加是必需的，对冷暴露诱导的 UCP1 蛋白的诱导也是必需的(Knudsen et al., 2014)。

3.3.2.7 饮食因素对米色脂肪的激活作用

饮食因素在白色脂肪棕色化中也发挥重要调节作用。甲硫氨酸摄入限制的大鼠(Hasek et al., 2010)和小鼠(Plaisance et al., 2010)，白色脂肪中 UCP1 被诱导，这种作用部分是通过增强 β 肾上腺素能受体信号通路来实现的(Plaisance et al., 2010)。辣椒素和一些非辛辣的辣椒素类似物(辣椒素酯类物质)通过激活 SNS，增强肾上腺髓质儿茶酚胺的分泌(Watanabe et al., 1987)，以及结合并激活瞬时受体电位通道[transient receptor potential (TRP) channel](Zhang et al., 2007; Ono et al., 2011)，也能诱导啮齿类动物白色脂肪的棕色化(Joo et al., 2010)。白藜芦醇能增加 MEF 来源的脂肪细胞的线粒体 DNA 含量、脂肪酸氧化和 UCP1 表达水平，提示具有促进棕色化作用(Mercader et al., 2011)。藻褐素或者藻褐素丰富的海藻提取物可以通过诱导腹部脂肪 UCP1 mRNA 和蛋白质的表达、诱导脂代谢相关基因的表达，刺激腹部白色脂肪棕色化，抵抗食物诱导的肥胖，减少腹部白色脂肪(Maeda et al., 2005; Maeda et al., 2007; Jeon et al., 2010; Hu et al., 2012)。藻褐素促进白色脂肪棕色化，但不影响棕色脂肪 UCP1 表达，提示具有作用部位的特异性。

3.3.2.8 microRNA 调节米色脂肪发育

冷暴露或肾上腺素能激活能诱导 miR-196a 表达。Hox 家族成员中 HOXC8 在人白色脂肪细胞的前体细胞高表达，能抑制棕色脂肪特异性关键基因表达(Moriet al., 2012)。*miR-196a* 基因位于 *HoxC8* 基因附近，与 HOXC8 的 mRNA 有保守的互补序列，能下调 HOXC8 的表达(Yekta et al., 2004)。这提示 miR-196a 可能通过抑制 HOXC8，促进白色脂肪细胞转换为米色细胞，而在脂肪组织特异性过表达 miR-196a 确实能诱导白色脂肪细胞中出现棕色样脂肪细胞(Mori et al., 2012)。

4. 脂肪组织与代谢

不同类型脂肪组织的脂肪细胞对代谢的作用各不相同(图版 X 图 1)，下面将分别介绍各自的功能。

4.1 白色脂肪组织与代谢

白色脂肪组织是以脂肪垫方式遍布全身，主要功能是参与能量贮存和动员。白色脂肪组织有很强的扩充能力，由脂肪细胞大小和数量增加所致过多的脂肪组织与肥胖和代谢综合征密切相关。此外，脂肪细胞还具有自分泌、旁分泌和内分泌的功能，例如，分泌瘦素和脂联素等。哺乳动物最大的脂肪垫位于腹内和皮下，在其他部位也存在脂肪组织,如脸部、四肢、骨髓等。不同部位的脂肪组织形态和功能存在差异(Prunet-Marcassus et al., 2006; Rosen and MacDougald, 2006; Gesta et al., 2007)。腹内的脂肪与Ⅱ型糖尿病和心血管疾病密切相关,而皮下的脂肪对代谢综合征则具有对抗作用(Bjorntorp, 1990; Lafontan and Berlan, 2003; Zeve et al., 2009)。乳腺和大腿的白色脂肪组对性激素敏感,颈部和上背部的白色脂肪组织对糖皮质激素敏感。正因为不同部位的白色脂肪组织具有不同特性,因此可以解释为什么有些部位脂肪组织堆积与严重的代谢紊乱有关,而有些则无关。

4.2 棕色脂肪对代谢的影响

4.2.1 能量消耗

棕色脂肪组织主要功能是产热御寒。多种哺乳动物在成年后体内仍保留一定量棕色脂肪组织。在成年人的颈部、锁骨上部和脊柱旁侧有活跃的棕色脂肪存在。在寒冷或过多能量摄入时,棕色脂肪细胞可以氧化脂肪酸以产热。人体内棕色脂肪组织的量与体重指数(body mass index, BMI)呈负相关,提示棕色脂肪组织在人体能量平衡调节中起重要作用。研究人员估计,50g 活跃的棕色脂肪组织就可以将人体静息能量消耗提高 20%。因此,刺激棕色脂肪细胞发育,可以用于对抗肥胖及相关代谢性疾病。

4.2.2 棕色脂肪消耗葡萄糖

棕色脂肪组织是重要的葡萄糖储存库,能够提高胰岛素敏感性和葡萄糖摄取。增加棕色脂肪组织量,葡萄糖摄取和消耗亦增加。将供体鼠的棕色脂肪组织移植到受体鼠的腹腔内,以增加棕色脂肪组织量,受体鼠葡萄糖耐量提高,胰岛素敏感性增加,体重减轻,脂肪体积下降。同样将正常饮食小鼠的 BAT 移植到 HFD 小鼠的腹腔内,能够完全逆转 HFD 诱导的胰岛素抵抗(Stanford et al., 2013)。而将正常饮食小鼠的棕色脂肪组织移植到 HFD 小鼠的肩胛间区也同样会逆转 HFD 诱导的肥胖和提高胰岛素敏感性(Liu et al., 2013)。另有研究表明,将胚胎的棕色脂肪组织移植到 STZ 诱导的Ⅰ型糖尿病小鼠皮下,能恢复其血糖(Gunawardana and Piston, 2012)。通过皮下微型渗透泵给与 CL316,243,能增加非肥胖大鼠的棕色脂肪组织体积,增加全身基础和胰岛素刺激葡萄糖消耗,但对体重却没有影响(de Souza et al., 1997)。

实验表明,成人冷暴露引起的葡萄糖摄取明显超过胰岛素刺激的骨骼肌葡萄糖摄取。冷暴露后棕色脂肪组织的葡萄糖摄取增加 12 倍,与全身能量消耗增加呈正相关,而胰岛素刺激的棕色脂肪组织葡萄糖摄取仅增加 5 倍(Orava et al., 2011)。在棕色脂肪组织中 GLUT4 表达比白色脂肪组织高(Orava et al., 2011)。冷暴露小鼠棕色脂肪组织上调的基因中,许多是参与葡萄糖摄取的基因(Yu et al., 2002)。GLUT1 和 GLUT4 在冷暴露小鼠的棕色脂肪组织的表达比其他组织更高(Bartelt et al., 2011)。冷暴露也能使 GLUT1 和 GLUT4 转运到细胞膜上,促进葡萄糖转运(Shimizu et al., 1998)。冷暴露能恢复肥胖小鼠的葡萄糖耐量,增加正常和肥胖小鼠的葡萄糖和脂肪酸摄取,而以棕色脂肪葡萄糖的摄取增加最明显(Bartelt et al., 2011; Nedergaard et al., 2011; Williams and Fisher, 2011)。这些结果说明：冷暴露时，棕色脂肪组织是重要的葡萄糖储存库，能够提高胰岛素敏感性和葡萄糖摄取(Vallerand et al., 1987; Skarulis et al., 2010)。棕色脂肪组织也可通过非胰岛素依赖方式摄取葡萄糖。选择性部分非经典 hedgehog 信号通路抑制剂能明显增加棕色脂肪组织和骨骼肌非胰岛素依赖的葡萄糖摄取，这种作用是通过 Smo-AMPK 轴激活来完成的(Teperino et al., 2012)。

4.2.3 清除血中甘油三脂

棕色脂肪组织还具有清除血中甘油三酯的作用(Bartelt et al., 2011)。脂肪酸是棕色脂肪组织的重要燃料或能源。棕色脂肪细胞首先利用自己储存的脂滴作为底物进行产热。早期的产热机制为交感神经释放的去甲肾上腺素促进甘油三酯水解为游离的脂肪酸，一些脂肪酸用于激活 UCP1，其他的脂肪酸则进入线粒体燃烧产热。但是棕色脂肪组织占体重很小一部分，储存的脂滴仅够维持短时产热。进一步的能量需要棕色脂肪组织以外的组织提供。棕色脂肪组织可将循环中的甘油三酯摄取并用于产热，因此能明显降低循环中的甘油三酯，对于预防和治疗代谢综合征具有潜在的应用价值(Bartelt et al., 2011; Nedergaard et al., 2011)。

4.2.4 内分泌功能

传统观念认为，棕色脂肪组织主要功能为氧化脂肪酸和葡萄糖以维持产热，而最近许多数据显示棕色脂肪组织也可能具有内分泌的作用。到目前为止，已知的棕色脂肪组织通过内分泌作用释放的是有活性的甲状腺激素。在棕色脂肪组织中存在 Dio2 能将 T4 转变为 T3，棕色脂肪产热被诱导后，Dio2 被明显激活(Silva and Larsen, 1983)。此外，在基础条件下，FGF21 主要在肝脏产生，而产热激活后，FGF21 在棕色脂肪组织表达增加，并由棕色脂肪细胞释放出来(Chartoumpekis et al., 2011; Hondares et al., 2011; Fisher et al., 2012)。BMP8b 也是重要的棕色脂肪组织释放因子，能增加棕色脂肪细胞对去甲肾上腺素的敏感性(Whittle et al., 2012)。棕色脂肪组织还能分泌 IGF-1(Gunawardana and Piston, 2012)、IL-6(Burysek and Houstek, 1997)、VEGF(Asano et al., 1999)等因子。这都提示棕色脂肪组织也可能具有内分泌的作用。

4.3 米色脂肪对代谢的影响

白色脂肪棕色化与棕色脂肪的激活一样，是预防和治疗肥胖及相关疾病，例如糖尿病、肝硬化等非常有用的策略。然而，现在很多关于白色脂肪棕色化研究都是在啮齿类动物进行的，但啮齿类动物的脂肪组织与人类有重要的不同，因此选择与人类脂肪组织相近的物种，如雪貂，对研究药物或饮食因素对白色脂肪棕色化的作用和代谢的影响将会具有更重要的意义。

5. 结　语

脂肪组织分为白色脂肪、棕色脂肪和米色脂肪。棕色脂肪组织和米色脂肪对代谢有益。白色脂肪中的皮下脂肪对代谢也有益，而内脏脂肪是胰岛素抵抗和Ⅱ型糖尿病的主要原因。因此选择性促进有利脂肪的发育，例如激活米色脂肪或棕色脂肪组织，抑制内脏脂肪发育将是治疗肥胖和肥胖相关疾病的新策略。

致谢

感谢国家自然科学基金的资助，我们对因篇幅原因未被引用的相关文献表示歉意。

参 考 文 献

Adams A C, Cheng C C, Coskun T, et al. 2012. FGF21 requires betaklotho to act in vivo. PLoS One, 7(11): e49977.

Ahima R S. 2006. Adipose tissue as an endocrine organ. Obesity (Silver Spring), 14 Suppl 5: 242S-249S.

Ahrens M, Ankenbauer T, Schroder D, et al. 1993. Expression of human bone morphogenetic proteins-2 or -4 in murine mesenchymal progenitor C3H10T1/2 cells induces differentiation into distinct mesenchymal cell lineages. DNA Cell Biol, 12(10): 871-880.

Akimoto T, Ushida T, Miyaki S, et al. 2005. Mechanical stretch inhibits myoblast-to-adipocyte differentiation through Wnt signaling. Biochem Biophys Res Commun, 329(1): 381-385.

Almind K and Kahn C R. 2004. Genetic determinants of energy expenditure and insulin resistance in diet-induced obesity in mice. Diabetes, 53(12): 3274-3285.

Almind K, Manieri M, Sivitz W I, et al. 2007. Ectopic brown adipose tissue in muscle provides a mechanism for differences in risk of metabolic syndrome in mice. Proc Natl Acad Sci U S A, 104(7): 2366-2371.

Alvarez R, de Andres J, Yubero P, et al. 1995. A novel regulatory pathway of brown fat thermogenesis. Retinoic acid is a transcriptional activator of the mitochondrial uncoupling protein gene. J Biol Chem 270(10): 5666-5673.

Apovian C M. and Gokce N. 2012. Obesity and cardiovascular disease. Circulation, 125(9): 1178-1182.

Arch J R, Ainsworth A T, Cawthorne M A, et al. 1984. Atypical beta-adrenoceptor on brown adipocytes as target for anti-obesity drugs. Nature, 309(5964): 163-165.

Arch J R. 2008. The discovery of drugs for obesity, the metabolic effects of leptin and variable receptor pharmacology: perspectives from beta3-adrenoceptor agonists. Naunyn Schmiedebergs Arch Pharmacol, 378(2): 225-240.

Arnsdorf E J, Tummala, P and Jacobs C R, et al. 2009. Non-canonical Wnt signaling and N-cadherin related beta-catenin signaling play a role in mechanically induced osteogenic cell fate. PLoS One, 4(4): e5388.

Asano A, Kimura K, Saito M. 1999. Cold-induced mRNA expression of angiogenic factors in rat brown adipose tissue. J Vet Med Sci, 61(4): 403-409.

Atit R, Sgaier S K, Mohamed O A, et al. 2006. Beta-catenin activation is necessary and sufficient to specify the dorsal dermal fate in the mouse. Dev Biol, 296(1): 164-176.

Attisano L, Wrana J L, Cheifetz S, et al. 1992. Novel activin receptors: distinct genes and alternative mRNA splicing generate a repertoire of serine/threonine kinase receptors. Cell, 68(1): 97-108.

Barbatelli G, Murano I, Madsen L, et al. 2010. The emergence of cold-induced brown adipocytes in mouse white fat depots is determined predominantly by white to brown adipocyte transdifferentiation. Am J Physiol Endocrinol Metab, 298(6): E1244-1253.

Barbera M J, Schluter A, Pedraza N, et al. 2001. Peroxisome proliferator-activated receptor alpha activates transcription of the brown fat uncoupling protein-1 gene. A link between regulation of the thermogenic and lipid oxidation pathways in the brown fat cell. J Biol Chem, 276(2): 1486-1493.

Bartelt A, Bruns O T, Reimer R, et al. 2011. Brown adipose tissue activity controls triglyceride clearance. Nat Med, 17(2): 200-205.

Beltowski J. 2003. Adiponectin and resistin--new hormones of white adipose tissue. Med Sci Monit, 9(2): RA55-61.

Bennett C N, Hodge C L, MacDougald O A, et al. 2003. Role of Wnt10b and C/EBPalpha in spontaneous adipogenesis of 243 cells. Biochem Biophys Res Commun, 302(1): 12-16.

Berger J and Moller D E. 2002. The mechanisms of action of PPARs. Annu Rev Med, 53: 409-435.

Berry R and Rodeheffer M S. (2013). "Characterization of the adipocyte cellular lineage in vivo." Nat Cell Biol 15(3): 302-308.

Bilkovski R, Schulte D M, Oberhauser F, et al. 2010. Role of WNT-5a in the determination of human mesenchymal stem cells into preadipocytes. J Biol Chem, 285(9): 6170-6178.

Bilkovski R, Schulte D M, Oberhauser F, et al. 2011. Adipose tissue macrophages inhibit adipogenesis of mesenchymal precursor cells via wnt-5a in humans. Int J Obes (Lond), 35(11): 1450-1454.

Billon N, Jolicoeur C, Raff M. 2006. Generation and characterization of oligodendrocytes from lineage-selectable embryonic stem cells in vitro. Methods Mol Biol, 330: 15-32.

Bjorntorp P. 1990. "Portal" adipose tissue as a generator of risk factors for cardiovascular disease and diabetes. Arteriosclerosis, 10(4): 493-496.

Boden G and Shulman G I. 2002. Free fatty acids in obesity and type 2 diabetes: defining their role in the development of insulin resistance and beta-cell dysfunction. Eur J Clin Invest, 32 Suppl 3: 14-23.

Bogacka I, Xie H, Bray G A, et al. 2005. Pioglitazone induces mitochondrial biogenesis in human subcutaneous adipose tissue in vivo. Diabetes, 54(5): 1392-1399.

Bordicchia M, Liu D, Amri E. Z, et al. 2012. Cardiac natriuretic peptides act via p38 MAPK to induce the brown fat thermogenic program in mouse and human adipocytes. J Clin Invest, 122(3): 1022-1036.

Bostrom P, Wu J, Jedrychowski M P, et al. 2012. A PGC1-alpha-dependent myokine that drives brown-fat-like development of white fat and thermogenesis. Nature, 481(7382): 463-468.

Bowers R R, Kim J W, Otto T C, et al. 2006. Stable stem cell commitment to the adipocyte lineage by inhibition of DNA methylation: role of the BMP-4 gene. Proc Natl Acad Sci U S A, 103(35): 13022-13027.

Burysek L, and Houstek J. 1997. beta-Adrenergic stimulation of interleukin-1alpha and interleukin-6 expression in mouse brown adipocytes. FEBS Lett, 411(1): 83-86.

Cabrero A, Alegret M, Sanchez R M, et al. 2001. Bezafibrate reduces mRNA levels of adipocyte markers and increases fatty acid oxidation in primary culture of adipocytes. Diabetes, 50(8): 1883-1890.

Calabro P, Golia E, Maddaloni V, et al. 2009. Adipose tissue-mediated inflammation: the missing link between obesity and cardiovascular disease? Intern Emerg Med, 4(1): 25-34.

Cannon B and Nedergaard J. 2004. Brown adipose tissue: function and physiological significance. Physiol Rev, 84(1): 277-359.

Cao W, Daniel K W, Robidoux J, et al. 2004. p38 mitogen-activated protein kinase is the central regulator of cyclic AMP-dependent transcription of the brown fat uncoupling protein 1 gene. Mol Cell Biol, 24(7): 3057-3067.

Carey A L, Formosa M F, Van Every B, et al. 2013. Ephedrine activates brown adipose tissue in lean but not obese humans. Diabetologia, 56(1): 147-155.

Carmona M C, Hondares E, Rodriguez de la Concepcion M L, et al. 2005. Defective thermoregulation, impaired lipid metabolism, but preserved adrenergic induction of gene expression in brown fat of mice lacking C/EBPbeta. Biochem J, 389(Pt 1): 47-56.

Cassard-Doulcier A M, Gelly C, Fox N, et al. (1993). Tissue-specific and beta-adrenergic regulation of the mitochondrial uncoupling protein gene: control by cis-acting elements in the 5'-flanking region. Mol Endocrinol, 7(4): 497-506.

Chambon P. 1996. A decade of molecular biology of retinoic acid receptors. FASEB J, 10(9): 940-954.

Chang J S, Fernand V, Zhang Y, et al. 2012. NT-PGC-1alpha protein is sufficient to link beta3-adrenergic receptor activation to transcriptional and physiological components of adaptive thermogenesis. J Biol Chem, 287(12): 9100-9111.

Chartoumpekis D V, Habeos I G, Ziros P G, et al. 2011. Brown adipose tissue responds to cold and adrenergic stimulation by induction of FGF21. Mol Med, 17(7-8): 736-740.

Chau M D, Gao J, Yang Q, et al. 2010. Fibroblast growth factor 21 regulates energy metabolism by activating the AMPK-SIRT1-PGC-1alpha pathway. Proc Natl Acad Sci U S A, 107(28): 12553-12558.

Chen Y, Siegel F, Kipschull S, et al. 2013. miR-155 regulates differentiation of brown and beige adipocytes via a bistable circuit. Nat Commun, 4: 1769.

Cheng S Y, Leonard J L, Davis P J. 2010. Molecular aspects of thyroid hormone actions. Endocr Rev, 31(2): 139-170.

Christodoulides C, Scarda A, Granzotto M, et al. 2006. WNT10B mutations in human obesity. Diabetologia, 49(4): 678-684.

Christy R J, Kaestner K H, Geiman, D E, et al. 1991. CCAAT/enhancer binding protein gene promoter: binding of nuclear factors during differentiation of 3T3-L1 preadipocytes. Proc Natl Acad Sci U S A, 88(6): 2593-2597.

Cinti S. 2001. The adipose organ: morphological perspectives of adipose tissues. Proc Nutr Soc, 60(3): 319-328.

Cinti S. 2005. The adipose organ. Prostaglandins Leukot Essent Fatty Acids, 73(1): 9-15.

Cinti S. 2009. Transdifferentiation properties of adipocytes in the adipose organ. Am J Physiol Endocrinol Metab, 297(5): E977-986.

Cinti S. 2011. Between brown and white: novel aspects of adipocyte differentiation. Ann Med, 43(2): 104-115.

Cinti S. 2012. The adipose organ at a glance. Dis Model Mech, 5(5): 588-594.

Clarke L, Bryant M J, Lomax M A, et al. 1997a. Maternal manipulation of brown adipose tissue and liver development in the ovine fetus during late gestation." Br J Nutr 77(6): 871-883.

Clarke S L, Robinson C E, and Gimble J M, et al. (1997b). CAAT/enhancer binding proteins directly modulate transcription from the peroxisome proliferator-activated receptor gamma 2 promoter. Biochem Biophys Res Commun, 240(1): 99-103.

Cnop M. 2008. Fatty acids and glucolipotoxicity in the pathogenesis of Type 2 diabetes. Biochem Soc Trans, 36(Pt 3): 348-352.

Cohen P, Levy J D, Zhang Y, et al. 2014. Ablation of PRDM16 and beige adipose causes metabolic dysfunction and a subcutaneous to visceral fat switch. Cell 156(1-2): 304-316.

Collins S, Yehuda-Shnaidman E, Wang H. 2010. Positive and negative control of Ucp1 gene transcription and the role of beta-adrenergic signaling networks. Int J Obes (Lond), 34 Suppl 1: S28-33.

Cousin B, Cinti S, Morroni M, et al. 1992. Occurrence of brown adipocytes in rat white adipose tissue: molecular and morphological characterization. J Cell Sci, 103 (Pt 4): 931-942.

Cypess A M, Chen Y C, Sze C, et al. 2012. Cold but not sympathomimetics activates human brown adipose tissue in vivo. Proc Natl Acad Sci U S A 109(25): 10001-10005.

Cypess A M, Lehman S, Williams G, et al. 2009. Identification and importance of brown adipose tissue in adult humans. N Engl J Med, 360(15): 1509-1517.

Dalamaga M, Diakopoulos K N, Mantzoros CS, et al. 2012. The role of adiponectin in cancer: a review of current evidence. Endocr Rev, 33(4): 547-594.

Dani C, Amri E Z., Bertrand B, et al. 1990. Expression and regulation of pOb24 and lipoprotein lipase genes during adipose conversion. J Cell Biochem, 43(2): 103-110.

de Souza C J, Hirshman M F, Horton E S. 1997. CL-316,243, a beta3-specific adrenoceptor agonist, enhances insulin-stimulated glucose disposal in nonobese rats. Diabetes, 46(8): 1257-1263.

DeBoer M D. 2013. Obesity, systemic inflammation, and increased risk for cardiovascular disease and diabetes among adolescents: a need for screening tools to target interventions. Nutrition, 29(2): 379-386.

del Mar Gonzalez-Barroso M, Pecqueur C, Gelly C, et al. 2000. Transcriptional activation of the human ucp1 gene in a rodent cell line. Synergism of retinoids, isoproterenol, and thiazolidinedione is mediated by a multipartite response element. J Biol Chem, 275(41): 31722-31732.

Ding X, Boney-Montoya J, Owen BM, et al. 2012. betaKlotho is required for fibroblast growth factor 21 effects on growth and metabolism. Cell Metab, 16(3): 387-393.

Drew J E. 2012. Molecular mechanisms linking adipokines to obesity-related colon cancer: focus on leptin. Proc Nutr Soc, 71(1): 175-180.

Elabd C, Chiellini C, Carmona M, et al. 2009. Human multipotent adipose-derived stem cells differentiate into functional brown adipocytes. Stem Cells, 27(11): 2753-2760.

Elsen M, Raschke S, Tennagels N, et al. 2014. BMP4 and BMP7 induce the white-to-brown transition of primary human adipose stem cells. Am J Physiol Cell Physiol, 306(5): C431-440.

Enerback S. 2010. Human brown adipose tissue. Cell Metab, 11(4): 248-252.

Esau C, Kang X, Peralta E, et al. 2004. MicroRNA-143 regulates adipocyte differentiation. J Biol Chem, 279(50): 52361-52365.

Fisher F M, Estall J L, Adams A C, et al. 2011. Integrated regulation of hepatic metabolism by fibroblast growth factor 21 (FGF21) in vivo. Endocrinology, 152(8): 2996-3004.

Fisher F M, Kleiner S, Douris N, et al. 2012. FGF21 regulates PGC-1alpha and browning of white adipose tissues in adaptive thermogenesis. Genes Dev, 26(3): 271-281.

Foltz I N, Hu S, King C, et al. 2012. Treating diabetes and obesity with an FGF21-mimetic antibody activating the betaKlotho/FGFR1c receptor complex. Sci Transl Med, 4(162): 162ra153.

Frontini A, and Cinti S. 2010. Distribution and development of brown adipocytes in the murine and human adipose organ. Cell Metab, 11(4): 253-256.

Fukui Y, Masui S, Osada S, et al. 2000. A new thiazolidinedione, NC-2100, which is a weak PPAR-gamma activator, exhibits potent antidiabetic effects and induces uncoupling protein 1 in white adipose tissue of KKAy obese mice. Diabetes, 49(5): 759-767.

Garg A. 2004. Regional adiposity and insulin resistance. J Clin Endocrinol Metab, 89(9): 4206-4210.

Gburcik V, Cawthorn W P, Nedergaard J, et al. 2012. An essential role for Tbx15 in the differentiation of brown and "brite" but not white adipocytes. Am J Physiol Endocrinol Metab, 303(8): E1053-1060.

Gemmell R T and Alexander G. 1978. Ultrastructural development of adipose tissue in foetal sheep. Aust J Biol Sci, 31(5): 505-515.

Gemmell R T, Bell A W, Alexander G. 1972. Morphology of adipose cells in lambs at birth and during subsequent transition of brown to white adipose tissue in cold and in warm conditons. Am J Anat, 133(2): 143-164.

Gesta S, Tseng Y H, Kahn C R. 2007. Developmental origin of fat: tracking obesity to its source. Cell, 131(2): 242-256.

Gilboa L, Nohe A, Geissendorfer T, et al. 2000. Bone morphogenetic protein receptor complexes on the surface of live cells: a new oligomerization mode for serine/threonine kinase receptors. Mol Biol, 11(3): 1023-1035.

Granneman J G, Li P, Zhu Z, et al. 2005. Metabolic and cellular plasticity in white adipose tissue I: effects of beta3-adrenergic receptor activation. Am J Physiol Endocrinol Metab, 289(4): E608-616.

Gregoire F M. 2001. Adipocyte differentiation: from fibroblast to endocrine cell. Exp Biol Med (Maywood), 226(11): 997-1002.

Gregor M F and Hotamisligil G S. 2011. Inflammatory mechanisms in obesity. Annu Rev Immunol, 29: 415-445.

Grujic D, Susulic V S, Harper M E, et al. 1997. Beta3-adrenergic receptors on white and brown adipocytes mediate beta3-selective agonist-induced effects on energy expenditure, insulin secretion, and food intake. A study using transgenic and gene knockout mice. J Biol Chem, 272(28): 17686-17693.

Guerra C, Koza R A, Yamashita H, et al. 1998. Emergence of brown adipocytes in white fat in mice is under genetic control. Effects on body weight and adiposity. J Clin Invest, 102(2): 412-420.

Guicheux J, Lemonnier J, Ghayor C, et al. 2003. Activation of p38 mitogen-activated protein kinase and c-Jun-NH2-terminal kinase by BMP-2 and their implication in the stimulation of osteoblastic cell differentiation. J Bone Miner Res, 18(11): 2060-2068.

Gunawardana S C and Piston D W. 2012. Reversal of type 1 diabetes in mice by brown adipose tissue transplant. Diabetes, Gunawardana, S.C., and Piston, D.W. 61(3): 674-682.

Guyton K Z, Xu Q, Holbrook N J. 1996. Induction of the mammalian stress response gene GADD153 by oxidative stress: role of AP-1 element. Biochem J, 314 (Pt 2): 547-554.

Hansen J B, Jorgensen C, Petersen R K, et al. 2004. Retinoblastoma protein functions as a molecular switch determining white versus brown adipocyte differentiation. Proc Natl Acad Sci U S A, 101(12): 4112-4117.

Hasek B E, Stewart L K, Henagan T M, et al. 2010. Dietary methionine restriction enhances metabolic flexibility and increases uncoupled respiration in both fed and fasted states. Am J Physiol Regul Integr Comp Physiol, 299(3): R728-739.

Hata K, Nishimura R, Ikeda F, et al. 2003. Differential roles of Smad1 and p38 kinase in regulation of peroxisome proliferator-activating receptor gamma during bone morphogenetic protein 2-induced adipogenesis. Mol Biol Cell. 14(2): 545-555.

Hausman D B, DiGirolamo M, Bartness T J, et al. 2001. The biology of white adipocyte proliferation. Obes Rev, 2(4): 239-254.

Heldin C H, Miyazono K, ten Dijke P, et al. 1997. TGF-beta signalling from cell membrane to nucleus through SMAD proteins. Nature, 390(6659): 465-471.

Larose M, Cassard-Doulcier A M, Fleury C, et al. 1996. Essential cis-acting elements in rat uncoupling protein gene are in an enhancer containing a complex retinoic acid response domain. J Biol Chem, 271(49): 31533-31542.

Larsen T M, Toubro S, van Baak M A, et al. 2002. Effect of a 28-d treatment with L-796568, a novel beta(3)-adrenergic receptor agonist, on energy expenditure and body composition in obese men. Am J Clin Nutr, 76(4): 780-788.

Lean M E. 1989. Brown adipose tissue in humans. Proc Nutr Soc, 48(2): 243-256.

Lee J Y, Takahashi N, Yasubuchi M, et al. 2012a. Triiodothyronine induces UCP-1 expression and mitochondrial biogenesis in human adipocytes. Am J Physiol Cell Physiol, 302(2): C463-472.

Lee P, Linderman J D, Smith S, et al. 2014. Irisin and FGF21 are cold-induced endocrine activators of brown fat function in humans. Cell Metab, 19(2): 302-309.

Lee Y H, Petkova A P, Mottillo E P, et al. 2012b. In vivo identification of bipotential adipocyte progenitors recruited by beta3-adrenoceptor activation and high-fat feeding. Cell Metab, 15(4): 480-491.

Leonardsson G, Steel J H, Christian M, et al. 2004. Nuclear receptor corepressor RIP140 regulates fat accumulation. Proc Natl Acad Sci U S A, 101(22): 8437-8442.

Li P, Zhu Z, Lu Y, Granneman J G. 2005. Metabolic and cellular plasticity in white adipose tissue II: role of peroxisome proliferator-activated receptor-alpha. Am J Physiol Endocrinol Metab, 289(4): E617-626.

Li X, Kim J W, Gronborg M, et al. 2007. Role of cdk2 in the sequential phosphorylation/activation of C/EBPbeta during adipocyte differentiation. Proc Natl Acad Sci U S A, 104(28): 11597-11602.

Li Y, Rankin S A, Sinner D, et al. 2008. Sfrp5 coordinates foregut specification and morphogenesis by antagonizing both canonical and noncanonical Wnt11 signaling. Genes Dev, 22(21): 3050-3063.

Liu X, Zheng Z, Zhu X, et al. 2013. Brown adipose tissue transplantation improves whole-body energy metabolism. Cell Res, 23(6): 851-854.

Long J Z, Svensson K J, Tsai L, et al. 2014. A Smooth Muscle-Like Origin for Beige Adipocytes. Cell Metab, 6;19(5):810-20

MacDougald O A and Lane M D. 1995. Transcriptional regulation of gene expression during adipocyte differentiation. Annu Rev Biochem, 64: 345-373.

Maeda H, Hosokawa M, Sashima T, et al. 2005. Fucoxanthin from edible seaweed, Undaria pinnatifida, shows antiobesity effect through UCP1 expression in white adipose tissues. Biochem Biophys Res Commun, 332(2): 392-397.

Maeda H, Hosokawa M, Sashima T, et al. 2007. Effect of medium-chain triacylglycerols on anti-obesity effect of fucoxanthin. J Oleo Sci, 56(12): 615-621.

Marinou K, Tousoulis D, Antonopoulos A S, et al. 2010. Obesity and cardiovascular disease: from pathophysiology to risk stratification. Int J Cardiol, 138(1): 3-8.

Massague J. 1998. TGF-beta signal transduction. Annu Rev Biochem, 67: 753-791.

Massaous J and Hata A. 1997. TGF-beta signalling through the Smad pathway. Trends Cell Biol, 7(5): 187-192.

McGregor R A and Choi M S. 2011. microRNAs in the regulation of adipogenesis and obesity. Curr Mol Med, 11(4): 304-316.

Mercader J, Palou A, Bonet M L. 2011. Resveratrol enhances fatty acid oxidation capacity and reduces resistin and Retinol-Binding Protein 4 expression in white adipocytes. J Nutr Biochem, 22(9): 828-834.

Mercader J, Palou A, Bonet M. 2010. Induction of uncoupling protein-1 in mouse embryonic fibroblast-derived adipocytes by retinoic acid. Obesity (Silver Spring) 18(4): 655-662.

Mercader J, Ribot J, Murano I, et al. 2006. Remodeling of white adipose tissue after retinoic acid administration in mice. Endocrinology, 147(11): 5325-5332.

Mochizuki N, Shimizu S, Nagasawa T, et al. 2000. A novel gene, MEL1, mapped to 1p36.3 is highly homologous to the MDS1/EVI1 gene and is transcriptionally activated in t(1;3)(p36;q21)-positive leukemia cells. Blood, 96(9): 3209-3214.

Mori M A, Thomou T, Boucher J, et al. 2014. Altered miRNA processing disrupts brown/white adipocyte determination and associates with lipodystrophy. J Clin Invest, doi: 10.1172/JCI73468.

Mori M, Nakagami H, Rodriguez-Araujo G, et al. 2012. Essential role for miR-196a in brown adipogenesis of white fat progenitor cells. PLoS Biol, 10(4): e1001314.

Morrison S F. 2011. 2010 Carl Ludwig Distinguished Lectureship of the APS Neural Control and Autonomic Regulation Section: Central neural pathways for thermoregulatory cold defense. J Appl Physiol (1985), 110(5): 1137-1149.

Mottillo E P, Bloch A E, Leff T, et al. 2012. Lipolytic products activate peroxisome proliferator-activated receptor (PPAR) alpha and delta in brown adipocytes to match fatty acid oxidation with supply. J Biol Chem, 287(30): 25038-25048.

Mudhasani R, Puri V, Hoover K, et al. 2011. Dicer is required for the formation of white but not brown adipose tissue. J Cell Physiol, 226(5): 1399-1406.

Mukherjee R, Davies P J, Crombie D L, et al. 1997. Sensitization of diabetic and obese mice to insulin by retinoid X receptor agonists. Nature, 386(6623): 407-410.

Nakamura K, Fuster J J, Walsh K. 2014. Adipokines: A link between obesity and cardiovascular disease. J Cardiol. 63(4):250-9.

Nedergaard J, Bengtsson T, Cannon B. 2007. Unexpected evidence for active brown adipose tissue in adult humans. Am J Physiol Endocrinol Metab, 293(2): E444-452.

Nedergaard J, Bengtsson T, Cannon B. 2011. New powers of brown fat: fighting the metabolic syndrome. Cell Metab, 13(3): 238-240.

Nishimura S, Manabe I, Nagasaki M, et al. 2007. Adipogenesis in obesity requires close interplay between differentiating adipocytes, stromal cells, and blood vessels. Diabetes ,56(6): 1517-1526.

Nohe A, Hassel S, Ehrlich M, et al. 2002. The mode of bone morphogenetic protein (BMP) receptor oligomerization determines different BMP-2 signaling pathways. J Biol Chem, 277(7): 5330-5338.

Obregon M J. 2014. Changing white into brite adipocytes. Focus on "BMP4 and BMP7 induce the white-to-brown transition of primary human adipose stem cells". Am J Physiol Cell Physiol, 306(5): C425-427.

Odegaard J I and Chawla A. 2013. Pleiotropic actions of insulin resistance and inflammation in metabolic homeostasis. Science, 339(6116): 172-177.

Ohno H, Shinoda K, Spiegelman B M, et al. 2012. PPARgamma agonists induce a white-to-brown fat conversion through stabilization of PRDM16 protein. Cell Metab, 15(3): 395-404.

Ohyama K, Das R, Placzek M. 2008. Temporal progression of hypothalamic patterning by a dual action of BMP. Development, 135(20): 3325-3331.

Ono K, Tsukamoto-Yasui M, Hara-Kimura Y, et al. 2011. Intragastric administration of capsiate, a transient receptor potential channel agonist, triggers thermogenic sympathetic responses. J Appl Physiol, 110(3): 789-798.

Orava J, Nuutila P, Lidell M E, et al. 2011. Different metabolic responses of human brown adipose tissue to activation by cold and insulin. Cell Metab, 14(2): 272-279.

Ouchi N, Parker J L, Lugus J J, et al. 2011. Adipokines in inflammation and metabolic disease. Nat Rev Immunol, 11(2): 85-97.

Pan D, Fujimoto M, Lopes A, et al. 2009. Twist-1 is a PPARdelta-inducible, negative-feedback regulator of PGC-1alpha in brown fat metabolism. Cell, 137(1): 73-86.

Pankov Y A. 1999. Adipose tissue as an endocrine organ regulating growth, puberty, and other physiological functions. Biochemistry (Mosc), 64(6): 601-609.

Patel P and Abate N. 2013. Role of subcutaneous adipose tissue in the pathogenesis of insulin resistance. J Obes, 2013: 489187.

Petrovic N, Walden T B, Shabalina I G, et al. 2010. Chronic peroxisome proliferator-activated receptor gamma (PPARgamma) activation of epididymally derived white adipocyte cultures reveals a population of thermogenically competent, UCP1-containing adipocytes molecularly distinct from classic brown adipocytes. J Biol Chem, 285(10): 7153-7164.

Picard F, Gehin M, Annicotte J, et al. 2002. SRC-1 and TIF2 control energy balance between white and brown adipose tissues. Cell, 111(7): 931-941.

Pico C, Bonet M L, Palou A. 1998. Stimulation of uncoupling protein synthesis in white adipose tissue of mice treated with the beta 3-adrenergic agonist CGP-12177. Cell Mol Life Sci, 54(2): 191-195.

Plaisance E P, Henagan T M, Echlin H, et al. 2010. Role of beta-adrenergic receptors in the hyperphagic and hypermetabolic responses to dietary methionine restriction. Am J Physiol Regul Integr Comp Physiol, 299(3): R740-750.

Poirier P, Giles T D, Bray G A, et al. 2006. Obesity and cardiovascular disease: pathophysiology, evaluation, and effect of weight loss. Arterioscler Thromb Vasc Biol, 26(5): 968-976.

Pope M, Budge H, Symonds M E. 2014. The developmental transition of ovine adipose tissue through early life. Acta Physiol (Oxf), 210(1): 20-30.

Prunet-Marcassus B, Cousin B, Caton D, et al. 2006. From heterogeneity to plasticity in adipose tissues: site-specific differences. Exp Cell Res, 312(6): 727-736.

Puigserver P, Vazquez F, Bonet M L, et al. 1996. In vitro and in vivo induction of brown adipocyte uncoupling protein (thermogenin) by retinoic acid. Biochem J, 317 (Pt 3): 827-833.

Puigserver P, Wu Z, Park C W, et al. 1998. A cold-inducible coactivator of nuclear receptors linked to adaptive thermogenesis. Cell, 92(6): 829-839.

Qian S W, Tang Y, Li X, et al. 2013. BMP4-mediated brown fat-like changes in white adipose tissue alter glucose and energy homeostasis. Proc Natl Acad Sci U S A, 110(9): E798-807.

Qiang L, Wang L, Kon N, et al. 2012. Brown remodeling of white adipose tissue by SirT1-dependent deacetylation of Ppargamma. Cell, 150(3): 620-632.

Qiu Y, Nguyen K D, Odegaard J I, et al. 2014. Eosinophils and type 2 cytokine signaling in macrophages orchestrate development of functional beige fat. Cell, 157(6): 1292-1308.

Rabelo R, Reyes C, Schifman A, et al. 1996. A complex retinoic acid response element in the uncoupling protein gene defines a novel role for retinoids in thermogenesis. Endocrinology, 137(8): 3488-3496.

Ragheb R, Shanab G M, Medhat A M, et al. 2009. Free fatty acid-induced muscle insulin resistance and glucose uptake dysfunction: evidence for PKC activation and oxidative stress-activated signaling pathways. Biochem Biophys Res Commun, 389(2): 211-216.

Rao R R, Long J Z, White J P, et al. 2014. Meteorin-like Is a Hormone that Regulates Immune-Adipose Interactions to Increase Beige Fat Thermogenesis. Cell, 157(6): 1279-1291.

Reaven G M. 2011. Insulin resistance: the link between obesity and cardiovascular disease. Med Clin North Am, 95(5): 875-892.

Redman L M, de Jonge L, Fang X, et al. 2007. Lack of an effect of a novel beta3-adrenoceptor agonist, TAK-677, on energy metabolism in obese individuals: a double-blind, placebo-controlled randomized study. J Clin Endocrinol Metab, 92(2): 527-531.

Ribet C, Montastier E, Valle C, et al. 2010. Peroxisome proliferator-activated receptor-alpha control of lipid and glucose metabolism in human white adipocytes. Endocrinology, 151(1): 123-133.

Rim J S and Kozak L P. 2002. Regulatory motifs for CREB-binding protein and Nfe2l2 transcription factors in the upstream enhancer of the mitochondrial uncoupling protein 1 gene. J Biol Chem, 277(37): 34589-34600.

Ringholm S, Grunnet Knudsen J, Leick L, et al. 2013. PGC-1alpha is required for exercise- and exercise training-induced UCP1 up-regulation in mouse white adipose tissue. PLoS One, 8(5): e64123.

Rodeheffer M S, Birsoy K, Friedman J M. 2008. Identification of white adipocyte progenitor cells in vivo. Cell, 135(2): 240-249.

Rosen E D and MacDougald O A. 2006. Adipocyte differentiation from the inside out. Nat Rev Mol Cell Biol, 7(12): 885-896.

Rosen E D and Spiegelman B M. 2000. Molecular regulation of adipogenesis. Annu Rev Cell Dev Biol, 16: 145-171.

Rosen E D, Walkey C J, Puigserver P, et al. 2000. Transcriptional regulation of adipogenesis. Genes Dev, 14(11): 1293-1307.

Roth Flach R J, Matevossian A, Akie T E, et al. 2013. beta3-Adrenergic receptor stimulation induces E-selectin-mediated adipose tissue inflammation. J Biol Chem, 288(4): 2882-2892.

Rubio A, Raasmaja A, Maia A L, et al. 1995a. Effects of thyroid hormone on norepinephrine signaling in brown adipose tissue. I. Beta 1- and beta 2-adrenergic receptors and cyclic adenosine 3',5'-monophosphate generation. Endocrinology, 136(8): 3267-3276.

Rubio A, Raasmaja A, Silva J E. 1995b. Thyroid hormone and norepinephrine signaling in brown adipose tissue. II: Differential effects of thyroid hormone on beta 3-adrenergic receptors in brown and white adipose tissue. Endocrinology, 136(8): 3277-3284.

Saito M, Okamatsu-Ogura Y, Matsushita M, et al. 2009. High incidence of metabolically active brown adipose tissue in healthy adult humans: effects of cold exposure and adiposity. Diabetes, 58(7): 1526-1531.

Sanchez R C, Ibanez C, Klaassen J. 2014. [The link between obesity and cancer]. Rev Med Chil, 142(2): 211-221.

Sanchez-Gurmaches J, Hung C M, Sparks C A, et al. 2012. PTEN loss in the Myf5 lineage redistributes body fat and reveals subsets of white adipocytes that arise from Myf5 precursors. Cell Metab, 16(3): 348-362.

Santos A, Bakker A D, de Blieck-Hogervorst J M, et al. 2010. WNT5A induces osteogenic differentiation of human adipose stem cells via rho-associated kinase ROCK. Cytotherapy, 12(7): 924-932.

Schulz T J, Huang P, Huang T L, et al. 2013. Brown-fat paucity due to impaired BMP signalling induces compensatory browning of white fat. Nature, 495(7441): 379-383.

Schulz T J, Huang T L., Tran T T, et al. 2011. Identification of inducible brown adipocyte progenitors residing in skeletal muscle and white fat. Proc Natl Acad Sci U S A, 108(1): 143-148.

Scime A, Grenier G, Huh M S, et al. 2005. Rb and p107 regulate preadipocyte differentiation into white versus brown fat through repression of PGC-1alpha. Cell Metab, 2(5): 283-295.

Seale P, Bjork B, Yang W, et al. 2008. PRDM16 controls a brown fat/skeletal muscle switch. Nature, 454(7207): 961-967.

Seale P, Kajimura S, Yang W, et al. 2007. Transcriptional control of brown fat determination by PRDM16. Cell Metab, 6(1): 38-54.

Sears I B, MacGinnitie M A, Kovacs L G, et al. 1996. Differentiation-dependent expression of the brown adipocyte uncoupling protein gene: regulation by peroxisome proliferator-activated receptor gamma. Mol Cell Biol, 16(7): 3410-3419.

Sell H, Berger J P. Samson P, et al. 2004. Peroxisome proliferator-activated receptor gamma agonism increases the capacity for sympathetically mediated thermogenesis in lean and ob/ob mice. Endocrinology, 145(8): 3925-3934.

Shan T, Liang X, Bi P, et al. 2013. Myostatin knockout drives browning of white adipose tissue through activating the AMPK-PGC1alpha-Fndc5 pathway in muscle. FASEB J, 27(5):1981-9

Shepherd P R, Gnudi L, Tozzo E, et al. 1993. Adipose cell hyperplasia and enhanced glucose disposal in transgenic mice overexpressing GLUT4 selectively in adipose tissue. J Biol Chem, 268(30): 22243-22246.

Shimizu Y, Satoh S, Yano H, et al. 1998. Effects of noradrenaline on the cell-surface glucose transporters in cultured brown adipocytes: novel mechanism for selective activation of GLUT1 glucose transporters. Biochem J, 330 (Pt 1): 397-403.

Shoelson S E, Lee J, Goldfine A B, et al. 2006. Inflammation and insulin resistance. J Clin Invest, 116(7): 1793-1801.

Siersbaek M S, Loft A, Aagaard M M, et al. 2012. Genome-wide profiling of peroxisome proliferator-activated receptor gamma in primary epididymal, inguinal, and brown adipocytes reveals depot-selective binding correlated with gene expression. Mol Cell Biol, 32(17): 3452-3463.

Silva J E and Rabelo R. 1997. Regulation of the uncoupling protein gene expression. Eur J Endocrinol, 136(3): 251-264.

Silva J E, and Larsen P R. 1983. Adrenergic activation of triiodothyronine production in brown adipose tissue. Nature, 305(5936): 712-713.

Skarulis M C, Celi F S, Mueller E, et al. 2010. Thyroid hormone induced brown adipose tissue and amelioration of diabetes in a patient with extreme insulin resistance. J Clin Endocrinol Metab, 95(1): 256-262.

Sottile V and Seuwen K. 2000. Bone morphogenetic protein-2 stimulates adipogenic differentiation of mesenchymal precursor cells in synergy with BRL 49653 (rosiglitazone). FEBS Lett, 475(3): 201-204.

Spiegelman B M. 1998. PPAR-gamma: adipogenic regulator and thiazolidinedione receptor. Diabetes, 47(4): 507-514.

Stanford K I, Middelbeek R J, Townsend K L, et al. 2013. Brown adipose tissue regulates glucose homeostasis and insulin sensitivity. J Clin Invest, 123(1): 215-223.

Sun K, Kusminski C M, Scherer P E. 2011a. Adipose tissue remodeling and obesity. J Clin Invest, 121(6): 2094-2101.

Sun L, Xie H, Mori M A, et al. 2011b. Mir193b-365 is essential for brown fat differentiation. Nat Cell Biol, 13(8): 958-965.

Takada I, Mihara M, Suzawa M, et al. 2007. A histone lysine methyltransferase activated by non-canonical Wnt signalling suppresses PPAR-gamma transactivation. Nat Cell Biol, 9(11): 1273-1285.

Takashima Y, Era T, Nakao K, et al. 2007. Neuroepithelial cells supply an initial transient wave of MSC differentiation. Cell, 129(7): 1377-1388.

Tang Q Q and Lane M D. 1999. Activation and centromeric localization of CCAAT/enhancer-binding proteins during the mitotic clonal expansion of adipocyte differentiation. Genes Dev, 13(17): 2231-2241.

Tang Q Q and Lane M D. 2012. Adipogenesis: from stem cell to adipocyte. Annu Rev Biochem, 81: 715-736.

Tang Q Q, Gronborg M, Huang H, et al. 2005. Sequential phosphorylation of CCAAT enhancer-binding protein beta by MAPK and glycogen synthase kinase 3beta is required for adipogenesis. Proc Natl Acad Sci U S A, 102(28): 9766-9771.

Tang Q Q, Jiang M S, Lane M D. 1999. Repressive effect of Sp1 on the C/EBPalpha gene promoter: role in adipocyte differentiation. Mol Cell Biol, 19(7): 4855-4865.

Tang Q Q, Otto T C, Lane M D. 2004. Commitment of C3H10T1/2 pluripotent stem cells to the adipocyte lineage. Proc Natl Acad Sci U S A, 101(26): 9607-9611.

Tang W, Zeve D, Suh J M, et al. 2008. White fat progenitor cells reside in the adipose vasculature. Science, 322(5901): 583-586.

ten Dijke P, Yamashita H, Sampath T K, et al. 1994. Identification of type I receptors for osteogenic protein-1 and bone morphogenetic protein-4. J Biol Chem, 269(25): 16985-16988.

Teperino R, Amann S, Bayer M, et al. 2012. Hedgehog partial agonism drives Warburg-like metabolism in muscle and brown fat. Cell, 151(2): 414-426.

Teruel T, Hernandez R, Benito M, et al. 2003. Rosiglitazone and retinoic acid induce uncoupling protein-1 (UCP-1) in a p38 mitogen-activated protein kinase-dependent manner in fetal primary brown adipocytes. J Biol Chem, 278(1): 263-269.

Tontonoz P, Kim J B, Graves R A, et al. 1993. ADD1: a novel helix-loop-helix transcription factor associated with adipocyte determination and differentiation. Mol Cell Biol, 13(8): 4753-4759.

Townsend K L, Suzuki R, Huang T L, et al. 2012. Bone morphogenetic protein 7 (BMP7) reverses obesity and regulates appetite through a central mTOR pathway. FASEB J, 26(5): 2187-2196.

Trajkovski M, Ahmed K, Esau C C, et al. 2012. MyomiR-133 regulates brown fat differentiation through Prdm16. Nat Cell Biol, 14(12): 1330-1335.

Tran K V, Gealekman O, Frontini A, et al. 2012. The vascular endothelium of the adipose tissue gives rise to both white and brown fat cells. Cell Metab, 15(2): 222-229.

Tseng Y H, Cypess A M, Kahn C R. 2010. Cellular bioenergetics as a target for obesity therapy. Nat Rev Drug Discov, 9(6): 465-482.

Tseng Y H, Kokkotou E, Schulz T J, et al. 2008. New role of bone morphogenetic protein 7 in brown adipogenesis and energy expenditure. Nature, 454(7207): 1000-1004.

Uldry, M., Yang, W., St-Pierre, J, et al. 2006. Complementary action of the PGC-1 coactivators in mitochondrial biogenesis and brown fat differentiation. Cell Metab, 3(5): 333-341.

Vallerand A L, Perusse F, Bukowiecki L J. 1987. Cold exposure potentiates the effect of insulin on in vivo glucose uptake. Am J Physiol, 253(2 Pt 1): E179-186.

Valmaseda A, Carmona M C, Barbera M J, et al. 1999. Opposite regulation of PPAR-alpha and -gamma gene expression by both their ligands and retinoic acid in brown adipocytes. Mol Cell Endocrinol, 154(1-2): 101-109.

van Marken Lichtenbelt W D, Vanhommerig J W, Smulders N M, et al. 2009. Cold-activated brown adipose tissue in healthy men. N Engl J Med, 360(15): 1500-1508.

Vernochet C, Peres S B, Davis K E, et al. 2009. C/EBPalpha and the corepressors CtBP1 and CtBP2 regulate repression of select visceral white adipose genes during induction of the brown phenotype in white adipocytes by peroxisome proliferator-activated receptor gamma agonists. Mol Cell Biol, 29(17): 4714-4728.

Vessby B. 2003. Dietary fat, fatty acid composition in plasma and the metabolic syndrome. Curr Opin Lipidol, 14(1): 15-19.

Villanueva C J, Vergnes L, Wang J, et al. 2013. Adipose subtype-selective recruitment of TLE3 or Prdm16 by PPARgamma specifies lipid storage versus thermogenic gene programs. Cell Metab, 17(3): 423-435.

Villarroya F, Iglesias R, Giralt M. 2007. PPARs in the Control of Uncoupling Proteins Gene Expression. PPAR Res, 2007: 74364.

Virtanen K A, Lidell M E, Orava J, et al. 2009. Functional brown adipose tissue in healthy adults. N Engl J Med, 360(15): 1518-1525.

Vosselman M J, van der Lans A A, Brans B, et al. 2012. Systemic beta-adrenergic stimulation of thermogenesis is not accompanied by brown adipose tissue activity in humans. Diabetes, 61(12): 3106-3113.

Walden T B, Timmons J A, Keller P, et al. 2009. Distinct expression of muscle-specific microRNAs (myomirs) in brown adipocytes. J Cell Physiol, 218(2): 444-449.

Wang E A, Israel D I, Kelly S, et al. 1993. Bone morphogenetic protein-2 causes commitment and differentiation in C3H10T1/2 and 3T3 cells. Growth Factors 9(1): 57-71.

Wang Q A, Tao C, Gupta R K, et al. 2013. Tracking adipogenesis during white adipose tissue development, expansion and regeneration. Nat Med, 19(10): 1338-1344.

Wang Z and Nakayama T. 2010. Inflammation, a link between obesity and cardiovascular disease. Mediators Inflamm, 2010: 535918.

Watanabe T, Kawada T, Yamamoto M, et al. 1987. Capsaicin, a pungent principle of hot red pepper, evokes catecholamine secretion from the adrenal medulla of anesthetized rats. Biochem Biophys Res Commun, 142(1): 259-264.

Weyer C, Tataranni P A, Snitker S, et al. 1998. Increase in insulin action and fat oxidation after treatment with CL 316,243, a highly selective beta3-adrenoceptor agonist in humans. Diabetes, 47(10): 1555-1561.

Whittle A J, Carobbio S, Martins L, et al. 2012. BMP8B increases brown adipose tissue thermogenesis through both central and peripheral actions. Cell, 149(4): 871-885.

Williams K J and Fisher E A. 2011. Globular warming: how fat gets to the furnace. Nat Med, 17(2): 157-159.

Wilson-Fritch L, Burkart A, Bell G, et al. 2003. Mitochondrial biogenesis and remodeling during adipogenesis and in response to the insulin sensitizer rosiglitazone. Mol Cell Biol, 23(3): 1085-1094.

Wilson-Fritch L, Nicoloro S, Chouinard M, et al. 2004. Mitochondrial remodeling in adipose tissue associated with obesity and treatment with rosiglitazone. J Clin Invest, 114(9): 1281-1289.

Wrana J L, Attisano L, Wieser R, et al. 1994. Mechanism of activation of the TGF-beta receptor. Nature, 370(6488): 341-347.

Wright W S, Longo K A, olinsky V W, et al. 2007. Wnt10b inhibits obesity in ob/ob and agouti mice. Diabetes, 56(2): 295-303.

Wu J, Bostrom P, Sparks L M, et al. 2012. Beige adipocytes are a distinct type of thermogenic fat cell in mouse and human. Cell, 150(2): 366-376.

Xie H, Lim B, Lodish H F. 2009. MicroRNAs induced during adipogenesis that accelerate fat cell development are downregulated in obesity. Diabetes, 58(5): 1050-1057.

Xu X, Ying Z, Cai M, et al. 2011. Exercise ameliorates high-fat diet-induced metabolic and vascular dysfunction, and increases adipocyte progenitor cell population in brown adipose tissue. Am J Physiol Regul Integr Comp Physiol, 300(5): R1115-1125.

Xue B, Coulter A, Rim J S, et al. 2005. Transcriptional synergy and the regulation of Ucp1 during brown adipocyte induction in white fat depots. Mol Cell Biol, 25(18): 8311-8322.

Xue R, Wan Y, Zhang S, et al. 2014. Role of bone morphogenetic protein 4 in the differentiation of brown fat-like adipocytes. Am J Physiol Endocrinol Metab, 306(4): E363-372.

Ye L, Wu J, Cohen P, et al. 2013. Fat cells directly sense temperature to activate thermogenesis. Proc Natl Acad Sci U S A, 110(30): 12480-12485.

Yekta S, Shih I H, Bartel D P. 2004. MicroRNA-directed cleavage of HOXB8 mRNA. Science, 304(5670): 594-596.

Yellaturu C R, Deng X, Cagen L M, et al. 2009a. Insulin enhances post-translational processing of nascent SREBP-1c by promoting its phosphorylation and association with COPII vesicles. J Biol Chem, 284(12): 7518-7532.

Yellaturu C R, Deng X, Park E A, et al. 2009b. Insulin enhances the biogenesis of nuclear sterol regulatory element-binding protein (SREBP)-1c by posttranscriptional down-regulation of Insig-2A and its dissociation from SREBP cleavage-activating protein (SCAP).SREBP-1c complex. J Biol Chem, 284(46): 31726-31734.

Yoo H J and Choi K M. 2014. Adipokines as a novel link between obesity and atherosclerosis. World J Diabetes, 5(3): 357-363.

Yoshida T, Sakane N, Wakabayashi Y, et al. 1994. Anti-obesity and anti-diabetic effects of CL 316,243, a highly specific beta 3-adrenoceptor agonist, in yellow KK mice. Life Sci, 54(7): 491-498.

Yu X X, Lewin D A, Forrest W, et al. 2002. Cold elicits the simultaneous induction of fatty acid synthesis and beta-oxidation in murine brown adipose tissue: prediction from differential gene expression and confirmation in vivo. FASEB J, 16(2): 155-168.

Yu Z K, Wright J T, Hausman G J. 1997. Preadipocyte recruitment in stromal vascular cultures after depletion of committed preadipocytes by immunocytotoxicity. Obes Res, 5(1): 9-15.

Yubero P, Barbera M J, Alvarez R, et al. 1998. Dominant negative regulation by c-Jun of transcription of the uncoupling protein-1 gene through a proximal cAMP-regulatory element: a mechanism for repressing basal and norepinephrine-induced expression of the gene before brown adipocyte differentiation. Mol Endocrinol, 12(7): 1023-1037.

Yubero P, Manchado C, Cassard-Doulcier A M, et al. 1994. CCAAT/enhancer binding proteins alpha and beta are transcriptional activators of the brown fat uncoupling protein gene promoter. Biochem Biophys Res Commun, 198(2): 653-659.

Zeve D, Tang W and Graff J. 2009. Fighting fat with fat: the expanding field of adipose stem cells. Cell Stem Cell, 5(5): 472-481.

Zhang J W, Tang Q Q Vinson C, et al. 2004. Dominant-negative C/EBP disrupts mitotic clonal expansion and differentiation of 3T3-L1 preadipocytes. Proc Natl Acad Sci U S A, 101(1): 43-47.

Zhang L L, Yan Liu D, Ma L Q, et al. 2007. Activation of transient receptor potential vanilloid type-1 channel prevents adipogenesis and obesity. Circ Res, 100(7): 1063-1070.

Zhou H, Mak W, Zheng Y, et al. 2008. Osteoblasts directly control lineage commitment of mesenchymal progenitor cells through Wnt signaling. J Biol Chem, 283(4): 1936-1945.

Zhou S, Eid K, Glowacki J. 2004. Cooperation between TGF-beta and Wnt pathways during chondrocyte and adipocyte differentiation of human marrow stromal cells. J Bone Miner Res, 19(3): 463-470.

Zingaretti M C, Crosta F, Vitali A, et al. 2009. The presence of UCP1 demonstrates that metabolically active adipose tissue in the neck of adult humans truly represents brown adipose tissue. FASEB J, 23(9): 3113-3120.

* 策划编辑：杜久林　中国科学院上海生命科学研究院神经科学研究所

以透明脑，观澄明心：斑马鱼神经功能研究进展

作　者：尚春峰　穆　宇　杜久林

中国科学院上海生命科学研究院神经科学研究所

- 1. 感觉信息加工和处理 / 143
- 2. 运动控制的规则 / 151
- 3. 学习和神经功能可塑性 / 154
- 4. 我国斑马鱼神经系统研究的进展 / 156
- 5. 新技术的发展和需求 / 158
- 6. 结语 / 164

摘要

斑马鱼是一种相对新颖的模式脊椎动物，具有脊椎动物保守的神经系统构造和丰富的行为模式。近年来随着在体电生理、光学成像、遗传工程等方法的建立和完善，幼龄斑马鱼因其脑部透明、结构简单的特点，日臻成为从突触、神经元、环路，到行为等多层次上，在全脑尺度上探究神经系统功能机制的理想动物模型。本文综述了近年来国际上利用斑马鱼，在感觉信息处理、运动控制、学习与神经可塑性等方向上所取得的重要成果，简要概括了我国科学家在相应领域的研究进展，并对新技术的开发提出了展望。随着研究思路的深化和实验手段的推陈出新，斑马鱼模式动物必将成为探索脑工作原理之利器，为神经科学研究带来更多的突破。

关键词

斑马鱼、神经环路、信息处理、视觉、嗅觉、运动、学习、可塑性、电生理、光学成像

从小到 1mm 长的线虫到大如超过 30m 长的蓝鲸，动物能迅速而灵活地响应环境变化，做出趋利避害的反应，这正是神经系统引人入胜之所在。神经系统负责接受和处理各种信息，评估周围环境和自身状态，做出有利于机体生存繁衍的行为和认知决策。对神经系统的研究，目的是希望能够阐明神经元及其组成的神经环路的结构和功能是如何实现复杂的认知和行为。经典的实验范式刻画了单个或一小群神经元的活动与动物个体行为之间的关联(Barlow, 1972; Parker and Newsome, 1998)，并通过操纵性实验探究其因果关系。然而，人类大脑包含约 8.6×10^{10} 个神经元(Azevedo et al., 2009)，即便是小小的秀丽隐杆线虫(*Caenorhabditis elegans*)，也有 302 个神经元，且各自具有独特的结构与连接。由于行为与认知往往涉及群体神经元、多个脑区的协同工作，对神经系统功能的研究也就需要大规模、并行记录多个神经元和脑区的活动，探究其相互间的结构和功能关联。因此，为了解析如此复杂的问题，我们需要选择合适的模式动物，其神经系统既有足够的代表性和复杂度，又有适当的可解性，能够充分利用目前可能的实验手段，对群体神经元甚至全脑大部分神经元的活动，在活体甚至清醒的动物上进行实时观察和系统分析，斑马鱼(*Danio rerio*)就是这样一种理想的选择。

斑马鱼为鲤科硬骨鱼类，原本游弋在喜马拉雅山南麓溪流池沼的浅层，与其他多种鱼类共同生活。斑马鱼可接受多种外界感觉信息，具有捕食猎物、逃避天敌，以及营社群生活等丰富的行为。与高等脊椎动物类似，其神经系统包含有端脑、间脑、中脑、小脑、后脑和脊髓等基本结构。1~2 周龄的斑马鱼，已具有成年动物的多种复杂行为和脑功能，但其大脑相对较小，以 1 周龄为例，约为 $300m^3$(背腹轴)×$500m^3$(左右轴)×$800m^3$(头尾轴)，含有数万个左右的神经元。在 10~20 倍物镜下，能观察到其大脑全貌，通过透明的皮肤可清楚分辨单个神经元的形态，从而可以实现在全脑尺度上对神经元活动进行在体电生理或光学成像的记录(Ahrens et al., 2013; Li et al., 2012; Mu et al., 2012; Wei et al., 2012)。记录神经活动时，斑马鱼头部通过琼脂糖凝胶束缚在培养皿中，释放其尾部

以观察感觉刺激所引起的运动，或者通过记录脊神经电活动判断其运动。此外，过去近 30 年的遗传发育生物学研究积累了大量的斑马鱼突变体和转基因品系；同时日新月异的现代基因工程技术可以实现对特定基因的编辑，使我们可以通过光学或化学手段标记特定的神经元类型和群体，对它们的神经活动进行特异性的记录和操纵。这些技术和资源的积累为利用斑马鱼开展神经科学研究提供了坚实的基础。基于此，在中国科学院 2012 年启动的科技先导 B 类专项"脑功能联结图谱的研究"和美国 2013 年启动的脑科学研究计划 "Brain Research Through Advanced Innovative Neurotechnologies（BRAIN）Initiative"中，斑马鱼也是其中的一个重要实验模型。本文将重点介绍近年来利用斑马鱼模型，国际上神经科学研究所取得的一些重要进展、及其所带来的启迪，并对未来的研究方向和技术更新提出期待和展望。

1. 感觉信息加工和处理

生活在复杂多变的自然界中，接受和处理各种重要的环境信息是动物生存、繁衍所必需的。斑马鱼具有视、听、嗅、味等多种感觉系统。近年来，在斑马鱼视觉和嗅觉系统方面所取得的研究进展，为揭示感觉信息加工和处理的机制提供了新的线索。

1.1 视觉系统研究

1.1.1 视网膜功能

视网膜是视觉系统接受视觉信号的第一站，因其简单而清晰的结构一直受到神经科学家的青睐。由于技术限制，既往工作多在离体标本上进行。近年来在斑马鱼上的研究工作，使得我们有机会窥见在体动物视网膜信息处理的机制。

在视网膜神经环路中，双极细胞（bipolar cell, BC）是第二级神经元。这些细胞接受光感受器细胞的输入，并将视觉信息传递给神经节细胞（retinal ganglion cell, RGC），其对光反应特性是视觉信息加工和处理的重要一环。Leon Lagnado 实验室应用光学成像方法，在活体斑马鱼视网膜上对双极细胞的功能进行了系统性的研究。GCaMP 是一种广泛使用的 Ca^{2+} 指示蛋白，是通过把环状变构的绿色荧光蛋白（GFP）、钙调蛋白（calmodulin）和肌球蛋白轻链激酶上结合 Ca^{2+} 的 M13 功能域融合而成，其荧光强度随 Ca^{2+} 浓度升高而增强。由于神经元活动增强伴随有细胞内 Ca^{2+} 浓度的升高，因而在表达了 GCaMP 的神经元上就可以通过荧光信号的变化来反映神经活动的改变，从而实现用光学成像方法记录神经元的活动。在 GCaMP 基础上，Dreosti 等（2009）开发了一种定位到突触前的 Ca^{2+} 指示蛋白 SynGCaMP2，其具有很好的线性和较宽的动态范围。将 SynGCaMP2 特异性地表达在双极细胞中，可以在活体斑马鱼上监测这些神经元轴突终末的神经活动。以前普遍认为双极细胞不产生动作电位，而是以去极化或超极化的方式来编码信息。Dreosti 等(2011)利用 SynGCaMP2 进行的光学成像实验发现,不论是给光反应(ON)还是撤光反应(OFF)、瞬时活动还是持续活动,双极细胞的轴突终末都表现出类似于动作电位的"全或无"式的快速 Ca^{2+} 活动,且这种活动受到视觉输入的调节。Odermatt 等(2012)在双极细胞中表达了另一种荧光蛋白 SypHy,用光学成像方法观察突触递质囊泡在细胞膜上的融合过程。通过给予不同强度的光刺激并记录相应的突触反应,他们发现,不论是呈现给光反应的突触还是撤光反应的突触,约有一半表现出非线性的特点,当光刺激强度逐渐增加时,其反应经过一个极小值和一个极大值后方达到平台期。分析表明,这种非线性突触具有更大的动态范围,可以更为有效地传递视觉信息。进一步地,Nikolaev 等(2013)研究了光刺激强度随时间周期性变化时双极细胞活动的适应性。发现给光反应和撤光反应的突触均可以按照对这种高时间对比度刺激的反应特点分为几种类型,有活动逐渐增加的,也有活动逐渐降低的。其中,初始反应较强的突触其活动会逐渐降低,而初始反应较弱的突触其活动则会逐渐增强。这两类反应性质不同的突触位于视网膜内网状层的不同亚层,其功能差异可能源自不同的神经环路的调节。其中,活动逐渐增强的突触与无长突细胞(amacrine cell,AC)的突起在空间位置上重叠,提示后者可能是调节的来源。利用 Ca^{2+} 指示蛋白 GCaMP3 测量无长突细胞对视觉刺激的反应,发现其活动逐渐降低,这可能导致向同一亚层内双极细胞的反馈性抑制性输入的下降,从而使得双极细胞的反应逐渐增强,这一推断也得到了药理实验结果的支持。这一系列工作,首次在活体模型上揭示了视网膜神经环路处理视觉信息的复杂性和丰富性。

中国科学院神经科学研究所杜久林研究组的张荣伟等采用在体电生理记录方法,系统地分析了视网膜视神经节细胞的功能发育(Zhang et al., 2010; Zhang et al., 2013)。运用全细胞膜片钳记录方法,可以细致地分析神经元的突触输入,区分自发活动和诱发反应中的不同突触输入、不同电流成分的相对贡献。作者细致测量了神经节细胞在不同发育阶段的 Cl^- 平衡电位,并考察了其与对光反应之间的关系。发现 Cl^- 平衡电位在发育过程中逐渐向超极化方向移动,从而使得抑制性神经递质 GABA 引起的反应由发育早期的兴奋性向发育晚期的抑制性转变,这种转变的时间节点与神经节细胞对光反应的出现有很好的对应关系(Zhang et al., 2010)。在同一神经节细胞上,不同亚细胞部位的 Cl^- 平衡电位也存在差异。作者在同时具有 ON 和 OFF 反应的神经节细胞上检验了这种差异的功能意义(Zhang et al., 2013)。通过记录不同树突及胞体部位的反应,测量局部 Cl^- 的平衡电位,发现在 ON 反应和 OFF 反应的树突之间,以及树突与胞体之间,均存在 Cl^- 平衡电位的差异,而这种差异主要由 K^+/Cl^- 共转运体 2(K^+/Cl^- co-transporter,KCC2)的差异表达所造成。这些工作为神经节细胞的形态发育研究提供了功能解释。

1.1.2 视觉朝向和方向选择型的神经环路机制

为了实现有效的信息编码,很多神经细胞会选择性地对感觉输入的某一特征(如运动信息)作出反应,这一过程称为特征检测(feature detection)。既往的研究已经在昆虫和哺乳类的大脑中发现了很多特征检测神经元,但由于实验手段的限制,通常只能记录单个或者一小群神经元的活动,其环路机制的研究进展缓慢。光学成像方法可以同时观察成像区域内群体神经元的形态和活动,有助于阐释特征检测的环路机制。斑马鱼的视觉中枢——视顶盖(optic tectum)位于斑马鱼中脑背部，分为内侧的胞体层(PVN)和外侧的纤维层(neuropil)。视顶盖主要接受视网膜神经节细胞的轴突投射，并将视觉信息传递至下游的多个脑区，在视觉相关行为中起关键作用。一方面，视顶盖中有为数众多的神经元对视觉刺激的反应具有方向选择性或朝向选择性(Niell and Smith, 2005)，另一方面，视顶盖空间位置和相对清晰的投射结构便于光学成像，因此，该脑区很适合研究特征检测的神经环路机制。

根据投射关系，可以推测视顶盖神经元的方向和朝向选择性有两个可能的来源：来自投射神经元神经节细胞，或由视顶盖内部的抑制性突触传递所导致(Ramdya and Engert, 2008)，支持局部环路处理的重要作用。这两种机制都得到了实验证明。Martin Meyer 实验室的工作关注神经节细胞投射到视顶盖的突触前结构(Nikolaou et al., 2012)。通过在神经节细胞中特异表达定位于突触前的 SynGCaMP3，他们得以观察视顶盖区所接受的这些投射的突触神经活动，发现具有不同方向选择性和朝向选择性的突触输入在空间上相互分离。这样，分别接受这些输入的视顶盖神经元就会产生方向选择性和朝向选择性。该研究还发现，大多数突触前的活动都偏好垂直或水平的朝向，和上偏后、下偏后或向前的运动方向，这可能与斑马鱼生活的视觉环境与行为特性相关。该实验室进一步研究了视顶盖内部环路对于方向和朝向选择性的贡献(Hunter et al., 2013)。通过对大量细胞视觉反应的分析，发现顶盖神经元按偏好方向可以分为上、下、前、后四类，其中对向后运动的偏好明显偏离神经节细胞的反应性质，提示顶盖内神经环路对方向信息进行了进一步的处理。此外，他们还研究了视觉剥夺对神经节细胞投射的影响，发现黑暗环境饲养改变了神经节细胞在视顶盖投射的朝向选择性分布，而方向选择性则不受影响(Lowe et al., 2013)。

Florian Engert 实验室的工作表明，局部抑制性突触传递对于视顶盖神经元的方向选择性是必要的(Ramdya and Engert, 2008)，且这些神经元所接受的抑制性输入确实具有方向选择性(Grama and Engert, 2012)。当用运动光条刺激斑马鱼时，电生理记录的结果显示，阈下反应中的抑制性成分较兴奋性成分具有更强的方向选择性，其方向偏好性与细胞阈上反应(即动作电位反应)是相反的。通过对突触输入成分的细致分析发现，对于神经元偏好的方向，抑制性突触输入的潜伏期更长；而对于相反方向的刺激，抑制性输入的潜伏期更短。该工作提示可能存在一种类似于 Reinhardt 运动检测的环路机制，视顶盖神经元通过不对称方式接受邻近中间神经元的抑制性输入，从而获得方向选择性。目前的研究尚未发现符合这一假说的中间神经元的类型。

Johann Bollmann 实验室通过遗传学操作得到了特异标记不同神经节细胞的若干斑马鱼品系，光学成像记录发现其中两个品系各自特异标记视顶盖的一种类型的抑制性神经元，这两种神经元对视觉刺激具有不同的方向选择性，分别偏好自头至尾和自尾至头方向的运动(Gabriel et al., 2012)。由于这两类神经元的树突在视顶盖中的分布存在明显的层次差异，联系前述的 Martin Meyer 实验室的工作，提示它们可能通过接受不同的视网膜输入获得了各自的方向选择性。作者进一步运用电生理和光学成像方法，发现这两类神经元接受的突触输入本身具有方向选择性，其中兴奋性成分的选择性与细胞的选择性相同，抑制性成分则偏好相反的方向。这些工作表明，发育过程中产生的特定投射关系使得顶盖神经元从上级输入获得方向选择性；进一步地，视顶盖来源的具有方向选择性的抑制性输入使得顶盖神经元的选择性更为显著。

1.1.3 视觉捕食行为的机制

斑马鱼视觉行为的一个主要功能是在捕食过程中识别并且追踪猎物。Florian Engert 实验室的 Bianco 等(2011)用不同大小的运动光点作为视觉刺激，发现 1 度视角的光点可以引起自由游动的斑马鱼汇聚双眼并朝向光点游动，类似于草履虫诱发的捕食行为。Trivedi 和 Bollmann(2013)通过闭环的方式，用小的方形图像刺激机械固定的斑马鱼，通过详尽的分析，发现斑马鱼的行为反应，包括眼球运动和摆尾，都类似于自由游动的斑马鱼对草履虫的反应，进一步支持了用这类人工刺激研究斑马鱼捕食行为神经机制的合理性。视顶盖向运动脑区发出的投射主要位于深部，与捕食行为相关。Herwig Baier 实验室的 Del Bene 等(2010)运用 Ca^{2+} 光学成像方法观察了该处神经纤维的活动。通过采用不同宽度的运动光条作为刺激，他们发现相对于视角 16°及以上的宽光条，窄小的刺激在该处可引起更强的反应。与此不同的是，接受神经节输入的顶盖浅层区域缺乏对光条宽度的选择性，提示顶盖内的神经信息处理对于深部纤维的偏好性至关重要。结合遗传筛选的手段，作者发现顶盖表层中有一群中间神经元，它们表现出相反的偏好性，对于 16°以上的运动光条有着更强的反应，提示这群神经元的抑制性输入可能造就了顶盖深部活动的选择性。进一步地，通过在这些神经元中特异表达 KillerRed，用光转化方法杀死这群神经元后，顶盖深部的神经活动就失去了选择性。同样地，通过定向表达破伤风毒素以抑制这些神经元的化学突触传递，则可降低斑马鱼的捕食能力。这些工作显示，具有生态学意义的视觉刺激，与恰当的实验设计相结合，可以在实验室中利用斑马鱼研究复杂行为的环路机制。

1.1.4 小胶质细胞对视觉功能的调节

神经元的特性和功能锻造于自身膜特性和环路连接，同时也受到神经系统中其他细胞的影响，比如少突胶质细胞(oligodentrocyte)构成的髓鞘保障了神经冲动传播的速率和保真度。小胶质细胞(microglia)是中枢神经系统中重要的免疫效应细胞，在病理状态下迅速激活，变成动态较强的阿米巴形态，迁移并参与到一系列免疫反应及组织修复过程中。在生理状态下，小胶质细胞处于"静息"状态，胞体位置固定，却拥有不断伸缩的众多突起。离体标本的小胶质细胞迅速转为激活态，而在体状态的静息态小胶质细胞不容易观察，因此，对于静息态小胶质细胞的生理功能所知甚少。杜久林研究组的李莹、杜旭飞等以斑马鱼为模式动物，发现了静息态小胶质细胞和神经元之间存在双向的功能调节(Li et al., 2012)。他们同时对小胶质细胞形态变化和神经元活动进行长时程在体记录，发现神经元活动升高，可吸引小胶质细胞向神经元靠近并与其胞体形成紧密接触，形成接触后神经元活动降低，这两组现象之间存在相关性。进而，他们通过局部的谷氨酸解笼锁实验人为局部改变神经元活动，证明神经元活动增强可吸引静息态小胶质细胞的突起靠拢，并且促进紧密接触的形成。这种紧密接触形成后被接触神经元的自发性电活动及视觉反应都被不同程度地削弱。该工作首次证明了神经元活动可以调控静息态小胶质细胞的运动,并揭示了小胶质细胞对神经元活动的稳态调节,为神经-免疫交叉领域提供了新的研究视角。

1.2 嗅觉系统研究

1.2.1 嗅球中的信息编码

嗅觉负责收集和处理长距离的化学信息,参与斑马鱼的觅食、求偶、逃生乃至洄游等重要行为。Rainer Friedrich 从 20 世纪 90 年代中期开始利用斑马鱼研究神经系统的功能,主要研究对象就是斑马鱼的嗅觉系统。他的早期工作主要集中在嗅球中的感觉信息的编码(Friedrich, 2006; Friedrich and Hyman, 2013; Friedrich and Korsching 1997, 1998)。这一方向在他实验室近年来的工作中得到了进一步延伸和拓展。

环境中的感觉信息丰富多样,并且经常发生连续的变化。动物的感知一方面能区别不同刺激对象,实现模式之间的去相关;另一方面又能够在复杂多变的背景噪声中保持稳定性,从不同的呈现方式中识别出同一对象,实现反应模式的均等化。也就是说,感觉系统可以把连续变化的感觉世界离散化,按照个体生存繁衍的需求分为若干小区间,只对跨区间的变化做出响应,即称为类型感知(categorical perception)(Freedman and Assad, 2011; Goldstone and Hendrickson, 2010)。Jörn Niessing 和 Rainer Friedrich 利用斑马鱼嗅球研究了这一现象(Niessing & Friedrich, 2010)。他们在实验中采用双光子成像方法,观察成年斑马鱼嗅球的主要神经元僧帽细胞(mitral cell)对氨基酸嗅觉刺激的反应。当施加单一种类的氨基酸时,发现僧帽细胞群体活动对刺激浓度的逐渐改变并不敏感,始终呈现一致的稳态反应。把两种不同的氨基酸进行混合,并且逐渐改变其混合比例,则发现僧帽细胞群体的稳态反应在某一混合比例附近突然发生切换,在群体神经元活动的多维向量空间中截然分开。这一发现表明斑马鱼嗅觉处理的早期区域嗅球中即存在着类型感知。进一步逐个细胞进行分析,发现稳态反应的突然变化是由一小部分僧帽细胞实现的,提示嗅球可通过不同僧帽细胞活动之间的组合实现对众多外界嗅觉对象的类型感知,这吻合于嗅觉编码多种化学信息的特点。

在此基础上,该实验室进一步研究了类型感知的环路机制。Martin Wichert 建立的理论框架,把高维空间中的模式去相关严格对应到神经元和神经环路的性质,发现神经元反应的非线性特性辅以神经元间交互连接的放大作用,就可以在随机网络中实现模式去相关;相应地,对嗅球中连接图谱的测绘则支持该脑区可能应用了这种机制(Wiechert et al., 2010)。对于类型感知的另一方面,模式均等化,朱培鑫等结合遗传学标记、光遗传学刺激和光/电生理学方法,研究了嗅球中表达启动子 *dlx4/6* 的一类中间神经元所起的作用(Zhu et al., 2013)。通过遗传学方法在这些神经元中特异性地表达光敏感阳离子通道 ChR2,他们可以通过蓝色激光来激活被标记的神经元。结合药理学,他们发现这些神经元除了发出抑制性化学突触到僧帽细胞,还通过电耦合通道与后者发生联系。当接受氨基酸嗅觉刺激时，dlx4/6 中间神经元自身产生兴奋性反应，且反应随刺激浓度线性增加。当刺激强度较低时，反应为阈下水平，不能产生抑制性突触传递，但可以通过电耦合通道把兴奋性活动传递到僧帽细胞，从而增强后者的活动。当刺激强度较高时，dlx4/6 神经元产生阈上反应，发放动作电位，在僧帽细胞上产生抑制性突触输入，这种抑制性作用超过电耦合通道的兴奋性作用，从而降低僧帽细胞的活动。通过这种机制，dlx4/6 神经元降低了僧帽细胞对高、低浓度嗅觉刺激所引起反应的差异，实现了反应模式的均等化，也扩大了信息传递的动态范围。

1.2.2 嗅觉信息编码的神经调制

除了僧帽细胞等主要的投射神经元，嗅球中还包含大量中间神经元。其中的嗅小球周细胞(juxtaglomerular cell 和 periglomerular cell)，包括前述的 dlx4/6 中间神经元在内，既含有主要的神经递质 GABA，还表达与之共释放的多巴胺(dopamine)，但是这些多巴胺对于嗅觉信息处理的意义并不清楚。Bundschuh 等(2012)用光遗传学方法激活这些神经元，在僧帽细胞上诱发出外向电流。通过进一步变换光刺激的时程，并用药理学方法分别阻断 GABA 能和多巴胺能的突触传递，发现长时间刺激诱发的外向电流中包含两种成份，GABA 和多巴胺分别介导了其中快的和慢的成份。作用机制研究发现，多巴胺不仅使僧帽细胞超极化，还提高其输入阻抗，从而降低僧帽细胞的背景活动，同时保持甚至提高其对嗅觉刺激的敏感性。Schärer 等(2012)则研究了多巴胺对嗅球输出的主要脑区——背侧前脑后部(Dp)的影响，发现该脑区接受来自下端脑腹侧(ventral subpallium)、视前区(preoptic region)和后侧结节室旁核(PTN of the posterior tuberculum)的多巴胺能投射。这些多巴胺能投射降低嗅束传来的抑制性突触传递，但对其中兴奋性突触传递没有影响，从而可以通过去抑制增强背侧前脑神经元的嗅觉反应，提高刺激的显著性。

1.2.3 高级脑区的嗅觉信息处理

1.2.3.1 前脑的嗅觉信息处理

对嗅球之外嗅觉系统高级脑区的研究近年来逐渐引起重视(Miura et al., 2012; Poo and Isaacson, 2009)。斑马鱼嗅球主要投射至背侧前脑后部、缰核和后侧结节(Kermen et al, 2013)，其中前两个脑区在近年来得到了较多关注。嗅球中主要的输出神经元是僧帽细胞，这些细胞的嗅觉反应在 20Hz 附近频段存在同步活动，可能代表了一种时域编码形式。当某一刺激激活的神经元在空间上弥散分布，可以通过这种同步活动组织起来。但是该动态特征跟嗅觉刺激的特征之间并没有明显的关联，因而其功能意义并不清楚。为了阐明这种同步活动在信息传递中的作用，Blumhagen 等(2011)采用了光遗传学方法控制嗅球僧帽细胞的活动，同时记录 Dp 的神经元的活动。他们把嗅球划分为若干个小区域(50m×50m 或 20m×20m)，每个区域包含几个到几十个神经元，利用数字微镜器件对激发光进行调制，通过改变每个微镜的偏转角度，只有投射到选定细胞区域的激发光才可投射到样品上,从而独立地激活每个区域的神经元。并可通过选择性地驱动任意一组僧帽细胞的同步活动,调节它们之间同步性的强弱。结果发现,Dp 神经元的确受到僧帽细胞活动的支配,但是仅限于低频范围,20Hz 的活动不论同步性多强,都不能传递到 Dp 神经元,也不能影响其对食物的嗅觉反应。进一步的实验表明,低频活动和嗅觉重要信息提取中表现出的上述选择性,都源自 Dp 神经元膜的被动特性。

1.2.3.2 缰核的嗅觉信息处理

斑马鱼嗅觉另一个重要的投射脑区是缰核(habenula)。该核团在脊椎动物脑中相对保守,与包括五羟色胺能系统、多巴胺能系统和乙酰胆碱能系统在内的多个神经调质核团存在交互联系,可以推测该脑区跟边缘系统关系密切。已有生理实验表明,该脑区神经元的活动跟表征负面情绪相关(Matsumoto and Hikosaka, 2009),但是感觉信息如何支配其活动并不清楚。近年来斑马鱼上的几项工作对这个问题进行了探讨。

从嗅球到缰核的投射,主要来自嗅球内侧的嗅小球群体,它们都终止于右侧缰核(Miyasaka et al., 2014)。右侧缰核与左侧缰核在核团大小、神经联系和化学分子表达上都存在很大差异。在 Stephen Wilson 实验室和 Emre Yaksi 实验室的联合研究中,Dreosti 等(2014)通过在发育早期的药理学操作调节 wnt 信号通路,获得左、右缰核形态发生反转的斑马鱼。在这些斑马鱼上,嗅觉输入依然投射到"右侧形态"的缰核,相应地视觉输入则投射到"左侧形态"的缰核,彼此空间位置发生了反转。当两侧缰核都表现为右侧或左侧形态时,缰核就只对嗅觉或者视觉输入有反应,提示缰核中左、右不对称的神经连接对于斑马鱼的感觉行为有重要意义。Yaksi 实验室的 Jetti 等(2014)研究了缰核环路中嗅觉信息的表征。通过变换多种嗅觉刺激,发现该处神经元可以根据对这些刺激的微弱选择性分为若干组,每组神经元在空间上聚集成簇。当没有嗅觉刺激时,这些神经元的自发活动表现出组内神经元之间的相似性。Suresh Jasuthasen 实验室 Krishnan 等(2014)的工作则研究了缰核神经元对同一刺激以不同浓度出现时的反应情况。通过对整个缰核的神经元进行分析,研究这一群体对斑马鱼生活中重要的胆酸刺激的反应,发现左、右缰核的腹侧部分反应显著,背侧部分则表现出左、右不对称性,与前述报道一致。当改变胆酸浓度时,缰核的反应强度随刺激强度增加,但没有明显的模式均等化现象。而尼古丁可以通过离子型乙酰胆碱受体使这种反应性质变为对中等浓度的选择性,恰好对应斑马鱼对中等浓度胆酸的偏好,提示缰核的生理功能可能受到调制。这些工作揭示了缰核结构与信息处理的复杂性,提示进一步明确其中细胞类型与突触连接的差异,对于理解缰核功能有着重要意义。

1.2.4 对伤害性物质的嗅觉反应

鱼类的嗅觉行为中,对伤害性物质的研究是一个由来已久的课题(Døving and Lastein, 2009; Speedie and Gerlai, 2009; von Frisch, 1942)。这种物质来自于破损皮肤的提取物,能诱发多种鱼类强烈的恐惧反应,但其主要作用成分一直不清楚。Mathuru 等(2012)使用生化分馏方法从中得到了多糖软骨素,发现这一成分可以有效地诱导斑马鱼冲窜、降速、下潜等主要恐惧反应。进一步的光学记录发现多糖软骨素可以在嗅球的背内侧后部引起反应，为进一步了解恐惧反应的神经机制提供了基础。

1.3　跨感觉模态的信息整合

在自然环境中，与动物生存繁衍相关的感觉对象复杂多变，远不是运动光条或者单一气味所能代表的，通常还处在更具动态性的背景中，这就需要动物个体尽可能多地收集信息以便恰当地进行识别与编码，跨模态的多感觉整合（multisensory integration）就是一种可能的策略（Partan and Marler, 2005）。已知斑马鱼的嗅觉可以调节其视觉功能，这种作用甚至在嗅球和视网膜的层次就已出现（Zucker and Dowling, 1987）。Leon Lagnado 实验室的 Esposti 等（2013）详细研究了其中的机制。通过在视网膜双极细胞表达 SyGCaMP2，并在神经节细胞表达 GCaMP3.5，他们对活体斑马鱼的视网膜神经通路的活动进行了光学成像，在给予视觉刺激的同时施加斑马鱼食物刺激（蛋氨酸）。发现，蛋氨酸刺激能够降低视网膜中 OFF 通路对刺激强度和时间对比度变化的反应，同时提高其敏感性，但对 ON 通路没有明显影响。这种调节作用是由多巴胺能神经传递所介导的。这与前人发现的结构基础相一致：嗅球到视网膜的终端神经（terminal nerve）投射到含有多巴胺的内网状层。值得进一步思考的是这种投射的特异性，食物刺激扩大了视觉通路的动态范围，可能有助于增强觅食能力，但这种调节仅局限于 OFF 通路的意义尚未可知。

除了觅食，还需要躲避天敌，动物才能维持生存繁衍。斑马鱼的逃跑行为中研究较多的是听觉刺激诱发的 C 形快速逃跑行为（fast-start C-shape escape），其神经环路清晰，包括毛细胞、听神经、后脑命令神经元 Mauthner 细胞、脊髓运动神经元（Eaton et al., 1977），Joseph Fetcho 实验室的经典工作探明了其中环路机制（Liu and Fetcho, 1999; O'Malley et al, 1996），但是在更复杂环境下的逃跑行为尚不清楚。杜久林实验室穆宇、李小泉等（Mu et al., 2012）发现，在听觉信号之前施加闪光刺激，使动物处于一种警觉状态，可以显著提高幼年斑马鱼的逃跑概率。采用相同刺激，在活体斑马鱼上进行的电生理记录研究了其中的环路机制，发现闪光刺激虽然不会引起 Mauthner 细胞发放动作电位，却可以强烈抑制听神经的自发放电，从而提高其信噪比；另一方面，闪光刺激提高了听神经到 Mauthner 细胞的突触传递效率。两种作用协调起来，有效地增强了 Mauthner 细胞的听觉反应，提高了 C 形逃跑的灵敏度和准确性。有意思的是，闪光刺激的这一作用也是由多巴胺能所介导的，激光损毁含有多巴胺能神经元的后侧下丘脑可阻断这一作用。这类由调质系统介导的跨模态整合可能代表了一种普遍的规律，神经调质系统通过相对粗疏但灵敏的方式感受机体内、外环境的变化，进而相应地对感觉-运动通路多个环节的神经信号传递进行调节，使得动物有选择性地处理和响应一些更有生存意义的刺激，从而更好地适应环境，有利于生存和进化。

2. 运动控制的规则

2.1 运动相关神经元的网络构筑

对斑马鱼神经系统的研究,早期集中在发育方面(Trevarrow et al., 1990; Woo and Fraser, 1995)。从 20 世纪 90 年代开始,对其生理功能的研究也逐渐展开,开启这一方向的包括 Joseph Fetcho 实验室。他们率先在斑马鱼上实现了神经活动的光学和电生理记录,研究了后脑 Mauthner 细胞及同源神经元如何介导了机械刺激诱发的 C 形逃跑,并且通过激光损毁特定神经元,建立了这种运动控制的因果联系(Liu and Fetcho, 1999; O'Malley et al., 1996)。在近年来的研究中,他们和其他实验室结合光遗传学、电生理学、药理学等手段,阐明了前运动神经元群在运动控制中的募集和组织过程。

Kaede 是来自珊瑚虫的一种荧光蛋白,在紫外光照射下从绿色荧光变为红色。McLean 和 Fetcho(2009)在脊髓中参与控制运动的兴奋性中间神经元内表达 Kaede,并在斑马鱼出生后逐日进行光转化,这样在光转化前分化出的中间神经元呈红色,之后分化的则呈绿色。通过这种方式,他们发现在发育过程中,控制大幅度快速运动的神经元更早分化,控制慢运动的神经元更晚分化。相应地,幼年斑马鱼的游动更早表现出快速成分,在电生理上也更早地记录到控制快速运动的神经元。

Fetcho 实验室后续的工作详细研究了后脑运动控制神经环路的结构和功能空间排布。Kinkhabwala 等(2011)在谷氨酸能兴奋性神经元和甘氨酸能抑制性神经元中分别表达红色和绿色荧光蛋白,并用免疫组化方法观察到若干重要转录因子在这两种神经元中的特异表达。研究发现,在冠状截面上,谷氨酸能神经元和甘氨酸能神经元从中线向外呈矢状交错排列,形成若干平行于背腹轴的条带,每条条带上表达特定的转录因子。通过对若干细胞形态的三维重构,发现同一条带中的神经元树突与轴突具有类似的形态和投射方向,同一条带神经元存在腹先背后的分化顺序。运用电生理学记录测量细胞的膜被动特性,发现同一条带内神经元的输入阻抗、轴突长度、所介导的运动速率均线性依赖于该神经元所处的背腹位置:腹侧神经元轴突更长,输入阻抗更低,自然能更快地传导神经冲动,与介导的更高运动速率相一致。在另一项工作中,鉴定了不同中间神经元在 Mauthner 细胞介导的逃跑行为中的作用(Koyama et al., 2011)。此前在金鱼上的工作已经发现,逃跑环路附近有多种中间神经元,它们以不同方式影响 Mauthner 细胞的活动。在绿色荧光蛋白标记甘氨酸能中间神经元的转基因斑马鱼上,Koyama 等依照突触连接位置差异,局部电转红色染料以标记不同群体的中间神经元,并成对记录所标记的中间神经元和 Mauthner 细胞,确认其突触连接类型,发现同一冠状截面上不同条带内,其神经元分别从特定方面调节 Mauthner 细胞的活动。进一步地,还发现从腹侧到背侧,运动神经元的活动水平递增,因而其树突伪足形态表现出更强的动态性,可能有助于建立前述的激活顺序(Kishore and Fetcho, 2013)。该系列工作从发育、形态、结构到功能对运动控制神经网络进行了全景式细致描绘。

2.2 运动神经元的激活顺序

脊髓运动神经元是控制肌肉、产生游泳运动的最后一级神经元。Bagnall 和 McLean(2014)利用在体膜片钳记录，在毫秒的精细时间尺度上，分析支配某一体节肌肉的运动神经元彼此间活动的同步性。发现控制同一体节肌肉的运动神经元，根据所控制肌肉的左右及背腹位置可分为四组，只有同组神经元才接受同步的突触输入(兴奋性和抑制性皆如此)，即这四组肌肉可以分别被控制。前人已知分别控制左、右侧肌肉是动物游动所必需的，他们进一步发现，背腹侧肌肉的分别控制对于斑马鱼调整身体姿势至关重要。

El Marina 实验室利用从成鱼脑分离出的脑干和脊髓标本，发现同一功能柱内的运动神经元按照所支配的肌肉类型分为四群，当刺激运动神经诱发运动时，这四群神经元会按照从腹到背的顺序，在从低频到高频的运动中渐次激活。机制研究发现，这一激活顺序可能与神经元活动的阈值和发放模式等生物物理特性相关，尤其是从腹侧到背侧，神经元胞体逐渐增大，这一顺序符合肌肉激活的大小原则：较小的运动神经元控制较弱的肌肉，在收缩力由弱渐强时更早被激活(Gabriel et al., 2011)。这个在离体标本上观察到的现象，被 Menelaou 和 McLean(2012)在活体幼年斑马鱼上的实验所证实。他们首先通过注入荧光染料，追踪了运动神经元到肌肉的投射，确认最大的运动神经元位于最背侧，主要支配快肌纤维，较小的运动神经元位于腹侧，支配慢肌纤维。在不同速率游动的斑马鱼上进行电生理记录，不仅发现这些运动神经元存在从腹侧到背侧的激活顺序，还发现沿此方向存在内在节律性逐渐降低、外来驱动逐渐增强的趋势，结合 Koyama 等的工作，提示斑马鱼运动控制中存在复杂的网络调控。

2.3 运动信息的编码方式

神经系统编码的一个基本问题是，究竟神经信息是由一小群神经元独立编码(即稀疏编码)，还是依靠大量神经元的活动的不同模式来编码(即群体编码)？Florian Engert 实验室的一系列工作研究了编码不同运动的神经元在空间上如何排布这一基本问题。

Douglas 等(2008)用光遗传学的手段研究了斑马鱼后脑的 Rohon-Beard 神经元和三叉神经节细胞，这些神经元负责把躯体感觉转换为运动输出。他们通过表达 ChR2，用光学刺激诱发这些神经元的动作电位发放，发现单个躯体感觉神经元上的少数几个动作电位即可诱发逃跑行为，提示可能存在稀疏编码方式。黄国华等用视觉刺激诱导了斑马鱼的趋光反应、视动反应和暗场下大幅转向行为，仔细分析了其运动过程中各时相的位移、转向等参数。发现，损毁后脑腹内侧向脊髓投射的神经元，在各种运动中都会特异地抑制转向行为，并且增加向前游动的行为，提示这些神经元负责普适性地把前向运动转化为转向(Huang et al., 2013)。Orger 等(2008)利用视动(optomotor)模型研究了负责不同行为的神经元是如何分布的。他们给予斑马鱼幼鱼以全场的运动光栅刺激，并高速拍摄斑马鱼运动和通过反馈控制来调整光栅刺激的方向，使斑马鱼始终感受到同一方向的刺激，避免了其自身运动和姿势变化的影响，从而得以诱发稳定的转向行为。结果发现，斑马鱼轨迹依赖于光栅方向和鱼体轴的夹角。利用双光子成像技术，记录后脑向脊髓投射的绝大部分神经元的活动，发现有不同的神经元分别与前游、左后转、右后转等运动相关，这些神经元在空间上彼此分隔。这一发现支持了稀疏编码机制。相比 Fetcho 实验室研究 Mauthner 细胞及相关环路如何介导逃避行为的工作，Engert 实验室的工作为从神经环路水平了解脊椎动物如何协调各种复杂运动奠定了基础。

关于运动神经元活动的有序结构是如何出现的，来自 Ehud Isacoff 实验室的 Warp 等 (2012) 进行了发育方面的研究。通过在脊髓神经元中表达 GCaMP3，作者得以用光学记录手段观察所标记的每个神经元的活动水平。随时间记录脊髓运动神经元群体的活动模式，发现在受精后 18~20h 时有一个迅速的转变，从零星散布的慢发放模式过渡到同侧神经元间正相关、对侧神经元间负相关的快发放模式。进一步地，光遗传学实验考察了细胞活动之间的关联。光敏感的 Cl^- 通道盐菌紫质 NpHR 可以在黄色激光照射下产生 Cl^- 电流从而使表达的神经元超极化，活动受抑。他们在所研究的运动神经元上同时表达 GCaMP3 和 NpHR，利用其激发光谱的分离，观察抑制某些神经元活动对其他神经元的影响。在此基础上，Warp 等 (2012) 发现，在受精 20h 后，抑制某一运动的神经元会造成同侧邻近的运动神经元活动降低，提示它们之间联系的加强。如果用光刺激抑制转变之前的零星发放，则会影响之后有序模式的建立。

2.4 眼动控制的神经机制

斑马鱼的重要视觉对象，如食物和天敌等，多为运动物体。为了追踪这些对象，斑马鱼需要保持运动刺激在视网膜上的位置，这是通过眼动反应(optokinetic response)和视动反应(optomotor response)来实现的。这类重要行为的神经基础在过去几年也变得更为清楚了。

在眼动反应中，面对运动的光栅，斑马鱼的眼睛会随之转动，直到不能进一步转动才迅速回到最初位置。这一运动的控制受到了多个实验室的关注。Herwig Baier 实验室在全脑大部分神经元中广泛表达 NpHR 从而可以通过光刺激来抑制神经元活动。通过逐步刺激不同的脑区，他们发现第五菱脑节是眼动反应所必需的。相应地，用 ChR2 实验进行充分性验证，也表明激活这个脑区足以引起眼动反应(Schoonheim et al., 2010)。为研究这一行为反应中感觉信息如何实现到运动信息的编码，Kubo 等 (2014) 用光遗传学方法发现，顶盖前区神经元的活动对于眼动发生是必要的，人为诱发顶盖前区活动也能诱导出眼动。作者进而用双光子成像方法记录整个顶盖前区的神经元对于水平转动光栅等刺激的反应，发现该区域可以根据神经元对刺激的偏好性，如平动或转动、转动的不同方向、单眼还是双眼输入，分为空间分离的若干脑区，并以此提出了一个有待实验证明的神经环路模型。

在眼动过程中，斑马鱼需要整合眼睛运动的速率信息来估计其当前位置，才能实现稳定的视觉追踪。位于后脑舌下前置核的神经元在眼动过程中表现出阶跃形式的活动，每次发生眼动后即改变放电频率到相应的水平，其后可维持长达数秒的活动平台期。形态上，这些神经元的树突接受编码眼动速率信息的输入。据此，该核团有可能实现了眼睛运动信息的整合，但其中环路机制尚不清楚。经典的线性吸引子模型(Seung, 1996)和多种交互网络模型(Anastasio, 1994; Goldman, 2009)建立在不同的神经环路基础上，均可以描述从速率到位置的整合过程，其间主要差异在于每次眼动发生后，整合神经元活动维持的时间常数不同。在线性吸引子模型中，该常数在所有整合神经元中保持一致；而交互网络模型中则存在多个不同的时间常数。因此，系统测量舌下前置核神经元群体的活动维持时间常数是明确整合环路机制的关键，而光学记录由于可以同时记录群体神经元的活动，在该问题的研究中可以起到事半功倍的作用。为此，David Tank 实验室的 Miri 等(2011)在清醒的幼年斑马鱼上进行光学记录，并同时观察其自发眼动。基于神经元活动与眼睛位置之间的时域相关性定位了舌下前置核中位置相关的神经元，并用光遗传学抑制方法确认这些神经元的活动是斑马鱼稳定注视所必需的。他们发现，这些整合神经元的活动维持时间常数存在明显的不一致性，从而否定了线性吸引子模型的假设。进一步地，他们利用光学记录得到的神经元空间位置信息，分析了成对神经元活动相关性与空间距离的关系，发现彼此间距越小的神经元对，其活动越相似，维持时间常数也更接近，从而支持了交互网络模型机制。神经活动相关性、维持时间常数沿背腹轴和头尾轴的变化则表明了拓扑投射的存在。参照 Fetcho 实验室的结果，提示发育过程和激活顺序上可能存在对应的差异。

3. 学习和神经功能可塑性

无论在发育过程中还是成年后，动物常常需要适应不同的生活环境，面对同样的刺激需要采取不同的行动，这就需要对旧有的信息处理机制和行为决策做出调整，也就是学习过程。神经系统的结构和功能的可塑性则是学习的神经基础。

经典的学习范式包括条件性学习和操作性学习，前者在环境中两个本来不相关的感觉刺激之间，后者则在动物自身行为和后果之间，建立联系。Erin Schuman 实验室的两项工作研究了条件性学习的神经基础。对于用琼脂糖凝胶固定的幼年斑马鱼(6~8 天)，Aizenberg 和 Schuman(2011)把运动光点刺激(条件性刺激)和躯体触碰的机械刺激(非条件性刺激)配对反复施加，显著增强了单独施加光刺激诱发原本由机械刺激引发的鱼尾摆动的概率。通过在鱼脑中注入化学合成的钙指示剂 Oregon Green BAPTA-1 AM，他们同时观察了斑马鱼小脑的群体神经元活动，发现分别存在对视觉刺激和机械刺激反应的神经元，二者在空间上相对分离；而反复施加配对刺激则显著增强视觉反应，提示小脑参与这一学习过程。激光局部损毁实验则证明小脑的作用在于学习而非已获得记忆的存储。值得注意的是，自发摆尾较少或者在配对刺激施加之前已有视觉反应的斑马鱼都被排除在实验之外，提示这种学习范式的稳健型和有效性尚待提高；同时，其神经环路机制也有待更细致地证实。在 Hinz 等(2013)的后续实验中，发现斑马鱼幼鱼学会把某个空间位置和同类的存在联系起来，从而产生对该位置的偏好性。这种偏好性甚至在较长时间(36h)内克服了斑马鱼幼鱼的趋光行为，说明形成了一种长期记忆，学习过程对蛋白质合成的依赖性也支持了这一结论。

  动物更偏向于学习对于个体生存繁衍有重要意义的感觉信息，逃避风险就是其中一种。来自 Hitoshi Okamoto 实验室的两项工作研究了恐惧学习的机制。Aoki 等(2013)发现前脑区域在操作性的主动回避学习中的作用。在实验中，成年斑马鱼身处分为两半的鱼缸中，需要在红光闪烁后从原本栖身的一侧转入另一侧以躲避电击伤害。光学记录结果显示，前脑背侧部分区域神经元原本对红光无反应，但在学习完成 24h 内出现了明显的反应性。损毁该脑区则影响已学会内容的提取，但是不妨碍完成新的学习任务，表明其红光反应可能涉及记忆的提取。进一步地，如果学习规则发生转换，前脑的活动区域相应地发生转移，提示其中可能存在着功能图谱。

  动物的学习能力也受到众多因素的影响。比如类似上述的逃避学习，如果动物事先经受了无法逃避的严重伤害，则不能学会逃避，表现出所谓习得性无助。背侧缰核作为情绪调节的关键脑区，被认为参与到学习能力的丧失当中。Agetsuma 等(2010)在背侧缰核外侧核(dHbL)中特异性表达破伤风毒素，阻断该处神经元的突触输出，这样的操作对斑马鱼的运动能力没有影响。但是，在重复进行条件性恐惧学习时，对照组斑马鱼会从初期的战栗反应转变到学会后的逃跑反应，而 dHbL 失活的斑马鱼则一直做出战栗反应，类似于习得性无助。相应地，来自 Jathusathan 实验室的 Lee 等(2010)在幼年斑马鱼上发现，通过光化学手段杀死缰核神经元，或者特异性表达神经毒素阻断其突触输出，均可以阻断斑马鱼的恐惧学习。这些工作提示，背侧缰核外核到脚间核的投射在恐惧反应因经验而改变的过程中起着关键作用。

  神经可塑性，即经验依赖的神经环路结构和(或)功能修饰，是动物学习的神经基础，其在突触层次研究得很深入，对突触传递长时程增强(long-term potentiation, LTP)的研究尤为详尽。然而，由于外周感受器如视网膜被认为需要把外界环境信息忠实地传递到中枢，其内部突触传递是否具有可塑性尚无定论。杜久林实验室的魏宏平、姚园园等，利用在体全细胞膜片钳记录和在体双光子成像等技术，在活体斑马鱼上研究 LTP 是否存在于发育期的视网膜(Wei et al., 2012)。结果发现，在受精后 3~6 天的斑马鱼上，高频电脉冲刺激能够在视网膜双极细胞到神经节细胞的兴奋性突触上诱导 LTP，而在 15~20 天的动物上则不能。通过进一步的功能实验，他们发现重复的光刺激也能在同一级突触上诱导 LTP，并且由电刺激和不同模式的光刺激诱导产生的 LTP 都能有效增加神经节细胞的对光反应。由此推测，一方面视网膜和其他脑区一样在发育过程中可能受到感觉经验的影响而产生可塑性变化；另一方面，在相对成熟后视网膜环路逐渐失去可塑性能力，从而稳定地编码视觉信息。

  来自 Florian Engert 实验室的 Valente 等(2012)同样关注发育过程中的学习。他们发现条件性学习和操作性学习的能力与斑马鱼发育程度有相关性。在条件性学习中，斑马鱼需要在视觉刺激与随后出现的电击之间建立联系；在操纵性学习中，斑马鱼需要移动自身位置逃避电击。通过连续训练，发现成功的条件性学习出现在受精后第四周。从不同批次中选取各个发育阶段的斑马鱼进行操纵性学习，发现第三周是学习成功实现的开始。考虑到 7 天大的斑马鱼即可进行视觉检测并有行为反应，该工作提示学习过程中需要多个脑区的协调。当然，这一结果也有可能受到了实验范式所限，未能充分发掘幼年斑马鱼的学习能力。值得注意的是，并非所有斑马鱼都在同一时间开始学会，也有少数个体在 10 天左右即表现出学习能力，提示可以通过预筛选，挑选出学习能力强的幼年个体进行相关的神经机制实验。

4. 我国斑马鱼神经系统研究的进展

我国已有数十家实验室采用斑马鱼进行神经科学研究，并取得了显著进步。其中，神经系统发育方面的研究尤为突出，涵盖了器官发生、神经调质系统发育、轴突发育与再生、脑血管网络发育等多个方向，概述如下。

清华大学生命科学院孟安明研究组利用斑马鱼模型，研究了 Grainyhead-like 2（*Grhl2*）基因影响听觉系统功能的分子机制（Han et al., 2011）。该基因在人类个体的突变会引起进行性听力损伤。利用 *Tol2* 转座子插入，他们获得了 *grhl2b* 突变的转基因斑马鱼。听觉器官的形态学分析、电生理及行为功能检测均发现这些斑马鱼幼鱼存在听觉功能缺陷。进一步的机制研究发现，正常的 grhl2b 蛋白可以通过结合在相应的增强子区域促进细胞连接蛋白 claudin-b 和 epcam 的表达，突变体中这两种蛋白表达降低。共注射 claudin-b 和 epcam 的 mRNA 可以恢复其蛋白水平，并能够有效缓解 *grhl2b* 突变所引起的听觉器官发育的缺陷，提示 grhl2b 调节听上皮细胞的完整性及分化，从而参与形成成熟的听器官和听觉感知。

神经调质系统参与甚至主导了很多基本的脑功能。以多巴胺能系统为例，它对运动、学习与记忆、情感与动机都有着重要的作用。然而，技术手段的限制使我们对调质系统的了解还所知甚少。例如，我们不十分清楚在同一核团内是否存在不同类型的多巴胺能神经元，它们是否介导不同的功能。同样，我们不了解不同多巴胺能神经元为什么对不同的感觉输入有不同反应特性。在哺乳动物中，也很难对单个多巴胺能神经元进行完整的形态学描述。能够稳定可靠地标记出释放调质的神经元对于研究其功能具有重要意义。北京大学生命科学院的张博研究组和林硕研究组应用以 *Tol2* 转座子为基础的"增强子诱捕"（enhancer-trap）技术，构建由不同增强子驱动的绿色荧光蛋白表达系。经过筛选，获得了可以标记单胺能神经元的转基因斑马鱼品系，即 ETVmat2:GFP。生化鉴定结果表明，该品系能够标记斑马鱼脑中的所有多巴胺能神经元核团，同时也标记五羟色胺能神经元核团，以及去甲肾上腺素能神经元核团（Wen et al., 2008）。这是首个能够清晰标记单胺能神经元的斑马鱼品系。在发育的分子机制方面，中国科学院健康科学研究所的乐卫东研究组及刘廷析研究组共同发现，已知对中脑多巴胺能神经元发育至关重要的孤核受体基因 *nr4a2*，是通过调节多巴胺能神经元的分化影响了胚胎发育阶段多巴胺能系统的形成（Luo et al., 2008）。而中山大学的谢富康研究组和李朝红研究组发现，食品添加剂中的苯甲酸钠可以下调酪氨酸羟化酶和多巴胺转运体的表达，从而造成多巴胺能神经元发育的缺陷(Chen et al., 2009)。

轴突生长是神经发育过程中的重要步骤，表现为长距离的投射、准确的目标定位、及高度的动态性。快速发育的斑马鱼是在体研究这些复杂过程的理想模式动物。南昌大学生命科学研究院徐洪研究组以斑马鱼视网膜神经节细胞轴突到对侧视顶盖的交叉投射为模型，利用 GAL4/UAS 表达系统、基因敲减，以及突变斑马鱼品系，发现神经节中的 cAMP 高浓度，促进了 Neuropilin-1a 的表达，从而增强了发育过程中轴突对中线处 Sema3D、Sema3E 等配体的敏感度，最终帮助建立神经节至对侧视顶盖的轴突投射(Dell et al., 2013)。复旦大学脑科学研究院彭刚研究组通过单细胞标记与嗅觉功能相关的端脑前背侧区神经元(anterior dorsal telencephalic, Adt)，发现 Adt 神经元的轴突具有三类投射方式，或者沿前联合(anterior commissure, AC)，或者沿视上束(supraoptic tract, SOT)，或者同时沿前联合及视上束进行投射。Adt 区局部导向受体的特异表达决定并协调了这三种投射：在背侧中区，有 Dcc 的高表达，该区的神经元受到 Dcc–Netrin 信号调节，主要沿 AC 进行投射；在腹侧外区，Robo2 表达较多，该区神经元受到 Robo2-slit 信号调节，主要沿 SOT 进行投射；而介于二者之间的区域，则出现了同时沿 AC 及 SOT 进行的轴突投射。此外他们还发现 Dcc-Netrin 信号还决定了 Adt 神经元轴突向腹侧的生长，当下调 Dcc 或者 Netrin 表达时，轴突投射位置更偏向背侧，甚至出现直接朝向背侧的多余轴突分支(Zhang et al., 2012)。中国科学院遗传与发育生物学研究所张建研究组通过研究运动神经元轴突生长，发现了一条调节脊髓运动神经元轴突发育的信号通路(Liu et al., 2012)。他们观察到 prdm14 基因较为特异地在脊髓运动神经元表达，突变或者敲减该基因使斑马鱼幼鱼个体的尾侧初级运动神经元轴突缩短，并出现运动功能障碍。进一步实验发现，prdm14 可以结合在 islet2 基因的启动子序列，并且过表达 islet2 可以挽救 prdm14 突变所造成的轴突生长缺陷。因此，该研究发现了 prdm14 这种运动神经元特异表达的基因，可以经由 islet2 调控脊髓运动神经元的轴突生长。

神经元轴突有效传递动作电位依赖于髓鞘的包裹，在中枢神经系统中，这是靠少突胶质细胞实现的。第二军医大学何成研究组与中国科学院神经科学研究所杜久林研究组合作，在少突胶质细胞中表达荧光蛋白，利用幼年斑马鱼通体透明的优点，对髓鞘发生进行了长时间的稳定动态观察(Liu et al., 2013)。发现，髓鞘发生是一个伴随着髓鞘形成与修剪的动态过程：开始阶段每个少突胶质细胞包裹多根轴突，并过度形成多个髓鞘片断；随着发育的进行，部分髓鞘片断发生修剪，其余髓鞘片断则逐渐延伸并稳定。髓鞘修剪过程是一个与所裹轴突直径相关的过程：髓鞘所包裹轴突直径越大，越不容易发生修剪。同时，同一少突胶质细胞来源的髓鞘间存在类似"竞争"的关系，即单个少突胶质细胞包裹多根轴突时，较细轴突上的髓鞘更容易被修剪，反之较粗轴突上的髓鞘更容易被保留。最后，通过基因敲减的方法，作者证明这种髓鞘修剪的过程依赖于膜锚定的神经调节蛋白 Neuregulin1(NRG1)，以及胶质细胞上的 ErbB 受体，提示，髓鞘修剪过程需要神经元轴突与少突胶质细胞髓鞘之间的相互作用。

相比哺乳动物，斑马鱼的中枢神经系统具有较强的再生能力，从而为神经损伤和再生研究提供了便利。我国科研工作者在这方面获得了一定的进展。汕头大学申延琴教授与 Melitta Schachner 教授所带领的研究团队，通过设计手术方法及衡量手术效率的行为学与形态学手段建立了有效的成年斑马鱼脊髓损伤模型(Fang et al., 2012)。在此基础上，他们系统检测了在脊髓损伤后表达上调的基因，包括手术后在中央管高表达的神经黏附分子 Contactin-2(Lin et al., 2012)、在创口尾侧室管膜细胞中高表达的穿窿体主蛋白(Pan et al., 2013)、在创口尾侧脊髓神经元与胶质细胞中高表达的骨架蛋白 Syntenin-a(Yu and Schachner, 2013)，并用基因敲减实验证明了这些蛋白在斑马鱼脊髓创伤后再生中的重要功能。中国科学技术大学生命科学学院胡兵研究组通过手术损伤成年斑马鱼视神经，并在术后检测从视网膜神经节细胞到视顶盖投射的重新建立(Zou et al., 2013)。发现，神经纤维投射的重新建立，并不依赖于新生神经元的产生，推翻了以前对这二者之间因果关系的推测。中国科学院营养科学研究所翟琦巍组和神经科学研究所杜久林组合作，利用单细胞电转方法在斑马鱼后脑神经元 Mauthner 细胞中表达烟酰胺核苷腺苷酰转移酶 2(nicotinamide mononucleotide adenylyltransferase2, Nmnat2)，发现 Nmnat2 对脊髓损伤后神经退化有保护功能，并且这种保护依赖于 Nmnat2 的 NAD 合成活性(Feng et al., 2010)。

斑马鱼透明而小巧的脑也为在体研究脑血管网络的发育提供了可能。杜久林研究组对大脑血管网络的发生过程做出了开创性的工作(Chen et al., 2012)。他们利用标记血管内皮细胞(vascular endothelial cell)的转基因斑马鱼，对脑部血管的发生进行长时间的实时追踪，并通过计算机图形学方法对所观察到的血管网络从几何特征和结构特征进行定量分析，揭示了脑部血管三维网络发育的基本过程。他们发现，斑马鱼脑血管发生过程中，不仅有血管生成，还有大量的血管修剪，二者的作用共同构建了斑马鱼脑部的血管网络。通过对中脑血管的精细观察并追踪每一根血管的发育命运，他们发现血管修剪多发生在环状结构中冗余的通路，提示修剪过程将会降低网络的复杂度，使血流更加有效。进一步网络拓扑结构的分析支持了这一点。鉴于血流相对较低的一支更容易被修剪，研究者探索了发育过程中的脑血管修饰与血流动力学之间的因果关系。通过人工阻断单节段血管血流、追踪单个上皮细胞命运、荧光共振能量转移依赖的分子功能成像，证明了血流动力学通过调节 Rac1 依赖的细胞迁移，从而导致血管修剪，造成发育过程中脑血管网络的结构简化和功能高效。这一研究结果，也暗合了宇宙中一个普遍存在的规律：无论是物理系统还是生物系统，无论是宏观的还是微观的，复杂网络的演变通常是一个在结构上从无序到有序、在功能上从低效到高效的自适应过程。

5. 新技术的发展和需求

虽然历史并不长久，用斑马鱼作为模型的神经系统功能研究已充分显现出蓬勃生机和潜力。斑马鱼模型的独特性，使得研究人员可以综合多种交叉手段，在全脑尺度上，在活体甚至清醒的标本上研究脊椎动物基本的感觉-运动与认知过程的突触机制，剖析其背后完整的神经环路。这一方面需要充分发挥现有实验手段的潜能，另一方面也将催生新的研究方法。

5.1 观察神经活动的探针

前述大量工作已经显示了光学成像方法在斑马鱼研究中的优势和重要性，这其中首要的是把神经活动的跨膜电位信号转化为光信号。蛋白质钙指示剂和化学合成的染料对细胞内 Ca^{2+} 浓度变化做出反应，间接达到了这一目的。由于可以通过遗传学方法进行体内表达，既避免了急性注射的损伤，还可以控制所要标记的细胞类型，因而蛋白质钙指示剂(如 GCaMP 系列)逐渐取代了传统的合成染料(如 Oregon Green BAPTA、Alexa 和 Rhod-2 等)。通过遗传工程的改造，GCaMP 系列钙指示剂目前已经升级到 GCaMP6，可以检测到单个动作电位相关的钙信号，大大超过以前使用较多的 GCaMP3 的灵敏度(Chen et al., 2013)。但即便如此，其信号变化的半衰期仍然是数百毫秒，远远不能反映发放神经冲动的毫秒级的信号变化。一种解决方法是通过去卷积运算，从与 Ca^{2+} 浓度变化相关的光信号中计算出动作电位发放的时间信息(Vogelstein et al., 2010)，为此需要所采用的钙指示剂具有较大的动态范围，可以反映从零星发放到成簇活动的动作电位的信息，同时在此范围内有较好的线性特性。染料 Rhod-2 是一种线性较好的钙指示剂(Yaksi and Friedrich, 2006)，但鉴于其灵敏度较低，且不能通过遗传学操纵进行特异表达，对 GCaMP 系列的进一步改造可能是更好的选择。

Ca^{2+} 作为重要的第二信使，参与多种细胞内生化过程，其浓度变化并不总是与神经活动相关，加上前述去卷积运算的要求，直接测量电压信号则是更优选择。斑马鱼标本的特点使得在体膜片钳记录包括成对记录并不困难，但面对记录群体神经元活动的要求时，还需要借助光学手段，这就要求开发出把跨膜电位信号直接转变为光学信号的指示剂。传统的电压敏感染料可以实现这一目的，但是缺乏细胞分辨能力，且同样不能通过遗传学操纵进行细胞标记。因此国际上有为数众多的实验室把发展电压敏感的蛋白质指示剂作为工作重点。最近，斯坦福大学 Michael Lin 实验室和 Mark Schnitzer 实验室分别报道了这样的工具，它们均可以线性地反映从静息膜电位以下到 0mV 的电压信号，并在数百赫兹的频率上跟随膜电位变化，达到了目前光学成像工具的极限(Gong et al., 2014; St-Pierre et al., 2014)。如果能顺利移植转移到斑马鱼体系中，这种新型工具必然会揭示出神经信息处理的时间细节，使我们得以了解众多神经元、神经核团之间的相互关系，同时有助于揭示群体神经元信息编码的机制。

观察活体动物上的神经活动，需要克服的一个问题是运动对光学信号的干扰。如果把斑马鱼脑视为刚体，图像处理可以解决这种干扰带来的主要问题。但是实际情况更为复杂，这就对神经活动的探针提出了新的要求。Cameleon 指示剂制备于钙调蛋白(calmodulin)和两种不同荧光蛋白的区段，Ca^{2+} 结合会拉近两种荧光蛋白的距离，从而发生能量转移，因而两种荧光强度的比值可以表征 Ca^{2+} 的浓度。由于发生运动时，两种荧光强度均会受到相同的影响，因而其比值不会发生变化。Fetcho 实验室采用其中的 YC2.1 研究了运动相关神经环路的活动。目前这一探针已发展到 YC3.60 和 YC-Nano140，可以记录 2 个以上动作电位相关的 Ca^{2+} 浓度变化(Horikawa et al., 2010)。

对神经环路功能的解析，很重要的是基于对其中不同神经元类型的了解，用它们各自的结构和功能特征来解释其生理功能。这一方面需要有恰当的遗传学、化学手段来识别、标记不同类型的神经元，另一方面需要从光学记录中对其进行区别。一种可能就是采用不同频段的荧光信号记录不同类型神经元的活动，Loren Looger 实验室通过对 GCaMP 和荧光蛋白 mRuby 进行改造，得到了蓝色、青色、黄色、红色的钙指示剂，为这个方向的技术发展打下了基础(Akerboom et al., 2013)。

5.2 新兴光学方法的应用

斑马鱼小巧透明的脑非常适合于光学成像，进行全脑尺度上细胞水平的神经功能成像。前面已经提到多项工作应用激光共聚焦或双光子显微镜，在单个核团甚至全脑实现了这种记录(Krishnan et al., 2014; Kubo et al., 2014; Portugues et al., 2014)。近年来复兴的光片照明显微镜(light sheet microscope 或 selective plan imaging microscope)则为这种策略提供了更好的手段(Huisken et al., 2004)。通过将激光聚焦至数微米厚的薄片，即所谓光片照明，结合垂直于光片方向上用 sCMOS 或 EMCCD 进行的逐帧采集，可以在最短 10ms 内采集一个光学层面，其成像速率快于点扫描 2~3 个数量级。如果实现样品相对照明光片和物镜在 z 方向的移动，就能够快速完成三维结构的扫描。由于只有成像层的细胞被照明，从而极大减少了光漂白的影响。利用此技术，Misha Ahrens 研究组及 Philipp Keller 研究组得以在 1.3s 内完成一次斑马鱼全脑扫描，对斑马鱼大脑 80%的神经元进行了 Ca^{2+} 成像，获得单细胞分辨率的全脑活动图谱(Ahrens et al., 2013)。Georges Debrégeas 实验室的 Panier 等(2013)则以 4Hz 的频率记录了斑马鱼脑中约 2.5 万个神经元的活动。该方法的建立大大鼓舞了 BRAIN Initiative 等计划的倡导者，Rafael Yuste 说："这相当惹人注目，说明记录动物全脑神经元的活动决非空谈"(Baker, 2013)。

光片照明显微镜在斑马鱼中的应用有几个方面尚需完善。首先，扫描时所产生的光刺激会穿过动物眼睛，可能会改变脑工作状态，影响动物对环境感觉输入的感知。为了解决这个问题，研究者考虑可以缩窄光片，从头尾方向照明，或者利用两个正交的扫描单元，在扫描空间上避开眼球所在的区域。但即便如此，也会因脑组织的散射使斑马鱼处于明亮背景中(GCaMP 等钙指示剂的激发光位于斑马鱼可视范围内)。因此，基于双光子的光片照明显微镜是一种解决策略。Scott Fraser、Emmanuel Beaurepaire 和 Willy Supatto 等实验室已经在这方面做出了可喜的尝试(Mahou et al., 2014; Truong et al., 2011)。其次，尽管斑马鱼脑颇为小巧，背腹深度仍有 300μm，从深部发射的荧光要经过上层脑组织才能被探测到，因而不可避免地会受到组织折射率不均匀性的影响，降低了图像质量。克服该问题的一种可能是结合光片照明显微镜与自适应光学手段。后者通过测量每个空间位置上的相位偏差，用空间光调制器或可变形反射镜加以补偿。最近已有工作表明自适应光学成像可以显著改善斑马鱼成像质量(Wang et al., 2014)，后续进展值得期待。另外，光片照明显微镜需要斑马鱼相对物镜移动从而改变每次成像的层面，惯性使得这一过程成为限速步骤。新兴的光场显微镜通过采集不同角度的荧光信号，用去卷积运算得到不同层次的信息，从而避免了调焦过程，可以实现 20Hz 的斑马鱼全脑成像(Prevedel et al., 2014)。但由于轴向分辨率与侧向分辨率之间的协调，目前这一方法还不能实现单细胞分辨率的成像，有待进一步完善。

5.3 遗传学操纵方法

遗传学操纵方法为神经科学研究提供了有力的工具(Crick, 1999)。斑马鱼可以每周产卵数百颗，且发育快速，很适合应用遗传学方法进行研究。然而，与具有长期积累的传统模式生物小鼠、果蝇、线虫相比，斑马鱼模式动物的遗传操纵工具与资源尚不够丰富。在最近几年中，随着越来越多研究者的加入，以及遗传工具的开发和经验的积累，这一情况已得到极大的改善。

传统的斑马鱼遗传操纵多基于正向遗传学(forward genetics)筛选，比如基于ENU(N-ethyl-N-nitrosourea)诱变的基因组局部定向突变技术(TILLING)，或者由假形反转录病毒介导的插入突变。延续这一趋势，以日本国立遗传研究所 Koichi Kawakami 实验室为代表，研究者借助 Tol2 转座酶介导的随机插入，利用增强子诱捕和基因诱捕(gene trap)方法将报告荧光蛋白 GFP 与工具分子 GAL4 随机插入到基因组中，经过大量筛选后建立资源库。Kawakami 实验室建立的资源库名为 zTrap，目前已经提供数百种不同的转基因品系供领域内研究者使用。同样应用 Tol2 转座系统和增强子诱捕方法，北京大学张博研究组和林硕研究组合作筛选得到了可以标记单胺能神经元的转基因斑马鱼品系 ETVmat2:GFP(Wen et al., 2008)。

正向策略有两方面的劣势，一是突变存在不定向性，工作量巨大；二是往往受到功能冗余基因的干扰，不容易获得可观察的突变表型(Huang et al., 2012)。因此，最近几年的技术开发逐渐转向反向遗传学(reverse genetics)，希望通过操纵特定目标基因来研究其功能。在 2008 年，研究者利用基于锌指核酸酶(zinc finger nuclease, ZFN)的方法成功敲除了斑马鱼中的靶基因(Doyon et al., 2008; Meng et al., 2008)，但是靶点选择的局限性使得这一方法推广不易。2011 年，张博研究组与林硕研究组开发了应用更广且更高效的方法，即基于转录激活因子类似效应物核酸酶(transcription activator-like effector nuclease, TALEN)的基因改造技术，并首次在斑马鱼上对内源基因进行了突变 (Huang et al., 2011)。TALEN 技术利用 TALE 蛋白与 DNA 进行配对，从而将核酸酶带到靶点对 DNA 进行切割。张博研究组与林硕研究组开发的"单元组装"(unit assembly)技术，对 TALE 蛋白进行组合，简化了操作流程，显著缩短了实验周期。随后，他们又进一步利用该技术，首次在斑马鱼上实现了基因敲入，将编码 EGFP 蛋白的 DNA 序列精确插入到基因组的靶位点上，并得到了稳定遗传的品系，但由于插入方式的原因，EGFP 不能正常表达(Zu et al., 2013)。这一技术还待进一步改进，使得在斑马鱼基因组上的定点修饰(包括基因敲除、敲入或条件性操控)成为可能。

最近出现的 CRISPER/CAS9 技术，采用 RNA 而非蛋白质识别切割位点，进一步简化了实验操作，加上该系统极高的切割效率，使之得到了极快的发展(Auer et al., 2014)。杜久林研究组李佳、张白冰等利用并改进了该技术，已成功地在斑马鱼模型上建立了高效的基因敲入技术，用于标记多种类型的神经元，并对其神经活动进行记录和操纵。

除了标记特定神经元类型，研究中还经常需要实现条件性表达或敲除，从而针对所研究的神经环路进行特异的操纵。由于斑马鱼脑部透明，还可建立特异调控区的条件性表达系统，例如，转入热激蛋白启动子及其下游基因后，可以通过双光子激光加热的方法，在单细胞水平驱动表达，或者利用光控的转录因子，在目标脑区进行激活表达。

5.4　病毒介导的跨突触标记

利用遗传操纵可以特异的观察和操控某一类具有相似基因表达特性的神经元群体，从而窥见不同神经元类型的反应差异，进而推断环路机制。然而，神经环路的关键在于其中的连接方式。一方面，同类型的神经元因为上下游联系不同其功能必然存在差异；另一方面，神经信息的存储也依赖于特定环路中的连接形式。以往对神经连接的了解依赖于各种局部刺激的功能实验和神经元形态的重构。近几年来，在小鼠模型上已成功建立了利用病毒进行跨突触标记的技术，可以在活体动物上追踪神经元的上下游投射关系，并在特定神经元的上游表达光学记录和光遗传学操纵工具，对相应的连接进行生理记录和操纵。

Rainer Friedrich 实验室最早利用病毒转染在斑马鱼中进行基因表达(Zou et al., 2014)。利用荧光蛋白作为报告基因，他们尝试了重组 1 型、2 型、5 型、6 型腺病毒相关载体(recombinant adeno-associated virus, rAAV)，慢病毒(lentivirus)，辛德毕斯病毒(Sindbis virus)和狂犬病病毒(rabies virus)，发现 rAAV 与慢病毒均不能有效感染斑马鱼，而 Sindbis 与 rabies 病毒则可以产生较强的感染。在幼鱼与成鱼中，Sindbis 病毒驱动的表达在感染后 12h 出现，并在 24h 达到峰值。与 *Huc* 启动子相比，Sindbis 病毒的表达强度更高。rabies 病毒的表达出现在感染后 3~4 天，并可持续 3 周左右。rabies 病毒可以跨过突触结构，进入到被感染神经元的突触前神经元中。在 Friedrich 实验室研究中使用的是改造后的 rabies 病毒，不具备跨突触的能力。为了验证和实现斑马鱼中 rabies 病毒是否能实现逆突触的追踪，杜久林研究组与中国科学院武汉物理与数学研究所徐富强研究组合作，正对这一技术进行探索。

为了保证使用的安全性，索尔克研究所的 Edward Callaway 实验室对 rabies 病毒进行了改造，将病毒基因序列中的糖蛋白置换为荧光蛋白，然后用禽白血病/肉瘤群病毒(avian sarcoma 和 leukosis virus, ASLV-A)的外壳糖蛋白对病毒进行包装。这种改装后的病毒被命名为 EnvA，它不能识别普通的哺乳动物细胞，而只能感染人工表达有 ASLV-A 受体(TVA)的神经元。被感染的表达 TVA 的细胞被称为"起点细胞"(starting cell)，如果在"起点细胞"同时人工表达正常的 rabies 病毒的外壳糖蛋白，就能帮助进入该神经元的病毒重新获得跨突触能力，从而标记突触前神经元。如果该神经元不具备 rabies 病毒的外壳糖蛋白，病毒只能留在该神经元中，从而实现安全可控的跨单级突触标记(Wickersham et al., 2006)。在斑马鱼中该策略是否可行还尚待验证，杜久林研究组和徐富强研究组首先检测了斑马鱼系统是否具备以下两个基本条件：ASLV-A 是否不感染普通的斑马鱼神经元；ASLV-A 是否感染表达 TVA 的斑马鱼神经元。通过注射质粒，在随机的少量神经元中表达 TVA 或者对照组质粒，他们发现这两点在斑马鱼系统中均成立，即 ASLV-A 能且只能感染表达有 TVA 的斑马鱼神经元。接下来，他们又通过在"起点细胞"中同时表达正常的 rabies 病毒外壳糖蛋白,发现的确能够实现形态学上的跨突触标记。目前,正在通过成对电生理记录,在功能上验证标记的神经元对是否存在单突触的联系。

5.5 行为学方法

基于斑马鱼表现出的丰富行为,已经建立了多种方法,用以研究其视动行为、逃跑行为、捕食行为、昼夜节律等的神经机制。然而,为了达到单细胞水平神经活动的记录精度,既往研究多在束缚标本上进行神经活动的记录,而行为实验则是独立进行的。这给行为神经机制的解释带来困难。其一,实验条件难以精确复制;其二,束缚动物不能发生运动,从而缺少运动的反馈输入和由运动造成的感觉信息的变化,使得这种范式得到的结果与自然行为有所差异;其三,束缚会改变动物的脑状态,从而影响实验结果。近几年出现的"虚拟现实"(virtual reality)系统,有助于解决这一难题。虚拟现实的特点在于根据动物的运动反馈控制视觉输入的变化,使被试动物能够与环境进行互动,从而有利于更完整地了解大脑如何处理"感觉-运动"信息。Florian Engert 实验室首先在斑马鱼中实现了这一系统(Ahrens et al., 2013a; Engert, 2013)。

Engert 实验室利用前文所述的视动运动,即动物在接受运动光栅刺激时,通过同向的游泳使身体与光栅保持相对静止。首先他们在头部固定的斑马鱼中,改变光栅运动速率,同时监测尾部的游泳行为,验证了斑马鱼幼鱼可以利用游泳的频率来适应不同流速。进而,他们结合 Fetcho 实验室对斑马鱼幼鱼假想游泳的研究的经验(Masino and Fetcho, 2005),记录两侧运动神经纤维的放电,并把这一运动信息反馈到光栅运动控制中,从而在麻痹斑马鱼上实现了闭环控制,可以实现在虚拟现实环境中的光学成像和电生理记录。对运动轨迹分析显示,虚拟现实中的斑马鱼运动与真实的运动几乎完全一致,即使经验丰富的研究者也无法单从记录到的运动轨迹区分虚拟现实与真实游泳。

斑马鱼脑小巧透明的优点体现在其幼年时期,但以往的行为学知识多来自成鱼的研究。在讨论具体某种行为时需要关注其可能存在的差异。比如前述的运动神经激活顺序的研究,体现出从幼鱼到成鱼的一致性;而斑马鱼的群体行为,则表现出幼年时对同窝的偏好而成年后的偏移(Gerlach and Lysiak, 2006)。中国科学技术大学胡兵研究组同时在斑马鱼的成鱼与幼鱼中,针对两类经典的视觉行为——眼动反应与视动反应进行了实验方法的探索,并将详细的描述发表在视频实验期刊库中(Zou et al., 2010)。眼动反应与视动反应都被作为行为范式筛选视觉功能缺陷的斑马鱼品系。在他们的工作中,不仅介绍了传统的利用斑马鱼幼鱼进行眼动反应,以及利用成鱼进行视动反应的检测方法,同时还详细描述了一种自主探索出的在成年斑马鱼中长时间检测眼动反应的实验方法。

6. 结　　语

斑马鱼作为一种新兴的脊椎模式动物，在神经系统功能研究中的应用急剧增多。其独特的优势可以充分发挥现有实验手段的极致，破除以前研究中受到的诸多限制，系统地研究神经科学中一些基本问题。综合国际发展现状和趋势，在基本的感觉运动环路中的研究已进行得颇为深入，感觉-运动的转换和认知相关行为方面的研究则尚在初期。在研究中，不仅要加深对于斑马鱼自然行为的了解(Patterson et al., 2013)，还应特别注意的是，之前在高等脊椎动物上的研究多局限于某一脑区或其局部，发挥斑马鱼的研究优势则需要充分考虑复杂网络的协作和全脑尺度上的把握，而非简单的层级模型，不宜照搬原有的实验范式。比如，从感觉输入到运动输出之间的神经信息加工和处理，可能受到动物个体内在与外在的多种扰动。内在的影响因素包括经验依赖的记忆、脑状态的改变、个体功能差异，或者消化、循环、免疫等跨系统的调节；而外在可包括昼夜/季节节律、群体或孤立的状态等。统筹考虑这些因素，设计简单可行的实验，充分发挥在体和全脑尺度的研究优势，斑马鱼模型必将在系统神经科学研究中扮演重要的角色。

实现中国科学院脑科学研究计划和美国 BRAIN Initiative 计划的核心内容，基本策略可概括三步曲：描述——记录单个神经元的形态和活动、神经元间的连接、群体神经元的活动；操控——上调或者下调环路中单个神经元的活动，检测它们在环路和认知行为中的功能；理解——根据前两个阶段的实验结果建立理论框架，深化对脑结构和功能的理解。目前国际范畴内，斑马鱼脑功能的研究已有一个很好的开端，特别是基于其特点的多种在体研究方法的逐步建立和成熟，这为充分发挥斑马鱼模型的优势来剖解全脑功能图谱提供了基础。

这个年轻而蓬勃的研究领域，尚需规范的理论框架，来整合越来越多的研究者的努力。这些理论，可能就在了解其全脑的过程中逐渐涌现。以斑马鱼神经系统为研究对象的愉悦之处在于，我们可同时把握住其小巧的"头"——多种形式的感觉输入，以及其流线形的"尾"——运动功能输出，以及"头""尾"之间扑朔迷离的神经系统工作原理。所有这些将会在脑结构与功能的剖析中，愈发清晰。对人类智力基础的认识，也很可能随着斑马鱼全脑功能的解析，迎来契机和曙光！

致谢

感谢顾珊烨博士对本文遗传学操纵部分的审读和修订意见。由于篇幅限制，本文未能覆盖相关领域的所有文献，特对其他作者表示歉意。

参 考 文 献

Agetsuma M, Aizawa H, Aoki T, et al. 2010. The habenula is crucial for experience-dependent modification of fear responses in zebrafish. Nature Neuroscience, 13:1354-1356.

Ahrens M B, Orger M B, Robson D N, et al. 2013. Whole-brain functional imaging at cellular resolution using light-sheet microscopy. Nature Methods, 10:413-420.

Aizenberg M, Schuman E M. 2011. Cerebellar-Dependent Learning in Larval Zebrafish. Journal of Neuroscience, 31:8708-8712.

Anastasio T J. 1994. The fractional-order dynamics of brainstem vestibulo-oculomotor neurons. Biological Cybernetics, 72:69-79.

Aoki T, Kinoshita M, Aoki R, et al. 2013. Imaging of Neural Ensemble for the Retrieval of a Learned Behavioral Program. Neuron, 78:881-894.

Auer T O, Duroure K, De Cian A, et al. 2014. Highly efficient CRISPR/Cas9-mediated knock-in in zebrafish by homology-independent DNA repair. Genome research, 24:142-153.

Azevedo F A C, Carvalho L R B, Grinberg L T, et al. 2009. Equal numbers of neuronal and nonneuronal cells make the human brain an isometrically scaled-up primate brain. The Journal of comparative neurology, 513:532-541.

Bagnall M W, McLean D L. 2014. Modular organization of axial microcircuits in zebrafish. Science, 343:197-200.

Baker M. 2013. Flashing fish brains filmed in action. Nature News. http://www.nature.com/news/flashing-fish-brains-filmed-in- action-1.12621 [2014-12-25]

Barlow H B. 1972. Single units and sensation: A neuron doctrine for perceptual psychology? Perception, 1:371-394.

Bianco I H, Kampff A R, Engert F. 2011. Prey capture behavior evoked by simple visual stimuli in larval zebrafish. Frontiers in systems neuroscience, 5:101.

Blumhagen F, Zhu P X, Shum J, et al. 2011. Neuronal filtering of multiplexed odour representations. Nature, 479:493-498.

Bundschuh S T, Zhu P, Schärer Y-P Z, et al. 2012. Dopaminergic modulation of mitral cells and odor responses in the zebrafish olfactory bulb. The Journal of neuroscience: the official journal of the Society for Neuroscience, 32:6830-6840.

Chen Q, Huang NN, Huang JT, et al. 2009. Sodium benzoate exposure downregulates the expression of tyrosine hydroxylase and dopamine transporter in dopaminergic neuronsin developing zebrafish. Birth Defects Research Part B: Developmental and Reproductive Toxicology, 86:85-91.

Chen Q, Jiang L, Li C, et al. 2012. Haemodynamics-driven developmental pruning of brain vasculature in zebrafish. PLoS biology 10:e1001374.

Chen T W, Wardill T J, Sun Y, et al. 2013. Ultrasensitive fluorescent proteins for imaging neuronal activity. Nature, 499:295-300.

Crick F. 1999. The impact of molecular biology on neuroscience. Philosophical Transactions of the Royal Society of London, Series B, Biological Sciences, 354:2021-2025.

Del Bene F, Wyart C, Robles E, et al. 2010. Filtering of visual information in the tectum by an identified neural circuit. Science, 330:669-673.

Dell A L, Fried-Cassorla E, Xu H, et al. 2013. cAMP-induced expression of neuropilin1 promotes retinal axon crossing in the zebrafish optic chiasm. The Journal of Neuroscience, 33:11076-11088.

Douglass A D, Kraves S, Deisseroth K, et al. 2008. Escape behavior elicited by single, Channelrhodopsin-2-evoked spikes in zebrafish somatosensory neurons. Current Biology, 18:1133-1137.

Døving K B, Lastein S. 2009. The alarm reaction in fishes - odorants, modulations of responses, neural pathways. Annals of the New York Academy of Sciences, 1170:413-423.

Doyon Y, McCammon J M, Miller J C, et al. 2008. Heritable targeted gene disruption in zebrafish using designed zinc-finger nucleases. Nature Biotechnology, 26:702-708.

Dreosti E, Esposti F, Baden T, et al. 2011. In vivo evidence that retinal bipolar cells generate spikes modulated by light. Nature Neuroscience, 14:951-952.

Dreosti E, Llopis N V, Carl M, et al. 2014. Left-Right Asymmetry Is Required for the Habenulae to Respond to Both Visual and Olfactory Stimuli. Current Biology, 24:440-445.

Dreosti E, Odermatt B, Dorostkar M M, et al. 2009. A genetically encoded reporter of synaptic activity in vivo. Nature Methods 6:883-889.

Eaton RC, Bombardieri RA, Meyer DL. 1977. The Mauthner-initiated startle response in teleost fish. Journal of Experimental Biology 66:65-81.

Esposti F, Johnston J, Rosa JM, Leung KM, Lagnado L. 2013. Olfactory Stimulation Selectively Modulates the OFF Pathway in the Retina of Zebrafish. Neuron 79:97-110.

Fang P, Lin J-F, Pan H-C, Shen Y-Q, Schachner M. 2012. A surgery protocol for adult zebrafish spinal cord injury. Journal of Genetics and Genomics 39:481-7.

Feng Y, Yan T, Zheng J, Ge X, Mu Y, et al. 2010. Overexpression of Wlds or Nmnat2 in mauthner cells by single-cell electroporation delays axon degeneration in live zebrafish. Journal of Neuroscience Research 88:3319-27.

Freedman DJ, Assad Ja. 2011. A proposed common neural mechanism for categorization and perceptual decisions. Nature Neuroscience 14:143-6.

Friedrich RW, Hyman SE. 2013. Neuronal computations in the olfactory system of zebrafish. Annual Review of Neuroscience, Vol 36 36:383-402.

Friedrich RW, Korsching SI. 1997. Combinatorial and chemotopic odorant coding in the zebrafish olfactory bulb visualized by optical imaging. Neuron 18:737-52.

Friedrich RW, Korsching SI. 1998. Chemotopic, combinatorial, and noncombinatorial odorant representations in the olfactory bulb revealed using a voltage-sensitive axon tracer. The Journal of neuroscience : the official journal of the Society for Neuroscience 18:9977-88.

Friedrich RW. 2006. Mechanisms of odor discrimination: neurophysiological and behavioral approaches. Trends in neurosciences 29:40-7.

Gabriel JP, Ausborn J, Ampatzis K, Mahmood R, Eklöf-Ljunggren E, El Manira A. 2011. Principles governing recruitment of motoneurons during swimming in zebrafish. Nature Neuroscience 14:93-9.

Gabriel JP, Trivedi CA, Maurer CM, Ryu S, Bollmann JH. 2012. Layer-specific targeting of direction-selective neurons in the zebrafish optic tectum. Neuron 76:1147-60.

Gerlach G, Lysiak N. 2006. Kin recognition and inbreeding avoidance in zebrafish, Danio rerio, is based on phenotype matching. Animal Behaviour 71:1371-7.

Goldman MS. 2009. Memory without feedback in a neural network. Neuron 61:621-34.

Goldstone RL, Hendrickson AT. 2010. Categorical perception. Wiley Interdisciplinary Reviews: Cognitive Science 1:69-78.

Gong Y, Wagner MJ, Zhong Li J, Schnitzer MJ. 2014. Imaging neural spiking in brain tissue using FRET-opsin protein voltage sensors. Nature Communications 5:3674-.

Grama A, Engert F. 2012. Direction selectivity in the larval zebrafish tectum is mediated by asymmetric inhibition. Frontiers in Neural Circuits 6.

Han Y, Mu Y, Li X, Xu P, Tong J, et al. 2011. Grhl2 deficiency impairs otic development and hearing ability in a zebrafish model of the progressive dominant hearing loss DFNA28. Human Molecular Genetics 20:3213-26.

Hinz FI, Aizenberg M, Tushev G, Schuman EM. 2013. Protein synthesis-dependent associative long-term memory in larval zebrafish. Journal of Neuroscience 33:15382-7.

Horikawa K, Yamada Y, Matsuda T, Kobayashi K, Hashimoto M, et al. 2010. Spontaneous network activity visualized by ultrasensitive Ca^{2+} indicators, yellow Cameleon-Nano. Nature Methods 7:729-32.

Huang KH, Ahrens MB, Dunn TW, Engert F. 2013. Spinal projection neurons control turning behaviors in zebrafish. Current Biology 23:1566-73.

Huang P, Xiao A, Zhou M, Zhu Z, Lin S, Zhang B. 2011. Heritable gene targeting in zebrafish using customized TALENs. Nat Biotech 29:699-700.

Huang P, Zhu Z, Lin S, Zhang B. 2012. Reverse genetic approaches in zebrafish. Journal of genetics and genomics = Yi chuan xue bao 39:421-33.

Hunter PR, Lowe AS, Thompson ID, Meyer MP. 2013. Emergent Properties of the Optic Tectum Revealed by Population Analysis of Direction and Orientation Selectivity. Journal of Neuroscience 33:13940-5.

Jetti SK, Vendrell-Llopis N, Yaksi E. 2014. Spontaneous activity governs olfactory representations in spatially organized habenular microcircuits. Current Biology 24:434-9.

Kermen F, Franco LM, Wyatt C, Yaksi E. 2013. Neural circuits mediating olfactory-driven behavior in fish. Frontiers in Neural Circuits 7:62-.

Kinkhabwala A, Riley M, Koyama M, Monen J, Satou C, et al. 2011. A structural and functional ground plan for neurons in the hindbrain of zebrafish. Proceedings of the National Academy of Sciences of the United States of America 108:1164-9.

Kishore S, Fetcho JR. 2013. Homeostatic regulation of dendritic dynamics in a motor map in vivo. Nature Communications 4.

Koyama M, Kinkhabwala A, Satou C, Higashijima S, Fetcho J. 2011. Mapping a sensory-motor network onto a structural and functional ground plan in the hindbrain. Proceedings of the National Academy of Sciences of the United States of America 108:1170-5.

Krishnan S, Mathuru AS, Kibat C, Rahman M, Lupton CE, et al. 2014. The Right Dorsal Habenula Limits Attraction to an Odor in Zebrafish. Current Biology:1-9.

Kubo F, Hablitzel B, Maschio MD, Driever W, Baier H, Arrenberg AB. 2014. Functional Architecture of an Optic Flow-Responsive Area that Drives Horizontal Eye Movements in Zebrafish. Neuron 81:1344-59.

Lee A, Mathuru AS, Teh C, Kibat C, Korzh V, et al. 2010. The habenula prevents helpless behavior in larval zebrafish. Current biology : CB 20:2211-6.

Li Y, Du X-f, Liu C-s, Wen Z-l, Du J-l. 2012. Reciprocal regulation between resting microglial dynamics and neuronal activity in vivo. Developmental cell 23:1189-202.

Lin J-F, Pan H-C, Ma L-P, Shen Y-Q, Schachner M. 2012. The cell neural adhesion molecule contactin-2 (TAG-1) is beneficial for functional recovery after spinal cord injury in adult zebrafish. PloS One 7:e52376.

Liu C, Ma W, Su W, Zhang J. 2012. Prdm14 acts upstream of islet2 transcription to regulate axon growth of primary motoneurons in zebrafish. Development 139:4591-600.

Liu KS, Fetcho JR. 1999. Laser ablations reveal functional relationships of segmental hindbrain neurons in zebrafish. Neuron 23:325-35.

Liu P, Du J-l, He C. 2013. Developmental pruning of early-stage myelin segments during CNS myelination in vivo. Cell research 23:962-4.

Lowe AS, Nikolaou N, Hunter PR, Thompson ID, Meyer MP. 2013. A Systems-Based Dissection of Retinal Inputs to the Zebrafish Tectum Reveals Different Rules for Different Functional Classes during Development. Journal of Neuroscience 33:13946-56.

Luo GR, Chen Y, Li XP, Liu TX, Le WD. 2008. Nr4a2 is essential for the differentiation of dopaminergic neurons during zebrafish embryogenesis. Molecular and Cellular Neuroscience 39:202-10.

Mahou P, Vermot J, Beaurepaire E, Supatto W. 2014. Multicolor two-photon light-sheet microscopy. Nature Methods 11:600-1.

Masino MA, Fetcho JR. 2005. Fictive swimming motor patterns in wild type and mutant larval zebrafish. Journal of Neurophysiology 93:3177-88.

Mathuru AS, Kibat C, Cheong WF, Shui G, Wenk MR, et al. 2012. Chondroitin fragments are odorants that trigger fear behavior in fish. Current biology : CB 22:538-44.

Matsumoto M, Hikosaka O. 2009. Two types of dopamine neuron distinctly convey positive and negative motivational signals. Nature 459:837-41.

McLean D, Fetcho J. 2009. Spinal interneurons differentiate sequentially from those driving the fastest swimming movements in larval zebrafish to those driving the slowest ones. Journal of Neuroscience 29:13566-77.

Meng X, Noyes MB, Zhu LJ, Lawson ND, Wolfe SA. 2008. Targeted gene inactivation in zebrafish using engineered zinc-finger nucleases. Nature biotechnology 26:695-701.

Miri A, Daie K, Arrenberg AB, Baier H, Aksay E, Tank DW. 2011. Spatial gradients and multidimensional dynamics in a neural integrator circuit. Nature Neuroscience 14:1150-9.

Miura K, Mainen ZF, Uchida N. 2012. Odor representations in olfactory cortex: distributed rate coding and decorrelated population activity. Neuron 74:1087-98.

Miyasaka N, Arganda-Carreras I, Wakisaka N, Masuda M, Sümbül U, et al. 2014. Olfactory projectome in the zebrafish forebrain revealed by genetic single-neuron labelling. Nature Communications 5:3639-.

Mu Y, Li X-qQ, Zhang B, Du J-lL. 2012. Visual input modulates audiomotor function via hypothalamic dopaminergic neurons through a cooperative mechanism. Neuron 75:688-99.

Niell CM, Smith SJ. 2005. Functional imaging reveals rapid development of visual response properties in the zebrafish tectum. Neuron 45:941-51.

Niessing J, Friedrich RW. 2010. Olfactory pattern classification by discrete neuronal network states. Nature 465:47-53.

Nikolaev A, Leung K-M, Odermatt B, Lagnado L. 2013. Synaptic mechanisms of adaptation and sensitization in the retina. Nature Neuroscience 16:934-41.

Nikolaou N, Lowe AS, Walker AS, Abbas F, Hunter PR, et al. 2012. Parametric Functional Maps of Visual Inputs to the Tectum. Neuron 76:317-24.

Odermatt B, Nikolaev A, Lagnado L. 2012. Encoding of Luminance and Contrast by Linear and Nonlinear Synapses in the Retina. Neuron 73:758-73.

O'Malley DM, Kao Y-H, Fetcho JR. 1996. Imaging the functional organization of zebrafish hindbrain segments during escape behaviors. Neuron 17:1145-55.

Orger MB, Kampff AR, Severi KE, Bollmann JH, Engert F. 2008. Control of visually guided behavior by distinct populations of spinal projection neurons. Nature Neuroscience 11:327-33.

Pan HC, Lin JF, Ma LP, Shen YQ, Schachner M. 2013. Major vault protein promotes locomotor recovery and regeneration after spinal cord injury in adult zebrafish. European Journal of Neuroscience 37:203-11.

Panier T, Romano Sa, Olive R, Pietri T, Sumbre G, et al. 2013. Fast functional imaging of multiple brain regions in intact zebrafish larvae using Selective Plane Illumination Microscopy. Frontiers in Neural Circuits 7:65-.

Parker J, Newsome WT. 1998. Sense and the single neuron: probing the physiology of perception. Annual review of neuroscience 21:227-77.

Partan SR, Marler P. 2005. Issues in the Classification of Multimodal Communication Signals. American Naturalist 166:231-45.

Patterson BW, Abraham AO, MacIver MA, McLean DL. 2013. Visually guided gradation of prey capture movements in larval zebrafish. Journal of Experimental Biology 216:3071-83.

Poo C, Isaacson JS. 2009. Odor representations in olfactory cortex: "sparse" coding, global inhibition, and oscillations. Neuron 62:850-61.

Portugues R, Feierstein CE, Engert F, Orger MB. 2014. Whole-brain activity maps reveal stereotyped, distributed networks for visuomotor behavior. Neuron 81:1328-43.

Prevedel R, Yoon Y-G, Hoffmann M, Pak N, Wetzstein G, et al. 2014. Simultaneous whole-animal 3D imaging of neuronal activity using light-field microscopy. Nature Methods

Ramdya P, Engert F. 2008. Emergence of binocular functional properties in a monocular neural circuit. Nature Neuroscience 11:1083-90.

Schärer Y-PZ, Shum J, Moressis A, Friedrich RW. 2012. Dopaminergic modulation of synaptic transmission and neuronal activity patterns in the zebrafish homolog of olfactory cortex. Frontiers in Neural Circuits 6.

Schoonheim PJ, Arrenberg AB, Del Bene F, Baier H. 2010. Optogenetic localization and genetic perturbation of saccade-generating neurons in zebrafish. Journal of Neuroscience 30:7111-20.

Seung HS. 1996. How the brain keeps the eyes still. Proceedings of the National Academy of Sciences 93:13339-44.

Speedie N, Gerlai R. 2009. Alarm substance induced behavioral response in zebrafish (Danio rerio). Behavioral Brain Research 188:168-77.

St-Pierre F, Marshall JD, Yang Y, Gong Y, Schnitzer MJ, Lin MZ. 2014. High-fidelity optical reporting of neuronal electrical activity with an ultrafast fluorescent voltage sensor. Nature Neuroscience

Trevarrow B, Marks DL, Kimmel CB. 1990. Organization of hindbrain segments in the zebrafish embryo. Neuron 4:669-79.

Trivedi Ca, Bollmann JH. 2013. Visually driven chaining of elementary swim patterns into a goal-directed motor sequence: a virtual reality study of zebrafish prey capture. Frontiers in Neural Circuits 7:86-.

Truong TV, Supatto W, Koos DS, Choi JM, Fraser SE. 2011. Deep and fast live imaging with two-photon scanned light-sheet microscopy. Nature methods 8:757-60.

Valente A, Huang K-HH, Portugues R, Engert F. 2012. Ontogeny of classical and operant learning behaviors in zebrafish. Learning & Memory 19:170-7.

Vogelstein JT, Packer AM, Machado Ta, Sippy T, Babadi B, et al. 2010. Fast nonnegative deconvolution for spike train inference from population calcium imaging. Journal of Neurophysiology 104:3691-704.

von Frisch K. 1942. Über einen Schreckstoff der Fischhaut und seine biologische Bedeutung. Zeitschrift für vergleichende Physiologie 29:49-145.

Wang K, Milkie DE, Saxena A, Engerer P, Misgeld T, et al. 2014. Rapid adaptive optical recovery of optimal resolution over large volumes. Nature methods 11.

Warp E, Agarwal G, Wyart C, Friedmann D, Oldfield CS, et al. 2012. Emergence of Patterned Activity in the Developing Zebrafish Spinal Cord. Current Biology 22:93-102.

Wei HP, Yao YY, Zhang RW, Zhao XF, Du JL. 2012. Activity-Induced Long-Term Potentiation of Excitatory Synapses in Developing Zebrafish Retina In Vivo. Neuron 75:479-489.

Wen L, Wei W, Gu W, Huang P, Ren X, et al. 2008. Visualization of monoaminergic neurons and neurotoxicity of MPTP in live transgenic zebrafish. Developmental biology 314:84-92.

Wickersham IR, Finke S, Conzelmann K-K, Callaway EM. 2006. Retrograde neuronal tracing with a deletion-mutant rabies virus. Nature Methods 4:47-49.

Wiechert MT, Judkewitz B, Riecke H, Friedrich RW. 2010. Mechanisms of pattern decorrelation by recurrent neuronal circuits. Nature Neuroscience 13:1003-U132

Woo K, Fraser SE. 1995. Order and coherence in the fate map of the zebrafish nervous system. Development 121:2595-2809.

Yaksi E, Friedrich RW. 2006. Reconstruction of firing rate changes across neuronal populations by temporally deconvolved Ca^{2+} imaging. Nature Methods 3:377-383.

Yu Y, Schachner M. 2013. Syntenin‐a promotes spinal cord regeneration following injury in adult zebrafish. European Journal of Neuroscience 38:2280-2289.

Zhang C, Gao J, Zhang H, Sun L, Peng G. 2012. Robo2–slit and Dcc–netrin1 coordinate neuron axonal pathfinding within the embryonic axon tracts. The Journal of Neuroscience 32:12589-12602.

Zhang R, Zhang S, Du J-l. 2013. KCC2-dependent subcellular ECl difference of ON-OFF retinal ganglion cells in larval zebrafish. Frontiers in Neural Circuits 7

Zhang R-w, Wei H-p, Xia Y-m, Du J-l. 2010. Development of light response and GABAergic excitation-to-inhibition switch in zebrafish retinal ganglion cells. The Journal of Physiology 588:2557-2569.

Zhu PX, Frank T, Friedrich RW. 2013. Equalization of odor representations by a network of electrically coupled inhibitory interneurons. Nature Neuroscience 16:1678-1686.

Zou M, De Koninck P, Neve RL, Friedrich RW. 2014. Fast gene transfer into the adult zebrafish brain by herpes simplex virus 1（HSV-1）and electroporation: methods and optogenetic applications. Frontiers in Neural Circuits 8

Zou S, Tian C, Ge S, Hu B. 2013. Neurogenesis of retinal ganglion cells is not essential to visual functional recovery after optic nerve injury in adult zebrafish. PloS One 8:e57280

Zou S-Q, Yin W, Zhang M-J, Hu C-R, Huang Y-B, Hu B. 2010. Using the optokinetic response to study visual function of zebrafish. Journal of visualized experiments: JoVE

Zu Y, Tong X, Wang Z, Liu D, Pan R, et al. 2013. TALEN-mediated precise genome modification by homologous recombination in zebrafish. Nature Methods 10:329-331.

Zucker CL, Dowling JE. 1987. Centrifugal fibres synapse on dopaminergic interplexiform cells in the teleost retina. Nature 330:166-168.

* 策划编辑：骆清铭　华中科技大学生物医学工程系

基于直接测量的哺乳动物全脑神经——血管网络精细结构及三维可视化

作　者：龚　辉[1,2]　李向宁[1,2]　袁　菁[1,2]　吕晓华[1,2]　李安安[1,2]
　　　　陈尚宾[1,2]　杨孝全[1,2]　曾绍群[1,2]　骆清铭[1,2]

1 华中科技大学-武汉光电国家实验室
2 华中科技大学生物医学工程系

▶ 1. 前言 / 172
▶ 2. 脑精细结构的标记 / 175
▶ 3. 精细结构获取方法 / 180
▶ 4. 脑的数字三维重建 / 186
▶ 5. 总结与展望 / 195

摘要

脑是人类和其他高等动物最重要、最复杂的器官。诸多脑的奥秘有待探索，认识脑、保护脑、创造脑对人类有着重大意义。目前，脑研究已成为全球最具有战略意义的科研主题。本文从研究脑的需求出发，简要阐明基于直接测量以神经元和毛细血管的真实尺度大小，实现哺乳动物全脑三维可视化的必要性和可行性。围绕标记、数据获取及可视化全脑神经-血管精细结构的目标，对相关标记技术和组织样本处理、光学成像、三维可视化等技术进行了综述和分析。结合作者课题组的工作，介绍了基于显微光学切片断层成像新技术获取小鼠全脑神经-血管精细结构的进展。最后，总结了全球脑研究计划的部署，展望了未来脑科学研究存在的挑战和发展方向。

关键词

脑研究、哺乳动物、全脑、真实尺度、精细结构、显微光学切片断层成像、三维可视化

1. 前　　言

1.1 脑研究的需求

众所周知，脑是人类和其他高等动物最精密、最复杂、最重要的器官，是感觉、语言、运动、学习、记忆、思维、情感等高级功能的中枢。迄今为止，我们对人脑或是其他哺乳动物的脑结构和功能还知之甚少。阐明脑的工作原理、揭示脑的奥秘是人类面临的最大挑战之一(郭爱克，2014)。另一方面，各种不断增长的神经系统疾病及其后遗症已经影响全球约十亿人口，给社会带来沉重负担(World Health Organization, 2006)。因此，认识脑、保护脑有着重要的科学意义和社会意义。目前，脑研究已成为自基因组计划以来国际上最重要的研究主题。

脑研究中最重要的研究目的之一就是理解脑的结构和功能，并阐明二者之间的关系(Lichtman et al., 2011)。毋庸置疑，结构是功能的基础，解析脑的真实结构是整个脑研究的重要环节。目前只是粗略地知道：人脑包含近千亿个神经元、上千亿个胶质细胞；每个神经元又通过树突和轴突与成百上千的神经元形成突触连接并进行信息传递，全脑百万亿计的突触连接又形成了高度复杂的神经网络(Helmstaedter, 2013)。脑结构的高度复杂性，一方面表现为神经细胞类型和数目众多，如神经系统中仅视网膜处就有约55种细胞(Mashland, 2001)。脑内究竟有多少种神经细胞至今仍是未明确的基本问题。另一方面还表现在神经连接环路的多样性。一般认为脑内有两种不同的神经连接环路，即神经元与其周围神经元建立的局部神经环路和神经元与远端脑区中神经元建立的长程神经环

路。复杂的脑功能往往需要局部神经环路和长程神经环路的协同作用才能完成。对于长程神经环路,可能从皮层一直投射至脊髓,几乎跨越全脑范围。即使特定于某一个神经元,其传入连接和传出投射的神经细胞就可能遍布全脑(Lichtman et al., 2011)。正因如此,对神经系统结构和功能的最终解析迫切需要从全脑精细结构研究开始。

此外,脑作为高能耗的器官,高度依赖于持续不断的能量供应。脑内密布着不同尺度大小的血管,动脉、小动脉、毛细血管、小静脉、静脉共同构成闭合而复杂的血管网络。血管网络为神经元和胶质细胞提供能量和营养,并把其产生的代谢物带走。脑血流供应中断几秒钟,神经元就会停止发放动作电位;若中断供血几分钟,神经元就会出现不可逆的坏死。当然,神经系统对血管网络和血流供应也存在调控作用;神经-血管耦合机制也是脑研究中的热点之一。神经与血管关系密切,通常被并称为神经-血管网络。当前,神经血管网络越来越被公认为是研究脑功能和疾病的靶点(Kleinfeld, 2011)。故而,获取哺乳动物全脑血管结构、特别是同时获取神经-血管网络精细结构是脑研究中的一项基础性重要任务,是全面理解脑功能和神经性疾病的重要保证。然而,人们对哺乳动物全脑范围内的细胞与血管网络的真实分布尚不清楚,特别缺乏从同一个脑内同时获取的神经-血管精细结构。究其原因在于成像技术难以兼顾厘米量级的全脑尺度大成像范围与单个神经细胞精细结构的分辨率。以图版XII图1所示小鼠脑为例:小鼠全脑在头尾、背腹、左右三个方向都是厘米量级;脑中的特定结构,如海马,为毫米量级;而一般神经元胞体的直径约为5~30μm,胞体发出的突起(树突或是轴突)直径往往小于1μm。此外,小鼠脑内较粗脑动脉血管直径可达100μm,而毛细血管的直径仅约为2~5μm。要获取全脑精细神经-血管结构,需要跨越4~5个数量级的几何尺度进行成像,这对现有的成像技术提出了极大的挑战。若要真实可视化脑内神经-血管的三维结构,图象采集时x、y、z三个方向的分辨率均应小于等于1μm(龚辉等,2014)。此处,我们提出"真实尺度"这一概念,特指成像分辨率达到或优于1μm、直接测量并能准确反映单个神经细胞形态(含突起)和毛细血管的研究水平。换言之,本文所述单神经元分辨率即指成像的体素分辨率至少达到1μm的真实尺度水平。

成像及其可视化产生的数据集可以将人眼无法分辨的结构信息放大,让我们能够直接、清楚地认识脑内结构,从而更好地理解相应的脑功能及其机制。因此尤为重要的是:以真实尺度获取不同类型神经元的精细形态、及其所处脑内的准确位置,解析其连接环路;以及为了避免个体差异,对同一个哺乳动物全脑同时获取所有神经细胞和血管的精细形态结构。目前,结合多种标记技术,实现在全脑范围、以真实尺度构建出感兴趣位置处的神经细胞和血管的精细形态,并给出神经连接通路和血管网络的可视化结果,已被国际上多个脑计划列为需要首先实现的目标。

1.2 脑研究进展

脑研究按照不同尺度可以分成三个层次:宏观尺度(macroscale)、介观尺度(mesoscale)、微观尺度(microscale)。宏观尺度主要包含不同脑区连接的研究,通常使用磁共振成像(magnetic resonance imaging, MRI)；介观尺度主要包含单个或一群神经细胞及其投射的研究，通常使用光学显微镜成像；而微观尺度主要包含突触连接层次的研究，一般需要借助电子显微镜(electron microscope, EM)成像(Bohland et al., 2009; Craddock et al., 2013)。近二十年脑科学研究发展迅猛，一方面已在分子、基因等微观水平，对神经元基因表达特异性、蛋白质功能，以及突触传递等方面取得了长足进步，对不同类型神经元的细胞工作机制有了深入认识。另一方面，得益于磁共振成像、正电子发射层析成像(positron emission tomography, PET)、脑电记录等先进技术的应用，对宏观水平的脑功能，如大脑皮层各脑区(如运动、感觉皮层等)的功能有较深入的理解；对多种神经性疾病的脑结构和功能有较全面的表征。然而，对于联系微观突触信息传递和宏观脑功能之间的介观层次的神经环路信息，还知之甚少。前期研究已知，脑功能不是由单个神经元或单一脑区独立产生，而是依赖于神经环路内的神经元集合、皮层功能柱或者局部脑区交互作用的结构，所以脑研究需要兼顾不同层次水平。虽然脑科学已在宏观尺度和微观尺度都取得了巨大的进步；但对于在宏观和微观之间的鸿沟，即介观尺度的认知几乎是空白，而那里可能正是大脑奥秘所在。目前，人们只对仅有302个神经元的线虫(*C. elegans*)的神经环路进行了较完整的研究，而对于拥有约10万个神经元的模式动物果蝇(*Drosophila*)脑的相关研究还只是刚刚起步(Chiang et al., 2011)。相比之下，研究高等哺乳动物拥有数以百亿计神经元的复杂神经环路，对传统技术而言是极大的挑战。

传统技术无法解决突起水平分辨和全脑范围探测这一矛盾。在宏观水平，磁共振成像被视为脑成像的"金标准"，但对组织样本只能实现数十微米的分辨能力，磁场强度、梯度场强度、成像时间、线圈灵敏度等都是制约分辨率的客观因素。目前，已有一些研究者开始用扩散磁共振成像(diffusion magnetic resonance imaging, dMRI)探索脑的细胞构筑和白质纤维束。dMRI能够测量脑组织中水分子的扩散运动信号，并以此推算关于细胞完整性和组织微观结构改变的相关信息，进而间接地反映出脑白质纤维束的物理和功能特性。dMRI为全脑神经纤维结构研究提供了独特的非侵入性活体检测手段，已成为脑成像研究领域中常用的方法之一，但仍然受限于分辨率。在微观水平，电子显微镜是显示脑组织精细结构的有力工具，其分辨率可达到1nm。但是，电子显微镜难以进行全脑的探测，据估算，仅成像$1mm^3$的样本就需1万人年工作量(Luo et al., 2008)。

光学显微镜(optical microscope)的分辨率介于MRI和EM之间，几乎可以观察到所有的神经元突起，而且适用于光学成像的神经标记技术非常丰富，对研究完整大脑的构筑和连接而言，光学显微成像技术更具有可行性和普适性。但是，传统的宽场光学显微镜不具备三维层析能力，又因组织对光存在着吸收和散射的影响，成像深度受到限制。为了解决高分辨率与大探测范围难以兼得的问题，近年来，开发了一系列新颖的光学成像技术且已被应用到脑研究中，诸如：光片照明显微成像、双光子序列断层成像和显微光学切片断层成像等(Osten et al, 2013)。这些成像技术分别结合一定的组织样品透明，或是机械切削，可以实现诸如小鼠全脑等样品的光学分辨率成像及三维可视化。此外，由于道德伦理和人脑尺度过大等原因，其他哺乳类动物(如小鼠)的脑研究往往先被选作研究对象。已经知道，小鼠与人类约有92%的基因同源，各种转基因技术更容易用于制备小鼠的神经性疾病模型。后文主要以小鼠为例来探讨哺乳动物全脑真实尺度的可视化；

涉及技术包括样品标记和处理、显微光学成像、海量数据三维可视化等。

2. 脑精细结构的标记

要理解大脑如何发挥功能，需要了解大脑内部神经元之间的环路连接，即神经元的输入、输出及电生理特性。因细胞和组织中水的含量较高，成像对比度较差，对大脑进行成像时，就需要对组织样品进行必要的组织处理，包括染色、示踪标记等，使其具有一定的对比度。同时因组织是浑浊介质，具有一定的光散射和吸收特性，光穿透深度会受到限制。对于大样本、厚组织，就需要先进行切片或化学透明处理等。以下从组织标记方法、组织处理技术的角度分别进行综述。

2.1 组织标记方法的发展

2.1.1 神经组织常规染色方法的发展

19世纪光学显微镜技术的发展、切片机的发明，以及神经组织染色方法的发展为现代神经解剖学奠定了重要基础，并促使人们对脑的复杂结构有了革命性的认识。1873年，意大利科学家Golgi发明了银染法，可染出神经元的完整形态，包括胞体和神经突起。西班牙科学家Cajal利用并改造了这种方法，发现了数十种神经细胞。这一类利用重金属盐对神经组织进行染色的方法统称为Golgi染色法(Sotelo, 2003)。1892年，德国生理学家Nissl发明了利用碱性染料对神经组织进行染色的方法(Windhorst and Johansson, 1999; Xiong and Gendelman, 2014)。Nissl染色包括一系列碱性染料，常用的如焦油紫、硫堇等，使脑内几乎所有细胞的细胞核和核周的块状物质(尼氏体)着色，可以用于研究脑的细胞构筑。但Nissl染色不能使神经细胞的树突和轴突着色，因此无法用于神经连接研究。Golgi染色能使完整的神经细胞结构，包括胞体、树突和轴突同时着色，从而能够研究神经元的连接关系(Keefer et al., 1976; Kiernan, 2008)。从19世纪中期一直到20世纪中期，关于神经系统结构认识主要是基于Golgi等常用组织染色方法。但这些传统方法只是应用于小块组织中，因此无法追踪长距离的突起连接。随着对神经系统了解的深入及技术的发展，研究者不断地改进Golgi、Nissl等染色方法(Friedland et al., 2006; Pilati et al., 2008)，使得这些方法目前仍然是神经生物学中经典的染色方法。华中科技大学骆清铭课题组改进了Golgi-Cox法，使之可用于全脑染色(Zhang et al., 2011)，该方法所得的结果可同时展示脑内不同位置处的单个神经元的形态，但由于神经突起非常密集，采用常规图像处理方法很难实现对纤细轴突和树突的分割与标识，也几乎不可能追踪神经环路中投射来源，因此不太适于精细结构的大范围研究(Lichtman et al., 2008)。

2.1.2 荧光标记技术

20世纪后期以来，神经生物学取得了迅猛的发展，尤其是在标记技术方面。陆续建立起各种适用于神经组织的染色方法，从神经束路追踪法，包括辣根过氧化物酶(HRP)法、菜豆凝集素顺行轴突追踪法、生物素葡聚糖胺顺行轴突追踪法、霍乱毒素B亚单位追踪法和生物胞素(biocytin)等；到化学神经解剖学方法，包括酶组织化学法、免疫组织化学法、免疫电镜技术、原位杂交组织化学法和神经系统功能活动形态定位法等(Keefer et al., 1976; Kiernan, 2008; Wu and Oertel, 1984; Xiong and Gendelman, 2014)。这些染色方法在神经系统结构或功能的研究中发挥了重要的作用。

荧光分子标记技术的引入和转基因技术的发展，为神经生物学研究的发展带来了新的契机。荧光分子标记技术可以分为两大类，传统的荧光染料标记和通过基因工程将荧光蛋白基因或荧光染料偶联的氨基酸序列转入细胞内的分子靶标技术。传统荧光标记技术又包括使用荧光染料标记特异抗体的免疫染色技术，以及通过荧光标记的化合物与细胞内靶分子特异反应的化学染色技术。荧光染料标记方法是通过化学的荧光染料对特定的细胞结构，如细胞核、细胞膜、细胞骨架等进行染色，可选择不同的荧光波长、光漂白速率、细胞穿透性等，达到识别特定细胞区域的目的，但该方法存在特异性不足的缺陷，如DNA特异的染料，一般都可标记RNA。相比之下，免疫组化和原位杂交可以特异性地标记神经元(Xiong and Gendelman, 2014)。免疫组化和原位杂交都是利用生物大分子(蛋白质和DNA)对生物组织进行染色，由于分子量巨大，很难到达组织深处，因此这些方法一般用于组织薄片。抗原和mRNA在细胞内的分布局限在特定的细胞结构，如胞体、特定突起或突起末梢，仅使用免疫组化和原位杂交的标记方法来染色完整神经元精细结构，尤其是对全脑水平的神经元染色较为困难。

染料注射技术是基于荧光染料的标记技术，通过显微注射等手段将染料注射到鼠脑内，实现对神经元的染色，通常使用具有特异性亲和的染料，如亲脂性荧光染料DiI、DiO等。这些特异性结合染料与神经元结合后，可以进入细胞并沿轴突或树突或细胞膜扩散，直至染到整个神经元；有些荧光染料还可以跨突触对上一或多级，或者下一或多级神经元进行染色。根据其传播方向的不同，这些染料分为顺行示踪剂和逆行示踪剂。2014年，Hongwei Dong等利用菜豆凝集素、霍乱毒素B亚基等示踪剂，结合切片技术构建了小鼠脑皮层连接图(Zingg et al., 2014)。由于荧光染料等分子探针会造成细胞损伤而被清除或直接被细胞代谢，荧光强度随扩散逐渐降低，且这些探针分子均具有假阳性，在精细结构研究中也有其局限性。

2.1.3 基于荧光蛋白的神经元特异标记技术

随着基因表达分析、高通量成像、电生理等技术的发展，21世纪以来研究人员们可以结合基因工程的分子靶标技术，使用荧光蛋白来标记特定基因产物，包括特定类型的神经元、血管内皮细胞等。如通过基因技术将分子靶标(如荧光蛋白)直接标记特定的基因产物，也可以通过在氨基酸序列上绑定具有细胞穿透性的荧光染料来标记靶蛋白。应用较多的是通过DNA重组等技术，将荧光蛋白基因转染到细胞内并结合到特定的靶标蛋白，用于检测特定类型的细胞结构和功能，这些新技术也可用于检测和调控神经突起和环路(Chung and Lee, 2009)及血管循环(Attwell et al., 2010)的结构和功能。常用的荧光蛋白包括绿色荧光蛋白、黄色荧光蛋白和红色荧光蛋白等。

野生型绿色荧光蛋白(wild type green fluorescent protein, wtGFP)最早于1962年由下村修在水母中发现。1992年马丁·查尔菲等成功在大肠杆菌和线虫中表达GFP，开始作为报告蛋白在生物学中应用(Chalfie et al., 1994)。随后钱永健等阐明了GFP的发光机制，通过对GFP改造，得到了XFP荧光蛋白，包括黄色、蓝色、青色荧光蛋白等(Cubitt et al., 1995; Shaner et al., 2005)。2008年，下村修、马丁·查尔菲和钱永健因在发现、应用和改造绿色荧光蛋白方面做出的突出贡献而分享了诺贝尔化学奖。与传统荧光染料相比，荧光蛋白具有明显优势：标记方法简单、性质较稳定、可遗传、能结合到靶标蛋白质的基因序列上共表达、可对真实状态的细胞进行实时定位观察等。荧光蛋白标记技术的发展，结合转基因技术和荧光显微镜、激光共聚焦显微镜技术的普遍应用，促进了人们对模式动物进行基因操作，进而实现对特定功能或特定蛋白质相关细胞和神经环路结构和功能的研究(Suzuki et al., 2007)。2000年，Guoping Feng等建立的Thy1-XFP品系的小鼠首次实现了荧光蛋白在神经系统特异性的表达(Zhao et al., 2010)，但是改造后的 *Thy1* 基因只能控制荧光蛋白在神经系统中表达，未实现对特定类型的神经元的标记。

由于细胞类型的特异性是由基因直接决定的，因此借助特异性表达的神经细胞标志物，结合荧光蛋白标记技术和Cre-loxp重组系统，即可实现对特定类型神经元的特异性标记。通过重组病毒基因，改造成病毒转染工具，可将含荧光蛋白的基因携带进入神经系统，与重组酶系统结合使用，可实现对特定类型细胞环路的标记。这类基因重组的病毒也被称为病毒型神经示踪剂。与非病毒型的神经示踪剂相比，病毒型神经示踪剂具有特异性、跨突触后荧光不会随之减弱、毒性较小等优点，为神经元网络拓扑结构研究提供了一种新的方法。2007年，Jean Livet等发展了脑虹技术，将GFP、YFP、RFP等多种荧光蛋白基因通过转到Cre小鼠的神经细胞中，清晰显示了不同神经细胞的连接关系(Livet et al 2007)。2008年Josh Huang实验室利用 *Cre* 重组酶基因敲除小鼠和病毒介导的基因转染技术，在小鼠脑中用遗传学的方法标记了具有特定功能的神经元类型(Kuhlman and Huang, 2008)。2014年Steven H. DeVires等结合Cre鼠系、重组腺相关病毒(AAV病毒)和狂犬病毒实现了视网膜中特定中间神经元的稀疏双标，实现了在一个较大区域中标记单个神经元(Zhu et al., 2014)。借助重组病毒在特定神经元上表达荧光蛋白，还可实现在体可视化突触水平的神经结构信息。

随着分子遗传学和基因重组技术的发展，还可对神经元胞体和突触蛋白的分布进行实时监测和定量分析，如Xu和Sudhof(2013)构建了整合有GFP和mCherry的腺病毒分别标记胞体和突触前蛋白，通过统计荧光蛋白的表达量，定量分析由病毒感染的神经元所投射的突触个数。该研究提供了一种可以借助于光学显微成像简单快速地获取神经元轴突和突触末梢分布区域的方法。

2.1.4 脑血管网络特异性标记技术

在血管网络构筑研究中，传统方法是通过冰冻切片或者石蜡切片将厚组织块切成薄片，再通过免疫组织化学的方法染色，标记特定的血管生物标志物，最后通过获取的二维图像进行三维重建(Gijtenbeek et al., 2005; McDonald and Choyke 2003; Wiederhold et al., 1976)。这种方法容易发生形变，造成三维重建的失真。为了获取连续完整的血管构筑信息，研究者们在血管中灌注荧光染料标记的植物凝集素、墨水等造影剂及血管铸型剂等进行标记。通常先通过静脉注射一定量的造影剂，待造影剂随血液循环一段时间，标记了目标组织的血管后，将动物处死，取出目标组织，通过切片成像或者光透明组织后整体成像。

针对脑血管整体网络的精细结构研究，主要是通过灌注造影剂后，利用 CT、扫描电镜、MRI 或光学显微镜等进行成像。2007 年，Dorr 等向小鼠脑内血管灌注硅橡胶，用微型 CT 对小鼠全脑进行成像；再对鼠脑进行 MRI 成像，用于对脑血管空间定位，但其血管分辨率约为 20μm，脑区分辨率为 62μm(Dorr et al., 2008)。2009 年，Brige P. Chugh 等通过向小鼠灌注荧光造影剂 microfil，利用微型 CT 全脑成像，实现了小鼠全脑大血管的成像，并对皮层不同区域的血流量进行了测量，其分辨率约为 16μm(Chugh et al., 2009)。这些成像技术分辨率较低，不能用于毛细血管等精细结构的分辨和重建，因此研究者尝试建立适于精细结构成像的标记和标本处理方法。2008 年，Bruno Weber 等通过向食蟹猴注射 Batson 树脂，用扫描电镜进行成像，对食蟹猴脑部视觉皮层的毛细血管进行了成像，分辨率达到了 3nm，但由于电镜的扫描范围有限，成像的总体积只有 100μm³(Weber et al., 2008)。2011 年 Mayerich 等用印度墨水灌注，获得了 500μm 见方的矢量化结果，分辨率可达 1μm(Mayerich et al., 2011)。这些研究的分辨率较高，但没有全脑范围的结果，无法从全局的角度分析血管网络的结构信息。

为了克服以上不足，人们尝试了其他办法，试图从整体的角度分析精细的血管网络结构。2008 年 Hisashi Hashimoto 等应用墨水灌注方式，用 Heidelberg 式滑动切片机，连续获取了 1000 个冠状面，以轴向 5μm 分辨率进行成像并做了三维绘制，初步展示大血管(Hashimoto et al., 2008)。2012 年，德国的研究者 Ali Ertürk 等利用植物凝集素的标记方法结合光透明技术，实现了对小鼠全脑血管的标记，分辨率为 2~3μm(Erturk et al., 2012)。2014 年，骆清铭课题组利用明胶-墨水灌注血管的方法结合自主研发的显微光学切片断层成像系统(micro-optical sectioning tomography, MOST)成功获取了小鼠全脑的血管网络连接图谱(Xue et al., 2014)。这些方法可以全脑范围内成像，并进行矢量化。但血管灌注须通过动脉、毛细血管才能到达静脉，因灌注液较黏稠而毛细血管管径较小，可能会引起成像时部分静脉缺失(Xie et al., 2012)。

因为 Nissl 染色可以标记细胞，同时反衬出血管，可同时实现血管追踪和脑区定位。2008 年，Mayerich 等利用 Nissl 染色和刀锋扫描成像技术实现了对小鼠嗅球血管的三维重建(Mayerich et al., 2008a; Mayerich et al., 2008b)。2013 年，骆清铭课题组改进了的 Nissl 染色法对小鼠全脑进行染色，结合自主研发的 MOST 系统，成功实现了以 1μm 的分辨率对小鼠全脑细胞和血管的同时可视化(Wu et al., 2014)。

2.2 组织处理技术的发展

由于大脑中神经细胞的分布非常密集，相邻神经纤维(树突或轴突)间距可能小于光学衍射极限，仅依靠光学成像方法很难分辨其走向和相互关系。如上文所述，近年来发展了一系列基于荧光分子的转基因技术，如转基因鼠、脑虹、光遗传技术等，采用多种荧光分别标记同一只脑内的不同类型神经细胞，可以通过多波长同时成像实现对不同神经元的投射路径的识别。标记技术的发展对于破解大脑神经连接网络的精细结构至关重要，在此基础上实现全脑精细结构可视化的必要条件是，组织处理方法能够保持荧光强度且适合于相应的成像方式。

2.2.1 常用组织处理技术

组织处理技术在生物组织学研究中有着广泛的应用，研究者们应用各种样本研究生物组织的显微结构已取得了丰硕的成果。根据研究目的不同，选择的组织处理方法也有所不同。针对不同的样本和成像系统要求，发展有多种组织学切片和包埋方法，切片技术包括振动切片、冰冻切片、石蜡切片、树脂超薄切片。为了保持生物组织的抗原性和原有状态，通常选择振动切片和冰冻切片；为了获取生物组织的超微精细结构，通常选择石蜡切片和树脂超薄切片。切片技术的不同，对组织标本的固定和包埋要求也不同。同时，常用的包埋方法所适用的样品类型和切片厚度范围存在很大差异，需针对样本的标记方法、成像条件等选择合适的组织处理方案。

2.2.2 适用于大样本组织处理技术的发展

为了获取高分辨率全脑图谱，近年来人们尝试研究和开发大标本全脑组织处理技术，根据不同的成像方式，发展了不同的样本预处理方法，如光透明技术和连续切片技术等。因为生物组织中水含量较高，而其折射率和生物大分子不匹配，使得生物组织具有高散射性，给光学成像造成困难。为此研究者们发展了光透明技术，使用有机溶剂使得组织内部成分的折射率相匹配，有效控制组织的光学特性，增加光在组织中的穿透深度。传统的光透明技术多用于光学诊断和光学治疗中，以消除皮肤对成像深度的影响。近年来研究者将这种技术应用于生物组织的三维成像(Liu et al., 2013; Zhu et al., 2013)，并逐步发展了适用于全脑光学成像的光透明技术。Erturk 等发展了一种基于有机溶剂的透明成像方法——3DISCO，结合光片照明成像技术，获得了小鼠全脑神经连接的三维重建结果(Erturk et al., 2012; Erturk et al., 2014)。2013 年，Karl Deisseroth 等发展的 CLARITY 技术通过电泳去除脑组织的脂类物质，使得全脑透明化(Chung et al., 2013)。但因其电泳设备的特殊性及光学参数的局限性，该方法难以用于多样本的对比及平行实验(Susaki et al., 2014)。在最近的研究中，亲水性的透明剂被证明可以有效地降低对荧光蛋白的损伤，其中 CUBIC 技术结合光片照明显微镜，对成年 GFP-M line 小鼠脑及生后 3 天的狨猴脑实现了快速的全脑成像(Susaki et al., 2014)。但是受物镜的工作距离及照明光片厚度的均匀性的限制，还难以对大样本实现像素 1μm 分辨率的成像。

为了获取高分辨率全脑精细结构信息,研究者们发展了适用于精细切片的样本包埋技术,包括石蜡包埋、火棉胶包埋、丙烯酰胺包埋、树脂包埋等。早在1949年,塑性包埋技术就被引入显微成像领域。研究者将生物组织包埋入硬度较高的树脂中,制备组织超薄切片,用于电子显微镜和光学显微镜成像观测,近年来,这种方法也在大样本组织处理中得到应用和发展。2007年,Ragan等针对石蜡包埋组织样品,采用双光子显微成像结合铣削加工样本的方式,获取了小鼠心脏的亚细胞分辨三维成像结果(Ragan et al., 2007)。由于石蜡包埋会对荧光造成淬灭,在上述方法的基础上,Ragan等(2012)又改用琼脂替代了石蜡,减少了包埋过程对组织荧光和变形的影响。而Allen脑科学研究所使用丙烯酰胺包埋技术,对AAV病毒结合EGFP标记小鼠的全脑进行成像,获得了介观层次的小鼠全脑连接组结果(Oh et al. 2014)。因为针对哺乳动物全脑的精细切削和高分辨率成像需要较长采集时间,这些样本包埋方法的长时间稳定性还有待证明。骆清铭课题组改进了高尔基染色法,实现了对小鼠全脑的同时染色,发展了全脑树脂包埋技术(Wang et al., 2012),结合自主研发的MOST系统,成功实现了体素1μm的分辨率对全脑神经元可视化(Li et al., 2010)。

然而,生物组织在塑化过程中需经过长时间脱水及有机溶剂处理,导致GFP荧光会发生淬灭,使特异性标记的样本应用受到局限。华中科技大学曾绍群、龚辉研究小组发现GFP在树脂包埋的过程中并没有变性;GFP严重的荧光淬灭来源于生色团的质子化;通过采用碱溶液进行化学激活处理树脂包埋的生物组织,可以恢复绝大多数的GFP分子荧光,从而可忠实地观察到被荧光蛋白标记的精细结构。基于这一发现,使用自主研发的荧光MOST技术,展示了如何获取密集的全脑神经网络数据(Yang et al., 2013; Xiong et al., 2014)。这一发现解决了塑性(树脂)包埋组织处理技术与荧光蛋白标记技术需兼容的难题,为实现宏观尺度生物组织的高分辨率荧光成像提供了切实可行的解决方案,将在诸如解析神经元形态、脑功能环路等一系列的基础科学研究中发挥重要作用。

3. 精细结构获取方法

光学方法为在介观水平获取脑组织信息提供了重要手段。光学显微成像技术可以达到亚微米水平的分辨率,因而可以清晰地对单个细胞及其亚细胞结构进行成像,同时可以分辨神经元的树突及轴突,可用于研究神经元的局部和长程连接关系。光学显微成像技术的优势在于可以与大量的光学染色和标记技术结合,特别是与基因标记技术结合,可以对大脑内部的神经元和神经胶质细胞进行特异性,获得局部神经元的精细形态结构;由投射神经元通过轴突的长程投射,了解脑区之间的相互联系;结合多色标记,跨突触病毒标记,以及突触特异性标记方法还可以明确神经通路的多级连接关系。可以说,光学显微成像技术和光学标记方法的结合是开展细胞分辨水平脑结构解析最为合适的技术。然而,由于生物组织的高散射特性,传统的光学显微成像方法,包括共聚焦成像技术,成像深度为50μm左右,限制了光学方法在脑科学研究中的应用。因此,为了让光学方法在脑研究中发挥更大的作用,实现前述真实尺度的脑精细结构和功能信息的三维可视化,近年来有多种新的光学成像机制和方法被相继发明出来,不断加速和深化对神经科学的研究。

3.1 局部厚组织光学成像

双光子显微成像、光学相干层析成像、光声成像等技术,因其自身成像机制的优势,有效地将生物组织中的光学成像深度扩展到了百微米甚至毫米量级,为局部脑结构和功能信息的三维可视化提供了新的技术手段。

双光子激发(Helmchen and Denk, 2005)是指一个荧光分子同时[在约 0.5 飞秒(fs)的时间内]吸收两个光子的能量,从而跃迁到激发态。由于荧光分子的双光子吸收截面远小于其单光子吸收截面,因此要实现双光子激发则要求激发光在空间和时间上高度聚焦(Denk et al., 1990)。在空间上,使用高数值孔径的物镜将光斑聚焦到数百纳米的尺寸;在时间上,采用脉冲宽度约 100fs 的激光,在聚焦的样品处激光的峰值功率密度就达到 10^{12}W/cm^2(Cahalan et al., 2002)。相比于单光子成像,双光子成像有两个主要的优点。首先,由于常用的荧光分子吸收光谱在可见光范围,而双光子激发使用的波长在近红外波段,该波段恰好是生物组织的"光学窗口",组织对该波段的光吸收和散射较弱,因此能够穿透更厚的组织,通常可达数百微米,并且带来较低的光毒性。其次,双光子荧光激发为一种非线性过程,因为激发效率与光子密度的平方成正比,只有在焦点附近,激发效率达到峰值,能够实现有效的荧光激发;而在非焦点位置,激发效率会迅速衰减,这使得双光子成像具有天然的光学层析效果。另一方面,即使在高散射特性的组织中,被散射的激发光子并不能达到足够高的光子密度而产生双光子激发荧光,因而不会产生焦外背景荧光干扰双光子成像的结果,能够保持良好的成像对比度。

双光子的成像深度决定了它适合在脑皮层以亚细胞分辨的能力研究神经细胞的结构(Xu et al., 2009)和功能(Liu et al., 2011),成为脑研究中特别是在体成像中重要的成像手段。对于脑成像而言,通常采用磨薄头骨或开颅窗等方法,以进行皮层表面的双光子成像。通过使用钙荧光探针等功能性荧光标记方法,既可以观察神经元的精细结构,也能对神经元或神经元网络的功能活动进行观察。它不仅可以观察快速的神经活动,也可以长时程监控神经细胞的动态变化。

与双光子不同,光学相干层析成像(optical coherence tomography, OCT)是一种通过探测后向散射光来测量组织中光学折射率不均匀性导致的散射信号,进行无标记成像的显微成像技术(Huang et al., 1991)。它利用干涉原理,实现对组织轴向结构信息的解析。其本质上是一个迈克尔逊干涉仪,宽带光源发出的光(探测光)一部分到达组织中,在组织中经过后向散射后与另一部分光(参考光)发生干涉,干涉信号被光电探测器记录。通过改变参考光的光程可实现样品深度方向的扫描成像。近年来为提高其成像速率,广泛采用了频域成像的技术,与上述时域技术不同的是,频域技术不需要进行轴向扫描,而是通过测量干涉光谱获得轴向的信息。成像系统采用的宽带光源的相干长度即为系统轴向分辨率,因此光源带宽越宽,系统轴向分辨率越高,一般在 1~15mm。成像深度在

2~3mm，为了实现轴向聚焦光斑尽可能的长，以获取较大范围的深度信息，通常样品前的物镜数值孔径会较小。由于横向分辨率与数值孔径成反比，因此，OCT横向分辨率较其他显微光学成像技术差。

近年来OCT技术在获得组织形态结构信息之外，还对一些生理参数进行成像和量化，相关技术也开始被应用到脑成像领域。OCT被证实可用于啮齿动物上单根髓质纤维成像(Arous et al., 2011)，还可测量大鼠躯体感觉皮层折射率(Binding et al., 2011)。利用偏振对组织双折射的敏感性,可以用偏振敏感OCT定位神经纤维束及微米量级神经纤维回路(fiber pathway)(Wang et al., 2011)。对神经活动中的散射变化进行测量，OCT还可以测量神经组织的功能活动(Chen et al., 2009)。利用多普勒原理，OCT还可以在高散射生物组织中对运动的成分进行高分辨率层析成像，已经成功用于对皮层血流及血管网络进行成像(Vakoc et al., 2009; Wang et al., 2011)。

光声成像技术是一种混合模式的光学成像技术，它测量的是组织内吸收系数的分布(Wang and Hu, 2012)。其原理是基于组织的光热效应，当脉冲激发光入射到生物组织，不同色团会吸收光，并产生热导致升温，在短时间内的组织内温度不均匀则会产生超声。光声成像系统结构较为简单，使用脉冲激光照射组织，再使用超声换能器探测产生的超声信号。光声成像有层析成像和显微成像两种模式，在层析成像模式中(Kruger and Liu, 1994)，超声换能器环绕样品探测不同角度的超声信号，并使用类似于超声层析成像的方法进行图像重建；而在显微成像模式中超声换能器在样品一侧做光栅扫描，类似于B超成像方式(Hoelen et al., 1998)。相比于传统的纯光学成像，光声成像由于使用了超声作为探测信号，而超声在组织中的散射比光要小至少一个数量级，因此能够在深层组织中获得优于纯光学成像的分辨率。光声成像的横向分辨率在使用光学聚焦照明时是与普通显微镜相同的，在使用宽场照明时是由超声聚焦的声焦点决定的，其纵向分辨率则是由超声探头的带宽决定的。此外，光声成像具有尺度可变的特性(Wang, 2009)，即其分辨率可以根据成像需要来进行调节，能够从光学衍射极限的分辨率调节到数百微米。光声成像中成像深度与分辨率是相互制约的因素，在不考虑分辨率时成像深度可达厘米量级；而在达到光学衍射极限的分辨率时，其成像深度也会被限制在1mm的范围内。

由于光声成像的对比度来源于光吸收，因此近年来被广泛应用于不同尺度上的在体脑血管结构和脑功能成像。Lihong Wang等率先将光声层析成像应用于在体无损的大鼠皮层血管结构成像(Wang et al., 2003)，并扩展到用于研究小鼠静息态下的脑功能连接中(Nasiriavanaki et al., 2014)。此外，Xiang等(2013)使用光声层析成像来研究大鼠癫痫发作的动态过程；Yang等(2007)使用光声层析成像研究小鼠脑损伤和恢复的动态过程；Deng等(2012)使用光声显微成像研究大鼠急性局灶性脑缺血后皮层的血液动力学参数变化；Liu等(2013)使用光学分辨的光声显微成像研究药物对脑皮层血管的影响。

3.2 脑深部光纤化成像

上述技术虽然可以在一定深度内获取生物组织的三维立体精细结构信息，但成像深度依然限制在百微米到几毫米以内，研究对象仅局限于脑皮层等浅表范围。为了进一步拓展对脑结构和功能信息的获取，人们开始借助去除部分脑组织或利用微光学探头等方式，开展脑深部组织光学成像的研究。去除部分脑组织可以较为容易的暴露深部脑区，但这种方式对实验动物损伤较大，导致其生理状态较其正常生理状态会不同，难以支持常规脑科学研究。与之对比，以光纤束或梯度折射率(gradient reflective index, GRIN)透镜作为光的传输介质，插入实验动物脑部的方法，可以最小创伤实现对传统显微镜无法到达的脑深部区域的神经活动及结构的高分辨率观察，因此，研制用于脑科学研究的小巧灵活的光纤探头更具有实用价值(Flusberg et al., 2005)。

光纤束由大约1万根光纤紧密排列组成，外径几百微米，光纤间相对位置固定，可以通过光纤束传输图像而无需在其远端进行扫描，因此可以设计外形小巧的微光学探头(Gmitro and Aziz, 1993)。为了以最小损伤插入实验动物脑内，光纤束探头外径需要尽可能小，因此，通常舍去前端成像物镜，以光纤束直接接触脑组织的方式成像(Rector and Harper, 1991; Zhang et al., 2012)。横向分辨率仅由单根光纤芯径决定，一般为2~3μm，可以满足细胞分辨的要求。光纤束法已成功与落射式宽场荧光成像(Zhang et al., 2012)、共聚焦扫描成像(Vincent et al., 2006; Hayashi et al., 2012)等技术结合，对颅骨下深至6mm的脑组织进行了细胞水平的观察。但无前端物镜导致系统轴向分辨率不高，存在背景荧光干扰，无法实现有效的光学层析。

另一种方式是以GRIN透镜作为微光学探头的前端物镜。GRIN透镜在垂直于光轴的平面上，其折射率由中心向四周逐渐变化。该器件一般应用于光通信领域，对光纤输出的光束进行准直。近年来，多个小组(Reed et al., 2002; Jung and Schnitzer, 2003; Gobel et al., 2004; Pillai et al., 2011)将这一器件引入微光学探头的研制，与单光子、多光子荧光成像技术结合，实现微米水平的分辨能力。GRIN透镜制成的微光学探头外径最小仅350μm，可以进入组织内几毫米甚至1cm深度，因此有能力对不同脑区神经结构和活动进行光学成像。例如，Schnizter小组利用GRIN透镜实现了对海马神经元和树突的长时程反复观察(Barretto et al., 2011)。但是，目前使用GRIN透镜的光纤成像系统的成像质量还需要进一步提高。GRIN透镜折射率分布的特点，GRIN透镜成像时边缘像差较视场中心更为严重。另外，由于GRIN透镜的光学传输特性是与折射率高度相关的，因此，成像时必然存在色差，当应用于多色成像时，这一缺陷也会限制GRIN透镜的成像效果。

3.3 全脑三维光学成像

由于脑科学研究需要在全脑范围获取神经连接的详细信息，而光学成像方法可以达到的成像深度还不能满足要求，因此，人们开始尝试将光学显微成像方法与组织学切片制备技术结合，获得连续切片的断层图像，以重建出脑组织的三维精细结构信息。然而，由于这一过程需要经过手动切片、切片制备、切片染色、光学成像、图像拼接等多个步骤才能获得一个切片的断层图像，过程非常耗时(Miyamichi et al., 2011)，而且手动切片难以保证数据的完整性，无法获得全脑的连续图谱数据。

因此，全脑范围光学显微成像方法迅速成为重要的研究方向。对于大部分传统光学显微成像方法而言，其所能达到的成像深度在几十微米以内，且会受到离焦的背景信号的严重影响。为了能够对厘米尺度的小鼠全脑样品进行成像，需要解决光穿透大样品问题，以及在成像中如何抑制离焦的背景光干扰。已有多个研究小组提出了不同的方法，其中代表性的技术包括光片照明显微成像技术、全光组织学，双光子序列断层成像技术和显微光学切片断层成像技术等。

光片照明显微成像技术(Huisken et al., 2004; Dodt et al., 2007)与传统的落射荧光成像技术不同，其照明光路与成像光路相互垂直。光片照明采取侧面照明的方式，照明光被调整为薄片式，对样品中的一个薄层进行照明或荧光激发，而成像光路则垂直于照明光片，对样品进行成像。每次成像只有光片所照明的薄层样品被激发，不存在传统落射荧光成像中的离焦背景荧光，同时也消除了离焦位置的光漂白和光损伤问题。光片照明方法具有较低的光漂白，且通过全场成像的方式可以得到较高的成像速率。通过移动样品可以得到样品的三维图像。光片照明显微成像技术应用于小鼠全脑成像，也需要解决光学穿透深度的问题，而这一问题是通过结合光透明技术实现的。近几年已经出现了多种光透明技术；包括使用光透明剂浸泡，实现全脑透明的 Scale(Hama et al., 2011)和 SeeDB(Ke et al., 2013)技术，以及通过电泳方法去除组织中的脂类物质而使得脑组织透明的 CLARITY 技术(Chung and Deisseroth, 2013)。一些显微镜公司也针对于这些光透明技术提供了高数值孔径、长工作距离的特殊物镜，其工作距离可以达到 4~8mm。

光片照明结合光透明技术尽管存在以上优点，但也存在一些不足。首先，迄今为止其成像范围还难以直接覆盖小鼠全脑；此外，即使进行了光透明处理，对于深部脑组织进行成像时，分辨率会显著下降，因此造成所获得的图像数据存在不同位置分辨率不同的情况，而这对于标准的图谱获取工作是十分不利的。为了将显微光学成像方法扩展到全脑成像，可以采用的另一思路是将显微光学成像和样品的逐层去除方法结合起来。通过样品的逐层去除，每次只需要对样品的表层或薄切片进行成像，从而避开了光穿透深度的限制。全光组织学，双光子序列断层成像技术和显微光学切片断层成像技术即采用这一实现思路。

Tsai 等(2003)提出的全光组织学法，完全使用光学方法实现成像和样品的逐层去除。全光组织学方法将双光子成像和组织的激光消融整合为一个系统。通过双光子成像系统对样品表面进行成像，样品固定于二维电动平台上，通过二维电动平台的移动，将多个成像视野拼接起来，得到样品表面的完整双光子成像结果。在完成双光子成像后，切换为激光消融光路，同样通过二维电动平台的移动，以及物镜在 z 方向的移动，逐个视野地去除样品表面已成像的部分。重复上述过程，可以得到样品的完整双光子成像结果。全光组织学方法是通过激光消融组织的方法去除已成像的组织，其系统及激光消融参数的设置相对复杂。多个研究小组采用了可实现自动切片-成像的方案，将基于机械切削的组织切片方法与光学成像方法结合，而无需经过手动切片、切片收集、制片、显微成像、图像配准等分离的流程。

Ragan 等(2012)提出的双光子序列断层成像技术通过将双光子成像和样品的振动切片结合起来。小鼠全脑用琼脂糖包埋，固定于三维电动平台上。每次只对样品表层进行双光子成像，通过三维电动平台的移动，将样品的一个断面分为多个区块，每个区块的大小对应于双光子成像的成像视野，从而可以自动地获得一个断面的完整图像。在结束断面图像的采集后,将样品移至切片位置,通过振动切片机去除表面已成像的部分。重复上述步骤,即可自动地获得小鼠全脑图像数据。由于双光子成像可以达到相对深的成像深度,因此,在成像时,可以对表层以下几十微米处进行成像,从而避免因振动切片造成的表面不平整对成像质量的影响。双光子序列断层成像技术已经被 Allen 脑科学研究所采用(Oh et al., 2014),通过结合 AAV 病毒 EGFP 标记特定脑区和细胞类型的轴突投射,按轴向每间隔 100μm 进行一次冠状面成像的方式,获得了 Allen 小鼠脑连接图谱。但是,该技术采用双光子点扫描成像方法,每个像素强度的采集是一个串行的过程,因此成像速率受到限制。上述 Allen 小鼠脑连接图谱的工作是通过间隔一定距离进行图像采集方式获得的,并没有得到小鼠全脑的连续完整的数据。如果要得到微米体素的完整数据,则每一只小鼠全脑的数据采集时间将达到 20 天以上。另外在建立脑图谱的过程中,由于每只小鼠只提供了部分脑区连接等的信息,故需要采集多个小鼠的全脑数据,才能构成完整脑的信息,因此长时间的图像采集不利于提高图像数据的获取效率。

经多年努力,骆清铭课题组发明了 MOST 技术(图版XII图 2),通过树脂包埋和精密切片技术,结合同步的显微图像获取,获得了体素分辨率为 1μm 的小鼠全脑连续三维结构图(Li et al., 2010)。通过树脂塑性包埋获得足够的硬度,再用金刚石刀进行超薄切片,这样的方式,可以保证获得 1μm 厚度的稳定切片。在切片的过程中,通过样品槽中水的压力,以及水循环系统的抽吸作用,使切片贴伏于金刚石刀具表面,并沿表面滑行,通过采用与切片运动同步的时间延迟积分成像方法,可以在切削的同时获取被切脑片的光学成像。光学成像采用了反射式的落射式成像,系统为模块化设计,运用该技术获得了小鼠全脑的高分辨率图像数据,成像质量高,且所得图像无需后期配准处理。在这一技术框架下,针对于不同标记的样品,研究团队研发了可用于高尔基或尼氏染色小鼠全脑的 MOST 系统,提出了通过尼氏染色的特点获得小鼠全脑血管网络的方法(图版XIII图 1),对神经元和血管网进行了空间共定位的定量分析(Wu et al., 2014)。该团队还基于共聚焦成像抑制背景荧光的方式,实现了可以对荧光样品进行成像的荧光 MOST 系统(fMOST),首次获得了神经元全脑长距离轴突投射通路的连续追踪(Gong et al., 2013)。同时基于双光子成像方式,该团队提出了将切片与成像过程分离的实现方式,即先使用双光子成像方式获取样品表面一定厚度的双光子成像图(z stack 图像),再通过机械切削去除样品表面已经成像的部分(Zheng et al., 2013)。

3.4 其他成像技术

X 射线层析成像(CT)是一种基于 X 射线吸收的三维成像技术(Kak and Slaney, 1988)。它是利用样品内部不同物质对 X 射线的吸收系数不同产生对比度,采集样品在不同角度的投影图像,再使用算法对样品的三维吸收分布进行重建,该技术已经被广泛应用于临床诊断。然而临床使用的 CT 分辨率低,软组织对比度差,无法满足高分辨率脑成像的需要。近年来发展的微型 CT 技术,使用微焦斑射线源提供的微米甚至纳米量级的焦斑,使用高分辨率的面阵列探测器,使得 CT 成像达到微米分辨率甚至纳米分辨率(Kiessling et al., 2004)。由于射线对软组织对比度不高,通常使用造影剂来改善成像质量。而使用高亮度的同步辐射光源(Chen et al., 2010)或者相衬技术(Pfeiffer et al., 2006)则可有效提高软组织的对比度。然而，受探测器视场的限制，在进行大范围成像时，微型 CT 往往无法获得较高的分辨率，难以同时实现大视场范围和高分辨率成像。Chugh 等(2009)使用微型 CT 研究小鼠的脑血容。Dorr 等(2007)用微型 CT 研究了小鼠全脑的三维血管结构。

Array tomography 是一种结合了电镜与荧光显微成像技术的方法(Micheva and Smith, 2007)。通过对树脂包埋的样品进行超薄切片(50~200nm)，并对所收集的切片、进行多次荧光染色(免疫荧光染色等方法)、洗脱和最后的重金属电镜染色，可以同时获得组织样品的精细结构信息和其内部的各种抗原、荧光蛋白等组分的三维空间分布信息。Array tomography 可用于获取小块脑组织的精细结构和分子组成等信息。但是因为要经过超薄切片、切片收集、多次染色、多次成像等步骤，因而非常耗时，难以应用于全脑大范围的成像。

4. 脑的数字三维重建

通过标记和成像，真实的组织结构被解析为蕴含有大量生物信息的连续图像，实现了数据的获取。但是，由此得到的图像仅仅是二维平面信息，要准确理解组织的三维结构，就需要对二维数字图像进行三维重建，建立适合于计算机表示和处理的立体模型。例如 20 世纪 90 年代发展起来的数字人研究，人们现在已经可以在计算机中模拟器官结构、运动过程、新陈代谢，甚至是遭受外来伤害的影响程度。但由于脑的高度复杂性，对其进行数字三维重建的研究一直都是难点和热点。随着标记和成像技术的发展，解析大脑的精度越来越高，获得的信息更加完整，已经具备了在全脑范围，以真实尺度构建出感兴趣位置处的神经元和血管的精细三维形态的基本条件，但也带来两方面亟待解决的问题。一是伴随高精度而来的海量图像数据，对于现有的数据存储、图像处理与分析、数据管理和共享等方式带来了全方位的挑战。二是脑内各结构单元形态、功能差异巨大，难以在原始数据中对这些复杂信息进行准确的辨识。

4.1 数字脑与大数据科学

4.1.1 从数字人到数字脑的发展历程

"数字人"是利用真实人体的数据在电脑里合成的三维人体模型，其起源可追溯到 1989 年美国国立医学图书馆(National Library of Medicine, NLM)发起的"可视人计划"(Visible Human Project, VHP)(Ackerman, 1998)。1994 年，VHP 完成了第一套人体断层图像数据集，包含 1871 个间隔 1mm 的断层。继美国的 VHP 计划之后，2000 年，韩国亚洲大学在韩国科技信息研究院的资助下完成了韩国可视人(Visible Korean Human,

VKH)的第一套数据(Kim et al., 2002)。中国虽然到2002年才发布了中国人体貌特征的数字人数据，但在南方医科大学和第三军医大学等的努力下，至今已经采集了10套以上的人体数据集，其中虚拟中国人(Virtual Chinese Human, VCH)男性1号是目前世界上三维空间分辨率最高、数据量最大的数字人数据集(Tang et al., 2010)。

三维重建能够更加直观地展现人体结构的三维立体形态，而断层图像的精细配准和分割是准确重建器官和组织的基础。美国的NLM早在VHP立项之初就非常重视在数字图像处理方面的研究，2000年VHP项目已经建立了全身骨骼、肌肉和心脏等部分器官的三维模型。美国徐榭教授课题组在VHP解剖图像中辨识并手工分割出1400余个组织器官(Kim et al., 2002)。NLM为了实现自动的器官轮廓分割，发起了一项开源软件项目Insight Segmentation and Registration Toolkit(ITK)，并在2002年的时候发布了第一个版本。ITK收集了几乎所有的图像分割和配准算法，具有免费、跨平台、效率高的特点，已成为生物医学图像处理和分析领域中重要的工具和标准。在国内，华中科技大学使用VCH数据集，实现了人体运动、消化、神经等七大系统中260个组织器官的精细分割和三维重建(Li et al., 2008)。在三维结构的基础上，更多的人体物理和生理参数将会加入到这个模型中，用于模拟更加复杂的过程，直至在计算机中建立完全真实的人体。美国华盛顿大学在1997年发起了"生理人"计划。徐榭教授等建立的VIP-Man人体辐射计算模型，可在计算机中对各种离子的辐射剂量分布进行分析(Xu et al., 2000)。华中科技大学在VCH数据集上实现了多粒子多角度超大能级外部辐射剂量计算(刘洋等，2009)，并正在开展基于虚拟现实技术的虚拟手术研究。与此同时，华中科技大学建立了数字化小动物数据获取和处理的整套技术方法和平台(分辨率0.02mm)，采集了国际上第一套完整成年大鼠、小鼠、青蛙等小动物的光学断层解剖数据集(Bai et al., 2006)，出版了大鼠断层解剖彩色图谱(刘谦和龚辉，2010)。

脑是人体最重要的器官，数字脑研究是数字人研究的自然延伸，是对更高目标的追求。传统的数字人研究中，由于图像获取的分辨率有限，也没有针对脑和神经组织的特殊标记，只能提供脑的基本宏观形态特征，例如沟回、脑室、灰质、白质的边界信息。因此，为了数字三维重建真实的脑，首先需要实施有效的标记和成像，相比数字人研究来说具有更大的挑战。对小型的哺乳动物，已经可以对脑进行单神经元分辨的解析，如华中科技大学获取的小鼠脑三维数据集第一次将约 $1cm^3$ 的小鼠脑分解成15 380个断层，结果发表在2010年的 *Science* 杂志上(Li et al., 2010)。2013年，同样是在 *Science* 上，德国和加拿大的研究人员发布了一套人脑的三维数字模型"BigBrain"，这项工作将一名65岁女性的大脑切成了7400多片，虽然精度远低于小动物脑的研究，但比现有数字人研究要高出一个数量级，已经能够在染色技术的帮助下，分辨数字图像上的不同的脑区(Amunts et al., 2013)。

从数字三维重建的角度看，数字人研究的基本原理和研究方法与数字脑研究较为一致，包括数据的处理、可视化、共享，以及图像信息分割和标识。数字人研究中已建立的研究工具也是开展数字脑研究的重要支撑，包括前面提到的ITK。

4.1.2 大数据——数字脑研究的机遇与挑战

在科学探索的推动下，研究人员一直在努力重建出更加精细和真实的脑。并且，为了研究更加复杂和高级的脑功能，重建的对象也逐渐从低等动物转向高等动物，甚至是人脑。因此，伴随脑体积的增大，数字脑的切片数量也越来越多，单张切片的尺寸和数字化精度也在不断提高，由此产生的数据量正在以难以想象的速度增长，逐渐超出了现有主流软、硬件的处理能力，成为典型的大数据（Big Data）问题。

大约15年前，第一套完整的数字人数据集仅有15GB，可以用4张DVD光碟存储。后来的 VCH 男1号的原始数据达到了540GB，也仅相当于目前一台普通计算机的存储能力。华中科技大学2010年公布的小鼠脑三维数据集已经达到了8TB（1TB=1024GB），已经相当于数万张 DVD 光碟的存储量。如果以相同的精度获得人脑的数据，预计数据量将增加1500倍，即12PB（1PB=1024TB）以上，仅能存储在大型数据中心。美国斯坦福大学、德国马普实验室已在探索使用电子显微镜重建脑中所有的突触网络，预计仅1mm^3 脑组织的解析数据就将超过 1PB，倘若在未来实现了完整人脑的三维重建，数据将达到惊人的 1ZB（1ZB=1024PB×1024PB），接近2010年全世界存储数据的总量。由此可见，不到20年时间，数字脑研究的数据计量单位已经从 GB 跨越到了 TB，约3~4个数量级，并即将进入 PB 时代。

脑科学中的大数据研究首先具有大数据研究的共性。如此工程化速度地产生大数据，对于目前现有的数据存储、图像处理与分析、数据管理和共享等方面带来了全方位的挑战，已成为限制脑科学研究发展的重要瓶颈问题，与其他科学大数据问题一起被广泛关注。2008 年，Nature 杂志率先出版了"Big Data"专刊，分析了大数据对当代科学的影响和意义。2011 年，Science 杂志推出"Dealing with Data"专刊，从多个科学领域介绍大数据所带来的技术挑战，其中就包括神经科学。

大数据概念与应用实践更多是开展于 IT 行业，与之相比，科学大数据的理论研究与实践还相对较少，特别是具有显著特征的脑科学大数据。脑由数以千亿计的神经元、胶质细胞，以及密布穿插其中的血管所组成。这些组织结构的尺度差异较大，典型神经元胞体的直径约为 10~30μm，由神经元胞体发出的突起仅有亚微米至 1μm 的直径，毛细血管和其他小细胞的直径约为 2~5μm（在第 1.3 小结有更详细的描述）。数据更加致密与饱和（Lichtman and Denk, 2011）。因此，神经元类型和形态，以及供血网络具有高度的复杂性，然而目前对于这类非结构化数据还缺乏有效的知识挖掘技术。

总而言之，取之不尽的实验数据是科学新发现的源泉，而传统的生物医学已经无法有效的分析和利用这些数据，大数据的出现必将改变人们认识脑的方式，产生脑科学领域的大发现。

4.2 海量数据的处理、可视化与共享在数字脑研究中的进展

根据不同的研究，存在研究层次、研究内容和研究对象的差异，脑的数字三维重建需要面对不一样的实施步骤和具体问题。德国马普研究所和美国冷泉港实验室的研究者，

将整个处理流程大致归纳为：以质量控制为目的的预处理，切片间和数据集间的配准，分割、量化和重建脑及其显微结构，在互联网上可视化和共享数据，以及与其他在线资源进行整合(Helmstaedter and Mitra, 2012)。其中，量化和重建脑及其显微结构将在下节作为一个专题介绍，其他部分在本节逐步介绍。

4.2.1 质量控制为目的的预处理

评价科学图像的质量指标有很多，但在针对脑组织的三维光学显微成像技术的论著中主要比较的是空间分辨率(resolution)、衬比度(contrast)和信噪比(signal to noise ratio, SNR)，以及上述指标间的均衡。对于大范围的高分辨率成像，直接从成像仪器中获得高质量图像的难度极高，需要对原始数据进行客观的预处理。因具体技术所遇到的问题及采取的预处理方法不尽相同，目前还没有通用的预处理方法(Osten and Margrie, 2013)。下面结合 3.2 节，以其中三家单位在预处理方面的工作为例进行介绍。

美国加利福尼亚大学圣迭戈分校 Kleinfeld 团队为全光组织学技术开发了一个包括 4 个步骤的预处理方法，以减弱噪声干扰，以及信号、背景不均匀对三维重建的影响(Tsai et al., 2003)。首先，使用一个 5×5 像素的低通滤波器去除图像背景里的白噪声和高斯噪声，滤波易造成的边界效应可通过引入具有边界信息的先验知识进行抑制。其次，使用高通滤波器校准单层图像内的不均匀性，具体用输入图像减去一个 81×81 像素的均值滤波结果。然后，利用非线性的灰度变换校准沿着轴向的层间图像的不均匀性，变换以每层图像像素最大值为变换基准。最后，利用一个 5×5 像素的双中值滤波器平滑信号区域的边界。通过上述处理，可以获得较好的毛细血管网的三维体绘制(volume rendering)效果，展示结果的重建体积约为 250μm³。全光组织学技术基于双光子荧光显微成像，上述预处理方法在此类技术中能解决一些共性问题。

德国马普研究所 Dodt 团队研发的 Ultramicroscopy 技术中，使用 Huygens 软件中的非盲反卷积算法对数据进行质量恢复，以提高数据的细节表现能力，这是共聚焦显微镜中常使用的方法(Dodt et al., 2007)。由于 Ultramicroscopy 技术使用光片(light sheet)从侧面照明样本，因样本无法理想的透明化导致照明不均匀，在原始图像上会产生的不均匀的横向条带状伪影。2010 年，该团队公开了一种伪影消除算法，先是利用非线性横向平滑，将横向条带状伪影从原始数据中提取出，再用原始数据除以提取出的伪影图像(Leischner et al., 2010)。虽然平滑操作有可能损失有效信息，但这种方法对于那些在数据中含有大量不均匀平行条纹干扰的研究(不仅仅是横向条纹)，具有很好的借鉴作用。

华中科技大学的 MOST 技术在预处理控制数据质量的研究中，较好地解决了因样本整体标记、连续 1μm 切片、反射照明等步骤带来的伪影问题(Li et al., 2010; Ding et al., 2013; Gong et al., 2013)。具体针对不同的神经细胞、血管标记技术或图像的特点，开发出不尽相同的预处理方法，归纳起来，有如下几个方面。对于周期性的条纹状伪影，可对图像沿条纹方向做均值投影，获得叠加有真实信息和条纹变化的曲线，再用移动中值滤波处理，获得真实信息的变化曲线，从而提取出条纹的变化，进而对原图进行恢复。对于在宏观上才能观察到的明暗区域变化，使用低通滤波提取不均匀的背景，对于局部小亮斑，使用形态学处理提取不均匀背景，原图除去背景即完成校准。对于层间图像的不均匀性，没有简单的求平均再加权，而是对原图进行初步的自动分割，准确提取出每一层图像背景的均值，再通过加权来进行校准。在早期研究中，曾使用过频域滤波去除周期性噪声，但从使用的效果上看，空间域滤波校准图像的同时能更好地保持原图细节（Li et al., 2010）。

从大数据计算的角度看，预处理的瓶颈主要是大量数据的读和写，其次才是计算的复杂性，分布式存储和分布式计算是提高效率的有效途径。此外，在预处理过程中，通过去燥、均匀化、去除背景等操作，可以大大减少原图中的无效信息含量，降低了后续处理的难度。

4.2.2 数据配准

对于数据配准是脑的数字三维重建的基本要求，用于校准数据点的空间位置，包含数据集内配准和数据集间配准这两个层次。

数据集内配准指的是断层数据集中同一水平层内不同小图间（mosaic），以及垂直方向上相邻大图间的无缝对齐，也就是将"像空间"中每个点的位置向其"物空间"中的真实位置对齐。数据配准的难度取决于数据采集技术和采集过程的质量控制，如果可以保证原始数据中每一张小图的精确位置，就可直接重建为三维数据，如果原始数据只是存在少量偏差，经刚性配准也可获得良好的效果，但对于在数据采集过程中样本发生过不规则变形或存在破损的数据，不仅配准难度大，准确性更无法保证。这类图像配准一直是计算机视觉领域需要解决的基本问题，其所涉及的具体技术方法在 Szeliski 等（2006）的综述文章中已有详尽的介绍，本文不再赘述。

数据集间配准具有更大的挑战，特别是脑三维数据集与标准脑图谱间的配准（Osten and Margrie, 2013），根据模型的数学性质，划分了刚性（rigid）、仿射（affine）、投影（projective）和非刚性（non-rigid）变换四个大类（Rueckert and Schnabel, 2011）。下面本文将介绍其中的三种，投影变换在脑三维数据配准中应用较少。

刚性变换仅仅包含旋转和平移两种矩阵变换（Oliveira and Tavares, 2014），非常适用于组织器官的空间方位校准，但不能解决因组织变形或个体差异所产生的不匹配。例如，MacKenzie-Graham 等（2004）使用 BrainSuite 工具对实现了参考脑标记空间到其他脑数据集的映射。Dorr 等（2008）使用一个已有的脑图谱作为参考脑，再将 40 个鼠脑数据集刚性配准到该参考脑上，然后对这些数据集进行平均，得到一个平均脑。

仿射变换是在刚性变换的基础上引入缩放和切变（shear transformation），也可以用矩阵变换来实现，它通过整体变形实现了尺寸差异较大样本之间的全局匹配，但不能处理局部形变的差异性。使用仿射变换配准不同脑三维数据集的研究有很多。例如，Sharief 等（2008）将多个鼠脑配准到一个参考脑上。Hawrylycz 等（2011）在构建 Waxholm 鼠脑图谱的过程中，将组织学数据集配准到对应的 MRI 数据集上。Gong 等（2013）将 fMOST 技术获得的全脑数据集配准到 Paxinos and Franklin 脑图谱上，得到神经元长程投射所途经的脑区信息。

非刚性变换进一步实现了局部变形，但实现的算法更加复杂，一般难以用矩阵变换来实现，可分为参数模型和非参数模型两类（Rueckert and Schnabel, 2011）。其中参数模型通常选取一些控制点来反映局部的变形程度，而非参数模型直接用每一个体素上描述变形。非刚性变换近年来发展很快，Klein 等(2009)对 AIR (automated image registration)等 14 种非刚性配准算法进行了评估和比较。为了提高全脑数据集非刚性配准的准确性，要求参考和目标的差异性较低，在非刚性变换之前，通常需要进行刚性变换(Dorr et al., 2008)或线性配准(Sharief et al., 2008; Hawrylycz et al., 2011)，甚至同时使用三种配准方法。例如，Aggarwal 等(2009)首先将 8 个鼠脑数据刚性配准到参考脑上，对这 9 个数据集求均值得到平均脑，再将 9 套数据集通过仿射变换线性配准到平均脑上，并再重复上述过程 4 次，获得最终的平均脑，最后再利用非刚性变换将 9 个原始数据集配准到最后获得的平均脑上。

从刚性配准、仿射配准到非刚性配准，配准的精度越来越高，但算法和时间复杂度也随之提高。特别是非刚性配准无法用矩阵计算实现，难以使用 GPU 等技术对其进行有效的加速，目前研究的对象大都是较低分辨率的全脑数据集。

4.2.3 海量数据集的三维可视化

随着成像技术的发展，人们可以获得更多、更复杂的图像数据，这些数据都需要依靠软件工具确保采集、管理、分析和可视化。因此，近二十年来相关的软件大量涌现，*Nature Methods* 杂志也分别在 2010 年和 2012 年对各种软件工具进行了综述(Walter et al., 2010; Eliceiri et al., 2012)。在生物成像领域使用的软件工具普遍具备三维数据可视化的能力，被广泛使用的包括 VSG Amira、Fiji/ImageJ(Abràmoff et al., 2004)、Bitplane Imaris 等。在脑成像领域，对可视化工具的需求直接体现在可处理的数据量上，如何用合理的代价实现 TB 级海量数据的可视化已成为重要的瓶颈问题。

这里不得不先提到三维数据可视化的艺术品——VTK(the Visualization Toolkit) (Schroeder et al., 1996)。VTK 是一个开源、跨平台、支持并行处理的图形应用二次开发包，目前的应用几乎涵盖了需要进行科学数据可视化的所有领域，包括前文介绍的数字人研究。VTK 从设计之初就关注于海量数据集的三维可视化，较好地兼容 MPI 等并行计算环境，是目前很多分布式并行可视化应用的基础，美国 Los Alamos 国家实验室就曾在高性能计算机上用其处理过接近 1PB 的数据。但是，VTK 可视化的数据量受到硬件条件的限制，在单台图形工作站上不能对超出内存量的数据进行可视化。

近年来，桌面级的海量数据可视化软件或模块已经开始出现，已经在一些脑成像研究中被应用，例如 Amira LDA、Octreemizer(Plate et al., 2002)、OpenGL Volumizer (Bhanirantka and Demange, 2002)，以及用于电镜研究脑连接的 Ssecrett(Jeong et al., 2010) 和 KNOSSOS[①]。利用这些工具，科学家能够在完整数据集中选择自己感兴趣的区域和分辨率的三维数据进行处理和可视化。多级分辨率(Multi-resolution)技术是上述软件实现对海量数据可视化的共同思路，它主要基于八叉树(Octree)的思想来实现(LaMar et al., 2000)。简单来说，就是通过对可视化所需数据的范围、分辨率进行估计，只载入有用数据进行可视化，从而大大减少了数据传输、缓存和渲染的压力。上述工作都由软件在后

---

① http://www.knossostool.org

台实现,而在使用者看来是在对完整的数据进行操作。

还有一些基于 WEB 的巨幅图像的可视化技术,同样采用了多级分辨率技术,其中最著名是 Google Map。现在的生物医学成像领域,海量数据的生产者很容易找到类似的技术来进行数据的分享,例如 Zoomify[②]、CATMAID(Saalfeld et al., 2009)和 Aperio ImageScope[③]等方式。在网页中导航、浏览二维图像,是目前最流行的数据共享方式,下节将介绍的三维脑数据共享的技术基础。

4.3.4 三维脑数据的共享与整合

许多数据库都变得太大而难以下载,这些数据库,对数据容量和管理提出了新的挑战。在脑科学领域,绘制数字化的脑图谱是三维脑数据最重要的数据共享形式和发展趋势。脑图谱是一个表达脑组织结构、脑功能结构、脑神经结构的模型,由数据库工具、图像处理工具等组成的软件系统。在脑图谱中,通过标记信息可以快速对海量数据进行索引,实现了图谱图像、结构名称索引和说明文字间的友好连接,大大提高了查找、传输和存储感兴趣数据的效率。近几年,在互联网上向公众开放的三维脑数据的图谱数据库如雨后春笋般的涌现,下面简要介绍几个规模大、影响大的网络脑图谱数据库。

美国加利福尼亚大学戴维斯分校在 BrainMaps.org 网站上刊载了已完成的高清晰全脑数据集(Mikula et al., 2007)。物种包括人、猕猴、狗、猫、小鼠,以及鱼类和鸟类等,可供阅览的图像超过 1.3 万张,数据总量已超过 140TB。通过词条索引、在线浏览、离线浏览工具,用户可以比较流畅的寻找和下载感兴趣的数据。另外,该网站还免费提供了一批大脑数据分析工具。

美国 Allan 脑科研究所开创了脑信息获取的"工业化大规模生产",虽然建立仅 10 年时间,但从 2007 年开始,该机构按照既定路线图已经先后发布了小鼠脑基因表达图谱、人脑基因表达图谱、小鼠脑皮层连接图谱,以及多种发育、疾病图谱,逐渐在实现其"大脑的逆向工程"的目标(Jones et al., 2009)。所有数据均能在 BrainMaps.org 上获得,主要以二维图像的形式浏览,但提供三维索引和三维数据浏览软件。

Neuromorpho.org 是世界上最大的神经元三维形态模型数据库,由美国乔治梅森大学的 Ascoli 教授创建和维护(Ascoli et al., 2007)。数据库收集了世界上 144 个研究团队获取和发表的 1.1 万个神经元数据,这些神经元数据来自于 24 个物种和 29 个脑区。每一个神经元都是经过识别的矢量数据,提供原始格式和标准 SWC 格式(Cannon et al., 1998)的下载,也可以用一个 Java 编写的浏览器或 GIF 格式动画进行三维浏览。值得一提的是,数据库中还包含了每一个神经元的原始信息、形态学定量分析结果、文献来源等,可用于检索。

众多脑共享数据库的出现,既给研究人员带来了大量的资料,但同时也增加了查找的负担。为了对数据、知识进行更好的整合,2005 年,美国 NIH 资助建立了神经科学信息框架(Neuroscience Information Framework, NIF)(Gardner et al., 2008)。NIF 就像是神

[②] http://www.zoomify.com
[③] http://www.leicabiosystems.com/index.php?id=8991

经科学领域的搜索引擎,科学家可以用它检索任何与神经科学相关的信息,例如数据集、工具、方法和材料等(Akil et al., 2011),详见 neuinfo.org。数据和知识的整合需要基于统一的术语库,大到对物种的分类,小到对神经元类型的命名,在 NIF 中是通过 NeuroLex,及伴随使用的 NIFSTD 进行约定(Bug et al., 2008; Larson and Martone, 2013)。

为了加强神经信息学领域的国际合作,早在 2000 年就由美国牵头建立了一个工作组(OECD-GSF-NI),也就是现在的国际神经信息学协调委员会(International Neuroinformatics Coordinating Facility, INCF),旨在建立一个有关神经系统所有数据的全球知识管理系统和网络协同研究环境,组织开展全球性脑科学科研大协作,具体的任务包括制定发展计划、工作指南、数据标准和共享规范等(De Schutter, 2009)。INCF 要求每个成员国都要用统一的标准建立神经信息学数据库网络,以提供能为世界共享的神经信息学资源(Bjaalie and Grillner 2007)。2011 年,受科技部委托,唐孝威院士等代表中国加入了 OECD-GSF-NI 工作组,参与国际合作。在国内,目前也建立有 INCF 的中国节点 www.brainbank.cn,以及全国神经信息学联络组。

4.3 脑内精细结构信息挖掘的进展

2007 年,图灵奖得主 Jim Gray 描绘了数据密集型科研第四范式,提出在与实验科学、理论分析和计算机模拟这三种经典科研范式相比,大数据科学将成为一种全新的科研范式(Hey et al., 2009)。对于脑科学来说,第四范式就是略过各种实验、理论和模型,直接通过数据挖掘技术,从海量的神经图像数据中取得新的科学发现,这自然离不开对脑内精细结构,如核团、胞体、血管、突起连接的信息挖掘(Akil et al., 2011)。

4.3.1 脑区、核团轮廓的识别

在组织染色的脑切片图像上识别脑区、核团的轮廓是脑图谱绘制中不可或缺的步骤。无论是传统印刷的图谱,还是数字脑图谱,几乎都采用人工方式识别脑区、核团。只是随着计算机技术的普及,开始在绘制数字脑图谱时使用 Illustrator、Photoshop、ImageJ 等绘图软件对轮廓线进行手工勾勒(Franklin and Paxinos, 1997; MacKenzie-Graham et al., 2004; Ali et al., 2005; Dong, 2008),这些轮廓线在三维空间中的连续性较差。随着脑断层数据集的规模越来越大,一方面有可能绘制出更加精细的图谱,但另一方面,在上万张图像上进行手工绘制的工作量同样巨大,需要发展自动的识别技术。

自动脑区分割研究主要集中在 MRI 脑图像上,已经实现了对脑外轮廓、白质轮廓和海马轮廓等高对比度区域的自动划分(Atkins and Mackiewich, 1998; Tang et al., 2000; Fischl et al., 2002)。2014 年,Selka 等撰写了一篇介绍如何从 MRI 图像中分割出肿瘤区域的综述,对这一领域的发展具有一定的概括性(Selka et al., 2014)。在组织学图像数据集上,除了对比度具有显著差异的脑外轮廓外,大部分脑区、核团轮廓对比度低、形状不规则,至今还没有能够有效识别这些轮廓线的自动方法。因此,在未来的一段时间内,人工识别脑区、核团还将成为急待解决的重要问题,相应地有必要重点发展一些计算机辅助技术,降低人工识别的工作量。

4.3.2 神经细胞的自动识别

神经细胞是脑的重要组成单元,在脑图像中尺寸大、信号强、基本呈椭球形,非常适合自动识别,但也面临着变化性、复杂性和细胞粘连等困难(Meijering, 2012)。其中,变化性是由不同的显微成像系统、染色方法、细胞类型和细胞密度所导致的图像特征多变,复杂性指的是分割对象的形态、尺度极其复杂,而且数量巨大,在细胞密度较大的区域出现细胞粘连的情况多,图像上难以区分彼此靠得很近的多个细胞。

针对变化性和复杂性,常用的细胞分割方法(Wu et al., 2010)包括:基于阈值的分割算法,基于全局阈值或者局部自适应阈值,在20世纪60年代就已被提出来;基于特征检测的分割算法,高斯或者拉普拉斯滤波、边缘检测等算法;基于形态学的算法,使用非线性的操作子,比如腐蚀、膨胀、开操作和闭操作。此外,还包括一些混合的分割方法,例如支持向量机和马尔科夫随机场等。Meijering在2012年特别发表综述,对细胞自动分割的发展历史进行了回顾(Meijering, 2012)。

在神经细胞的分割中,最大的挑战是粘连细胞的分割,当多个细胞被检测为一个细胞,会严重影响细胞密度计算、空间分布和形态分析结果的准确性(Li et al., 2008; Al-Kofahi et al., 2010)。针对这个问题,目前已经发展出了多种方法,包括分水岭算法、可变模型法、Hough变换法、边缘跟踪法、基于凹点的算法、图论切割和多尺度高斯拉普拉斯滤波算法、ITCN算法、迭代投影的算法等(Canny, 1986; Maragos, 1987; Fernandez et al., 1995; Garrido and de la Blanca, 2000; Zhang et al., 2004; Byun et al., 2006; Parvin et al., 2007; Al-Kofahi et al., 2010)。上述方法在应用于粘连细胞数据时各有所长,都在各自设定的应用场景下获得了不错的结果,但也不同程度地遇到了效率低、过分割、噪声敏感、适应性差、不易扩展到三维空间等问题。目前仍没有在哺乳动物全脑范围内实现所有神经细胞的计数、定位或轮廓的精确提取,最近一些工作很有希望接近这个目标(Frasconi et al., 2014; Quan et al., 2013)。

4.3.3 神经突起的自动识别

神经突起,包括轴突和树突,是数字化重建神经元形态和脑网络的重要组成部分。神经突起具有显著的纤维状结构特征,而且会形成完整的网络结构,理解脑网络对于人们理解认识大脑是至关重要。对它们进行准确的追踪识别是进行数字化三维重建的基础。

19世纪末,科学家就已经利用设备将显微镜的成像结果投影到纸张上,再对显微结构进行描绘,例如记录神经突起的走向。1965年,Glaser等设计了一个系统,由人与计算机交互,控制显微镜,并保存关键点的坐标,达到追踪神经纤维走向的目的(Glaser and Van der Loos, 1965)。为了减少人为的干预,Neurolucida(Glaser and Glaser, 1990)和NeuroZoom(Young et al., 1997)使用鼠标定义关键点进行追踪,甚至能拓展到三维空间。2010年,在Peng等构建的V3D可视化平台中,可手工选取纤维的端点,由计算机得到最优的路径(Peng et al., 2010)。Amira和Neurostudio(Rodriguez et al., 2008)软件平台也具有类似的交互追踪功能,甚至还提供了完全自动的追踪模块,但追踪效果对数据质量的依赖性较高。

在具体的算法上，按照数据处理的范围，将追踪算法分为全局和局部两大类(Meijering, 2010)。全局处理类方法需要用到原图全局信息进行分割，大部分方法均包含二值化、骨架化、矫正和图表示等操作(Dima et al., 2002; Rodriguez et al., 2006; Zhang et al., 2007; Vasilkoski and Stepanyants, 2009)。局部处理类方法的基本思想是根据起始点预测下一个点，与全局处理类方法相比，只用到当前位置局部邻域的信号，对图像质量的要求较低，计算资源的消耗较少。局部处理类方法包括矢量追踪(vector tracking)(Al-Kofahi et al., 2003; Al-Kofahi et al., 2008)、区域增长(region growing)(Rodriguez et al., 2009)、开放曲线蛇形算法(open-curve snake)(Wang et al., 2011)、超级椭球模型(superellipsoid model)(Tyrrell et al., 2007)等，通过对起点和终点进行约束，可以使得这些方法具有更好的鲁棒性。

大量的自动追踪技术的出现，有希望节省用人工识别突起(也包括血管)的大量时间，目前以人工分割出一个完整神经元形态就需要数小时到数天的努力工作。虽然商界和学术界创造出各种算法、程序，并声称成功解决了突起的自动追踪，但科学家在具体使用时，会发现这些方案并不那么适合自己。全自动的追踪技术适用的应用场景还非常有限，对大范围、高密度、长距离轴突的追踪质量还有待提高。为了推动这一领域的发展，2009~2010 年，美国的霍华德休斯医学研究所举办了第一届的自动神经突起追踪算法的竞赛(DIADEM challenge)，参赛单位超过 120 家，他们用各自的算法处理给定的 6 套不同类型的测试数据，从鲁棒性、自动化程度、效率和原创性等角度，对竞赛结果进行了综合评判(Brown et al., 2011; Gillette et al., 2011; Liu, 2011)。

5. 总结与展望

目前，脑研究已被许多国家视为科技战略的重中之重。2013 年，欧盟和美国分别启动脑研究计划。欧盟的"人脑计划"(Human Brain Project, HBP)，旨在解读超过上兆个脑神经细胞的联结，以研究人类情感、意识与思维。而这些复杂的运算，将通过超级计算机多段多层的模拟来实现。美国的"推进创新神经技术脑研究计划"(Brain Research through Advancing Innovative Neurotechnologies, BRAIN)拟研发新型脑研究技术，以更高的时空分辨率建立脑活动图谱。未来十年，这两大脑研究计划预计投入分别超过 10 亿欧元和 30 亿美元。在我国，《国家中长期科学和技术发展规划纲要(2006~2020)》中已将"脑科学与认知"列入基础研究科学前沿问题；2011 年以来，国家自然科学基金委员会、国家科技部、中国科学院分别启动了一系列脑研究的重大计划。

这些重大的脑研究计划极大地促进了脑研究的进展。美国已在介观层次鼠脑连接组方面实现了不同脑区结构连接的定量分析；德国和加拿大则在数字人脑的精细结构获取方面拔得头筹；日本先后发展了全脑透明的 Scale 和 CUBIC 技术并用于获取全脑精细结构……我国脑研究同样勇立世界科研潮头。目前，我国已在分子及细胞神经科学、发育神经科学、脑认知科学、脑疾病机制、计算神经科学等方面取得长足进步；在动物模型、光遗传、光学成像、人工智能、脑网络连接组等新技术和新方法方面有重要进展(详见《生命科学》2014年第6期脑科学研究专刊)。值得一提的是，我国骆清铭教授课题组发明了一系列MOST/fMOST技术；成功获取了荧光标记、Golgi染色、Nissl染色和印度墨水染色的小鼠全脑数据集，三维分辨率达到1μm；开发了适应于海量数据集的自动分析技术，展示了对小鼠全脑进行高分辨率成像和可视化的能力(图版XIII图2)。未来几年，有望完成小鼠全脑真实尺度神经-血管精细结构3D脑图谱。

"认识你自己"，脑研究的发展任重道远、充满挑战。纵观世界，目前仅实现了小鼠全脑真实尺度的精细结构成像；与单个细胞水平可视化完整人脑的梦想还非常遥远。已经知道，以单神经元分辨率采集一个单色标记小鼠脑的数据量都是TB级，单层冠状面图片的数据量是几百兆甚至数GB。如果进行多色成像或对灵长类动物全脑成像，其数据量更是成倍增长。推进脑研究前进的第一个挑战是：大样本、高分辨率、高通量光学成像。对于灵长类全脑(包含人脑)如何保持光学分辨率并实现高通量成像，这里面有许多技术问题亟待研究；同时，还有必要实现规范而工业化的流程产生图像数据。第二个挑战是：特定神经通路和特定样本的优化标记。对某一通路的稀疏标记、对不同成分的多色标记、对灵长类大脑的特异性标记，都是有待深化和改进的方向。第三个挑战就是：大数据的可视化和信息化。按照预期的工程化速度产生大数据，对于目前现有的数据存储、图像处理与分析、数据管理和共享等方式已带来了全方位的挑战。同时，由于神经元类型和形态的高度复杂性，面对已获取的大数据图像，不可能采用过去手工分割图像的做法，必须发展新的方法，以实现自动地对多层次、多类型、多形态、高密度的神经结构和功能图像的数字化处理，在此基础上实现单神经元分辨的全脑可视化和定量分析的目标。最后，如何对获取的脑数据加以整合和模拟，实现脑功能的认知，发展人工智能，研究人类智力，也是从理解脑到发展脑的飞跃将是更进一步的挑战。

挑战与机遇并存。目前，正是脑研究发展的黄金时期。不难预见，脑研究将凝聚物理、生物、化学、数学、信息和认知等多学科力量，催生一系列相关新原理、新方法和新技术，必将成为新兴产业的重要引擎。当前，我国政府已加大资金投入，强化统筹规划；研究团队正在团结协作、勇攀高峰。脑科学是我国的优势学科之一，我国脑研究有望在若干领域取得重大突破，为全人类更好的"认识脑、保护脑、创造脑"做出重要贡献。

致谢

本研究获得国家自然科学基金创新群体项目(No. 61121004)，国家自然科学基金(Nos. 30727002, 81127002)，国家重大科学仪器设备开发专项(No. 2012YQ030260)和985计划资助。

参 考 文 献

龚辉, 曾绍群, 李安安, 等. 2014. 单神经元分辨水平的小鼠全脑网络可视化. 生命科学, 26(6): 626-633.

郭爱克. 2014. 脑科学：机遇与挑战. 生命科学, 26(6): 543-544.

刘谦, 龚辉. 2010. 大鼠断层解剖彩色图谱. 武汉: 华中科技大学出版社.

刘洋, 张郭智, 孙文娟, 等. 2009. 基于可视中国人高分辨人体结构数据集的辐射剂量模拟研究进展. 激光生物学报, 18(5): 585:590.

Abràmoff M D, Magalhães P J, Ram S J. 2004. Image processing with ImageJ. Biophotonics International, 11(7): 36-43.

Ackerman M J. 1998. The visible human project. P IEEE, 86(3): 504-511.

Aggarwal M, Zhang J, Miller M I, et al. 2009. Magnetic resonance imaging and micro-computed tomography combined atlas of developing and adult mouse brains for stereotaxic surgery. Neuroscience, 162(4): 1339-1350.

Akil H, Martone M E, Van Essen D C. 2011. Challenges and opportunities in mining neuroscience data. Science, 331(6018): 708-712.

Ali A A, Dale A M, Badea A, et al. 2005. Automated segmentation of neuroanatomical structures in multispectral MR microscopy of the mouse brain. Neuroimage, 27(2): 425-435.

Al-Kofahi K A, Can A, Lasek S, et al. 2003. Median-based robust algorithms for tracing neurons from noisy confocal microscope images. IEEE T Inf Technol B, 7(4): 302-317.

Al-Kofahi Y, Dowell-Mesfin N, Pace C, et al. 2008. Improved detection of branching points in algorithms for automated neuron tracing from 3D confocal images. Cytom Part A, 73A(1): 36-43.

Al-Kofahi Y, Lassoued W, Lee W, et al. 2010. Improved automatic detection and segmentation of cell nuclei in histopathology images. IEEE T Bio-Med Eng, 57(4): 841-852.

Amunts K, Lepage C, Borgeat L, et al. 2013. BigBrain: An ultrahigh-resolution 3D human brain model. Science, 340(6139): 1472-1475.

Arous J B, Binding J, Leger J-F, et al. 2011. Single myelin fiber imaging in living rodents without labeling by deep optical coherence micrscopy. J Biomed Opt, 16: 116012.

Ascoli G A, Donohue D E, Halavi M. 2007. NeuroMorpho.Org: a central resource for neuronal morphologies. J Neurosci, 27(35): 9247-9251.

Atkins M S, Mackiewich B T. 1998. Fully automatic segmentation of the brain in MRI. IEEE T Med Imaging, 17(1): 98-107.

Attwell D, Buchan A M, Charpak S, et al. 2010. Glial and neuronal control of brain blood flow. Nature, 468(7321): 232-243.

Bai X, Yu L, Liu Q, et al. 2006. A high-resolution anatomical rat atlas. J Anat, 209(5): 707-708.

Barretto R P J, Ko T H, Jung I C, et al. 2011. Time-lapse imaging of disease progression in deep brain areas using fluorescence microendoscopy. Nat Med, 17(2): 223-228.

Bhanirantka P, Demange Y. 2002. OpenGL Volumizer: A toolkit for high quality volume rendering of large data sets. Proc IEEE Visualization and Graphics, 45: 53.

Binding J, Arous J B, Leger J-F, et al. 2011. Brain refractive index measured in vivo with high-NA defocus-corrected full-field OCT and consequences for twophoton microscopy. Opt Express 19: 4833.

Bjaalie J G, Grillner S. 2007. Global neuroinformatics: The international neuroinformatics coordinating facility. J Neurosci, 27(14): 3613-3615.

Bohland J W, Wu C Z, Barbas H, et al. 2009. A proposal for a coordinated effort for the determination of brainwide neuroanatomical connectivity in model organisms at a mesoscopic scale. Plos Comput Biol, 5(3): e1000334.

Brown K M, Barrionuevo G, Canty A J, et al. 2011. The DIADEM data sets: Representative light microscopy images of neuronal morphology to advance automation of digital reconstructions. Neuroinformatics, 9(2-3): 143-157.

Bug W J, Ascoli G A, Grethe J S, et al. 2008. The NIFSTD and BIRNLex vocabularies: building comprehensive ontologies for neuroscience. Neuroinformatics, 6(3): 175-194.

Byun J Y, Verardo M R, Sumengen B, et al. 2006. Automated tool for the detection of cell nuclei in digital microscopic images: Application to retinal images. Mol Vis, 12(105-07): 949-960.

Cahalan M D, Parker I, Wei S H, et al. 2002. Two-photon tissue imaging: Seeing the immune system in a fresh light. Nat Rev Immunol, 2(11): 872-880.

Cannon R C, Turner D A, Pyapali G K. 1998. An on-line archive of reconstructed hippocampal neurons. J Neurosci Meth, 84(1-2): 49-54.

Canny J. 1986. A computational approach to edge detection. IEEE T Pattern Anal, 6: 679-698.

Chalfie M, Tu Y, Euskirchen G, et al. 1994. Green fluorescent protein as a marker for gene expression. Science, 263(5148): 802-805.

Chen R C, Longo R, Rigon L, et al. 2010. Measurement of the linear attenuation coefficients of breast tissues by synchrotron radiation computed tomography. Phys Med Biol, 55(17): 4993-5005.

Chen Y, Aguirre AD, Ruvinskaya L, et al. 2009. Optical coherence tomography (OCT) reveals depth-resolved dynamics during functional brain activation. J Neurosci Methods, 178: 162-173.

Chiang A S, Lin C Y, Chuang C C, et al. 2011. Three-dimensional reconstruction of brain-wide wiring networks in Drosophila at single-cell resolution. Curr Biol, 21(1): 1-11

Chugh B P, Lerch J P, Yu L, et al. 2009. Measurement of cerebral blood volume in mouse brain regions using micro-computed tomography. Neuroimage, 47(4): 1312-1318.

Chung H J, Lee H K. 2009. Constructing a road map from synapses to behaviour. EMBO Rep, 10(9): 958-962.

Chung K, Deisseroth K. 2013. Clarity for mapping the nervous system. Nat Methods, 10(6): 508-513.

Chung K, Wallace J, Kim S Y, et al. 2013. Structural and molecular interrogation of intact biological systems. Nature, 497(7449): 332-337.

Craddock R C, Jbabdi S, Yan C G, et al. 2013. Imaging human connectomes at the macroscale. Nat Methods, 10: 524-539.

Cubitt A B, Heim R, Adams S R, et al. 1995. Understanding, improving and using green fluorescent proteins. Trends Biochem Sci, 20(11): 448-455.

De Schutter E. 2009. The international neuroinformatics coordinating facility: evaluating the first years. Neuroinformatics, 7(3): 161-163.

Deng Z L, Wang Z, Yang X Q, et al. 2012. In vivo imaging of hemodynamics and oxygen metabolism in acute focal cerebral ischemic rats with laser speckle imaging and functional photoacoustic microscopy. J Biomed Opt, 17(8).

Denk W, Strickler J H, Webb W W. 1990. Two-photon laser scanning fluorescence microscopy. Science, 248(4951): 73-76.

Dima A, Scholz M, Obermayer K. 2002. Automatic segmentation and skeletonization of neurons from confocal microscopy images based on the 3-D wavelet transform. IEEE T Image Process, 11(7): 790-801.

Ding W, Li A, Wu J, et al. 2013. Automatic macroscopic density artefact removal in a Nissl-stained microscopic atlas of whole mouse brain. J Microsc, 251(2): 168-177.

Dodt H U, Leischner U, Schierloh A, et al. 2007. Ultramicroscopy: three-dimensional visualization of neuronal networks in the whole mouse brain. Nat Methods, 4(4): 331-336.

Dong H. 2008. The Allen reference atlas: A digital color brain atlas of the C57Bl/6J male mouse. New York: John Wiley & Sons Inc.

Dorr A E, Lerch J P, Spring S, et al. 2008. High resolution three-dimensional brain atlas using an average magnetic resonance image of 40 adult C57Bl/6J mice. Neuroimage, 42(1): 60-69.

Dorr A, Sled J G, Kabani N. 2007. Three-dimensional cerebral vasculature of the CBA mouse brain: A magnetic resonance imaging and micro computed tomography study. Neuroimage, 35(4): 1409-1423.

Eliceiri K W, Berthold M R, Goldberg I G, et al. 2012. Biological imaging software tools. Nat Methods, 9(7): 697-710.

Erturk A, Becker K, Jahrling N, et al. 2012. Three-dimensional imaging of solvent-cleared organs using 3DISCO. Nat Protoc, 7(11): 1983-1995.

Erturk A, Lafkas D, Chalouni C. 2014. Imaging cleared intact biological systems at a cellular level by 3DISCO. J Vis Exp, 89: e51382.

Fernandez G, Kunt M, Zrÿd J P. 1995. A new plant cell image segmentation algorithm. Lecture Notes in Computer Science, 974: 229-234.

Fischl B, Salat D H, Busa E, et al. 2002. Whole brain segmentation: Automated labeling of neuroanatomical structures in the human brain. Neuron, 33(3): 341-355.

Flusberg B A, Cocker E D, Piyawattanametha W, et al. 2005. Fiber-optic fluorescence imaging. Nat Methods, 2(12): 941-950.

Franklin K B, Paxinos G. 1997. Mouse brain in stereotaxic coordinates.

Frasconi P, Silvestri L, Soda P, et al. 2014. Large-scale automated identification of mouse brain cells in confocal light sheet microscopy images. Bioinformatics, In press.

Friedland D R, Los J G, Ryugo D K. 2006. A modified Golgi staining protocol for use in the human brain stem and cerebellum. J Neurosci Methods, 150(1): 90-95.

Gardner D, Akil H, Ascoli G A, et al. 2008. The neuroscience information framework: a data and knowledge environment for neuroscience. Neuroinformatics, 6(3): 149-160.

Garrido A, de la Blanca N P. 2000. Applying deformable templates for cell image segmentation. Pattern Recogn, 33(5): 821-832.

Gijtenbeek J M, Wesseling P, Maass C, et al. 2005. Three-dimensional reconstruction of tumor microvasculature: simultaneous visualization of multiple components in paraffin-embedded tissue. Angiogenesis, 8(4): 297-305.

Gillette T A, Brown K M, Ascoli G A. 2011. The DIADEM metric: comparing multiple reconstructions of the same neuron. Neuroinformatics, 9(2-3): 233-245.

Glaser E, Van der Loos H. 1965. A semi-automatic computer-microscope for the analysis of neuronal morphology. IEEE T Bio-Med Eng, 1: 22-31.

Glaser J R, Glaser E M. 1990. Neuron imaging with Neurolucida—a PC-based system for image combining microscopy. Comput Med Imag Grap, 14(5): 307-317.

Gmitro A F, Aziz D. 1993. Confocal microscopy through a fiber-optic imaging bundle. Opt Lett, 18(8): 565-567.

Gobel W, Kerr J N D, Nimmerjahn A. 2004. Miniaturized two-photon microscope based on a flexible coherent fiber bundle and a gradient-index lens objective. Opt Lett, 29(21): 2521-2523.

Gong H, Zeng S Q, Yan C, et al. 2013. Continuously tracing brain-wide long-distance axonal projections in mice at a one-micron voxel resolution. Neuroimage, 74: 87-98.

Hama H, Kurokawa H, Kawano H, et al. 2011. Scale: a chemical approach for fluorescence imaging and reconstruction of transparent mouse brain. Nat Neurosci, 14(11): 1481-U1166.

Hashimoto H, Kusakabe M, Ishikawa H. 2008. A novel method for three-dimensional observation of the vascular networks in the whole mouse brain. Microsc Res Tech, 71(1): 51-59.

Hawrylycz M, Baldock R A, Burger A, et al. 2011. Digital atlasing and standardization in the mouse brain. Plos Comput Biol, 7(2):e1001065.

Hayashi Y, Tagawa Y, Yawata S, et al. 2012. Spatio-temporal control of neural activity in vivo using fluorescence microendoscopy. Eur J Neurosci, 36:2722-2732.

Helmchen F, Denk W. 2005. Deep tissue two-photon microscopy. Nat Methods, 2(12): 932-940.

Helmstaedter M, Mitra P P. 2012. Computational methods and challenges for large-scale circuit mapping. Curr Opin Neurobiol, 22(1): 162-169.

Helmstaedter M. 2013. Cellular-resolution connectomics: challenges of dense neural circuit reconstruction. Nat Methods, 10:501-507.

Hey T, Tansley S, Tolle K. 2009. The fourth paradigm: data-intensive scientific discovery. Washington: Microsoft Research.

Hoelen C G A, de Mul F F M, Pongers R et al. 1998. Three-dimensional photoacoustic imaging of blood vessels in tissue. Opt Lett, 23(8): 648-650.

Huang D, Swanson E A, Lin C P, et al. 1991. Optical coherence tomography. Science, 254(5035): 1178-1181.

Huisken J, Swoger J, Del Bene F, et al. 2004. Optical sectioning deep inside live embryos by selective plane illumination microscopy. Science, 305(5686): 1007-1009.

Jeong W K, Beyer J, Hadwiger M, et al. 2010. Ssecrett and neurotrace: Interactive visualization and analysis tools for large-scale neuroscience data sets. IEEE Comput Graph, 30(3): 58-70.

Jones A R, Overly C C, Sunkin S M. 2009. Innovation the Allen brain atlas: 5 years and beyond. Nat Rev Neurosci, 10(11): 821-U878.

Jung J C, Schnitzer M J. 2003. Multiphoton endoscopy. Opt Lett, 28(11): 902-904.

Kak A C, Slaney M. 1988. Principle of computerized tomographic imaging. IEEE Press.

Ke M T, Fujimoto S, Imai T. 2013. SeeDB: a simple and morphology-preserving optical clearing agent for neuronal circuit reconstruction. Nat Neurosci, 16(8): 1154-U1246.

Keefer D A, Spatz W B, Misgeld U. 1976. Golgi-like staining of neocortical neurons using retrogradely transported horseradish peroxidase. Neurosci Lett, 3(5-6): 233-237.

Kiernan J A. 2008. Histological and histochemical methods: theory and practice. Cold Spring Harbor Laboratory Press.

Kiessling F, Greschus S, Lichy M P, et al. 2004. Volumetric computed tomography (VCT): a new technology for noninvasive, high-resolution monitoring of tumor angiogenesis. Nat Med, 10(10): 1133-1138.

Kim J Y, Chung M S, Kwang W S, et al. 2002. Visible Korean human: another trial for making serially-sectioned images. Stud Health Technol Inform, 85: 228-233.

Klein A, Andersson J, Ardekani B A, et al. 2009. Evaluation of 14 nonlinear deformation algorithms applied to human brain MRI registration. Neuroimage, 46(3): 786-802.

Kruger R A, Liu P. 1994. Photoacoustic ultrasound: pulse production and detection of 0.5% Liposyn. Med Phys, 21(7): 1179-1184.

Kuhlman S J, Huang Z J. 2008. High-resolution labeling and functional manipulation of specific neuron types in mouse brain by Cre-activated viral gene expression. PLoS One, 3(4): e2005.

LaMar E, Hamann B, Joy K I. 2000. Multiresolution techniques for interactive texture-based volume visualization. Proc of the conference on Visualization, 355-361.

Larson S D, Martone M E. 2013. NeuroLex.org: an online framework for neuroscience knowledge. Front Neuroinform, 7: 18.

Leischner U, Schierloh A, Zieglgansberger W, et al. 2010. Formalin-Induced fluorescence reveals cell shape and morphology in biological tissue samples. Plos One, 5(4).

Li A A, Liu Q, Zeng S Q, et al. 2008. Construction and visualization of high-resolution three-dimensional anatomical structure datasets for Chinese digital human. Chinese Sci Bull, 53(12): 1848-1854.

Li A, Gong H, Zhang B, et al. 2010. Micro-optical sectioning tomography to obtain a high-resolution atlas of the mouse brain. Science, 330(6009): 1404-1408.

Li G, Liu T, Nie J, et al. 2008. Segmentation of touching cell nuclei using gradient flow tracking. J Microsc 231, (1): 47-58.

Lichtman J W, Denk W. 2011. The big and the small: challenges of imaging the brain's circuits. Science, 334(6056): 618-623.

Lichtman J W, Livet J, Sanes J R. 2008. A technicolour approach to the connectome. Nat Rev Neurosci, 9(6): 417-422.

Liu X, Quan T, Zeng S, et al. 2011. Identification of the direction of the neural network activation with a cellular resolution by fast two-photon imaging. J Biomed Opt 16(8): 080506.

Liu Y, Yang X, Gong H, et al. 2013. Assessing the effects of norepinephrine on single cerebral microvessels using optical-resolution photoacoustic microscope. J Biomed Opt, 18(7).

Liu Y, Yang X, Zhu D, et al. 2013. Optical clearing agents improve photoacoustic imaging in the optical diffusive regime. Opt Lett, 38(20): 4236-4239.

Liu Y. 2011. The DIADEM and beyond. Neuroinformatics, 9(2-3): 99-102.

Livet J, Weissman T A, Kang H, et al. 2007. Transgenic strategies for combinatorial expression of fluorescent proteins in the nervous system. Nature, 450(7166): 56-62.

Luo L, Callaway E M, Svoboda K. 2008. Genetic dissection of neural circuits. Neuron, 57(5): 634-660.

MacKenzie-Graham A, Lee E F, Dinov I D, et al. 2004. A multimodal, multidimensional atlas of the C57BL/6J mouse brain. J Anat, 204(2): 93-102.

Maragos P. 1987. Tutorial on advances in morphological image processing and analysis. Opt Eng, 26(7): 267623-267623-.

Mayerich D M, Abbott L, Keyser J. 2008. Visualization of cellular and microvascular relationships. IEEE Trans Vis Comput Graph, 14(6): 1611-1618.

Mayerich D, Abbott L, McCormick B. 2008. Knife-edge scanning microscopy for imaging and reconstruction of three-dimensional anatomical structures of the mouse brain. J Microsc, 231(Pt 1): 134-143.

Mayerich, D, Kwon J, Sung C, et al. 2011. Fast macro-scale transmission imaging of microvascular networks using KESM. Biomed Opt Express, 2(10): 2888-2896.

McDonald D M, Choyke P L. 2003. Imaging of angiogenesis: from microscope to clinic. Nat Med, 9(6): 713-725.

Meijering E. 2010. Neuron Tracing in Perspective. Cytom Part A, 77A(7): 693-704.

Meijering E. 2012. Cell segmentation: 50 years down the road. IEEE Signal Proc Mag, 29(5): 140-145.

Micheva K D, Smith S J. 2007. Array tomography: A new tool for imaging the molecular architecture and ultrastructure of neural circuits. Neuron, 55(1): 25-36.

Mikula S, Trotts I, Stone J M, et al. 2007. Internet-enabled high-resolution brain mapping and virtual microscopy. Neuroimage, 35(1): 9-15.

Miyamichi K, Amat F, Moussavi F, et al. 2011. Cortical representations of olfactory input by trans-synaptic tracing. Nature, 472(7342): 191-196.

Nasiriavanaki M, Xia J, Wan H L, et al. 2014. High-resolution photoacoustic tomography of resting-state functional connectivity in the mouse brain. Proc Natl Acad Sci U S A, 111(1): 21-26.

Oh S W, Harris J A, Ng L, et al. 2014. A mesoscale connectome of the mouse brain. Nature, 508(7495): 207-214.

Oliveira F P M, Tavares J M R S. 2014. Medical image registration: a review. Comput Method Biomec, 17(2): 73-93.

Osten P, Margrie T W. 2013. Mapping brain circuitry with a light microscope. Nat Methods, 10(6): 515-523.

Parvin B, Yang Q, Han J, et al. 2007. Iterative voting for inference of structural saliency and characterization of subcellular events. Ieee T Image Process, 16(3): 615-623.

Peng H C, Ruan Z C, Long F H, et al. 2010. V3D enables real-time 3D visualization and quantitative analysis of large-scale biological image data sets. Nat Biotechnol, 28(4): 348-U375.

Pfeiffer F, Weitkamp T, Bunk O, et al. Phase retrieval and differential phase-contrast imaging with low-brilliance X-ray sources. Nat Phys, 2(4): 258-261.

Pilati N, Barker M, Panteleimonitis S, et al. 2008. A rapid method combining Golgi and Nissl staining to study neuronal morphology and cytoarchitecture. J Histochem Cytochem, 56(6): 539-550.

Pillai R S, Lorenser D, Sampson D D. 2011. Deep-tissue access with confocal fluorescence microendoscopy through hypodermic needles. Opt Express, 19(8): 7213-7221.

Plate J, Tirtasana M, Carmona R, et al. 2002. Octreemizer: a hierarchical approach for interactive roaming through very large volumes. Proc of the symposium on Data Visualisation.

Quan T W, Zheng T, Yang Z Q, et al. 2013. NeuroGPS: automated localization of neurons for brain circuits using L1 minimization model. Sci Rep, 3: 1414-1-7.

Ragan T, Kadiri L R, Venkataraju K U, et al. 2012. Serial two-photon tomography for automated ex vivo mouse brain imaging. Nat Methods, 9(3): 255-258.

Ragan T, Sylvan J D, Kim K H, et al. 2007. High-resolution whole organ imaging using two-photon tissue cytometry. J Biomed Opt 12(1): 014015.

Rector D, Harper R. 1991. Imaging of hippocampal neural activity in freely behaving animals. Behav Brain Res, 42(2): 143-149.

Reed W A, Yan M F, Schnitzer M J. 2002. Gradient-index fiber-optic microprobes for minimally invasive in vivo low-coherence interferometry. Opt Lett 27(20): 1794-1796.

Rodriguez A, Ehlenberger D B, Dickstein D L, et al. 2008. Automated three-dimensional detection and shape classification of dendritic spines from fluorescence microscopy images. PLoS One, 3(4).

Rodriguez A, Ehlenberger D B, Hof P R, et al. 2006. Rayburst sampling, an algorithm for automated three-dimensional shape analysis from laser scanning microscopy images. Nat Protoc, 1(4): 2152-2161.

Rodriguez A, Ehlenberger D B, Hof P R, et al. Wearne 2009. Three-dimensional neuron tracing by voxel scooping. J Neurosci Meth, 184(1): 169-175.

Rueckert D, Schnabel J A. 2011. Medical image registration. Biomedical Image Processing, Springer, 131-154.

Saalfeld S, Cardona A, Hartenstein V. 2009. CATMAID: collaborative annotation toolkit for massive amounts of image data. Bioinformatics, 25(15): 1984-1986.

Schroeder W, Martin K, Lorensen B. 1996. The visualization toolkit: an object-oriented approach to 3-D graphics. New Jersey: Prentice Hall.

Selka M R, Thakare M, Chilke B. 2014. Review on detection and segmentation of brain tumor using watershed and thresholding algorithm. IORD Journal of Science and Technology, 1(2):11-14.

Sharief A A, Badea A, Dale A M, et al. 2008. Automated segmentation of the actively stained mouse brain using multi-spectral MR microscopy. Neuroimage, 39(1): 136-145.

Sotelo C. 2003. Viewing the brain through the master hand of Ramon y Cajal. Nat Rev Neurosci, 4(1): 71-77.

Susaki E A, Tainaka K, Perrin D, et al. 2014. Whole-brain imaging with single-cell resolution using chemical cocktails and computational analysis. Cell, 157(3): 726-739.

Suzuki T, Matsuzaki T, Hagiwara H, et al. 2007. Recent advances in fluorescent labeling techniques for fluorescence microscopy. Acta Histochem Cytochem, 40(5): 131-137.

Szeliski R. 2006. Image alignment and stitching: a tutorial. Foundations and Trends in Computer Graphics and Vision, 2(1): 1-104.

Tang H, Wu E X, Ma Q Y, et al. 2000. MRI brain image segmentation by multi-resolution edge detection and region selection. Comput Med Imag Grap, 24(6): 349-357.

Tang L, Chung M S, Liu Q A, et al. 2010. Advanced features of whole body sectioned images: virtual Chinese human. Clin Anat, 23(5): 523-529.

Tsai P S, Friedman B, Ifarraguerri A I, et al. 2003. All-optical histology using ultrashort laser pulses. Neuron, 39(1): 27-41.

Tyrrell J A, di Tomaso E, Fuja D, et al. 2007. Robust 3-D modeling of vasculature imagery using superellipsoids. Ieee T Med Imaging, 26(2): 223-237.

Vakoc B J, Lanning R M, Tyrrell J A, et al. 2009. Three-dimensional microscopy of the tumor microenvironment in vivo using optical frequency domain imaging. Nat Med, 15(10): 1219-1223.

Vasilkoski Z, Stepanyants A. 2009. Detection of the optimal neuron traces in confocal microscopy images. J Neurosci Meth, 178(1): 197-204.

Vincent P, Maskos U, Charvet I, et al. 2006. Live imaging of neural structure and function by fibred fluorescence microscopy. EMBO Rep 7, 1154-1161.

Walter T, Shattuck D W, Baldock R, et al. 2010. Visualization of image data from cells to organisms. Nat Methods, 7(3): S26-S41.

Wang C, Yang Y, Ding Z, et al. 2011. Monitoring of drug and stimulation induced cerebral blood flow velocity changes in rat sensory cortex using spectral domain Doppler optical coherence tomography. J Biomed Opt, 16(4).

Wang H, Black A J, Zhu J, et al. 2011. Reconstructing micrometer-scale fiber pathways in the brain: multi-contrast optical coherence tomograhpy based tractography. Neuroimage, 58:984-992.

Wang L V, Hu S. 2012. Photoacoustic tomography: In vivo imaging from organelles to organs. Science, 335(6075): 1458-1462.

Wang L V. 2009. Multiscale photoacoustic microscopy and computed tomography. Nat Photonics, 3(9): 503-509.

Wang Q, Li A, Gong H, et al. 2012. Quantitative study on the hygroscopic expansion of spurr resin to obtain a high-resolution atlas of the mouse brain. Exp Biol Med (Maywood), 237(10): 1134-1141.

Wang X D, Pang Y J, Ku G, et al. 2003. Noninvasive laser-induced photoacoustic tomography for structural and functional in vivo imaging of the brain. Nat Biotechnol, 21(7): 803-806.

Wang Y, Narayanaswamy A, Tsai C L, et al. 2011. A broadly applicable 3-D neuron tracing method based on open-curve snake. Neuroinformatics, 9(2-3): 193-217.

Weber B, Keller A L, Reichold J, et al. 2008. The microvascular system of the striate and extrastriate visual cortex of the macaque. Cereb Cortex, 18(10): 2318-2330.

Wiederhold K H, Bielser W, Schulz U, et al. 1976. Three-dimensional reconstruction of brain capillaries from frozen serial sections. Microvasc Res, 11(2): 175-180.

Windhorst U, Johansson H. 1999. Modern techniques in neuroscience research. Springer Verlag.

World Health Organization. Neurological disorders: public health challenges. Geneva: World Health Organization, 2006

Wu J, He Y, Yang Z, et al. 2014. 3D BrainCV: simultaneous visualization and analysis of cells and capillaries in a whole mouse brain with one-micron voxel resolution. Neuroimage 87: 199-208.

Wu Q, Merchant F, Castleman K. 2010. Microscope image processing. Academic Press.

Wu S H, Oertel D. 1984. Intracellular injection with horseradish peroxidase of physiologically characterized stellate and bushy cells in slices of mouse anteroventral cochlear nucleus. J Neurosci, 4(6): 1577-1588.

Xiang L Z, Ji L J, Zhang T, et al. 2013. Noninvasive real time tomographic imaging of epileptic foci and networks. Neuroimage, 66: 240-248.

Xie B, Miao P, Sun Y, et al. 2012. Micro-computed tomography for hemorrhage disruption of mouse brain vasculature. Transl Stroke Res, 3(Suppl 1): 174-179.

Xiong H, Gendelman H E, 2014. Current laboratory methods in neuroscience research. Springer.

Xiong H, Zhou Z, Zhu M, et al. 2014. Chemical reactivation of quenched fluorescent protein molecules enables resin-embedded fluorescence microimaging. Nat Commun, 5: 3992.

Xu T, Yu X, Perlik A J, et al. 2009. Rapid formation and selective stabilization of synapses for enduring motor memories. Nature, 462(7275): 915-919.

Xu W, Sudhof T C. 2013. A neural circuit for memory specificity and generalization. Science, 339(6125): 1290-1295.

Xu X G, Chao T C, Bozkurt A. 2000. VIP-man: An image-based whole-body adult male model constructed from color photographs of the visible human project for multi-particle Monte Carlo calculations. Health Phys, 78(5): 476-486.

Xue S, Gong H, Jiang T, et al. 2014. Indian-ink perfusion based method for reconstructing continuous vascular networks in whole mouse brain. PLoS One, 9(1): e88067.

Yan Z, Hu B, Zhang Y, et al. 2013. Development of a plastic embedding method for large-volume and fluorescent-protein-expressing tissues. PLoS One 8(4): e60877.

Yang S H, Xing D, Lao Y Q, et al. 2007. Noninvasive monitoring of traumatic brain injury and post-traumatic rehabilitation with laser-induced photoacoustic imaging. Appl Phys Lett, 90(24).

Young W, Nimchinsky E, Hof P, et al. 1997. NeuroZoom software user guide and reference books. San Diego: YBM Inc.

Zhang B, Li A, Yang Z, et al. 2011. Modified Golgi-Cox method for micrometer scale sectioning of the whole mouse brain. J Neurosci Methods 197(1): 1-5.

Zhang B, Zimmer C, Olivo-Marin J C. 2004. Tracking fluorescent cells with coupled geometric active contours. Proc of IEEE International Symposium on Biomedical Imaging, 1(2): 476-479.

Zhang H, Yuan J, Fu L. 2012. Imaging Fourier transform endospectroscopy for in vivo and in situ multispectral imaging. Opt Express 20(21): 23349-23360.

Zhang Y, Zhou X B, Degterev A, et al. 2007. Automated neurite extraction using dynamic programming for high-throughput screening of neuron-based assays. Neuroimage, 35(4): 1502-1515.

Zhao S, Zhou Y, Gross J, et al. 2010. Fluorescent labeling of newborn dentate granule cells in GAD67-GFP transgenic mice: a genetic tool for the study of adult neurogenesis. PLoS One, 5(9).

Zheng T, Yang Z, Li A, et al. 2013. Visualization of brain circuits using two-photon fluorescence micro-optical sectioning tomography. Opt Express 21(8): 9839-9850.

Zhu D, Larin K V, Luo Q, et al. 2013. Recent progress in tissue optical clearing. Laser Photon Rev 7(5): 732-757.

Zhu Y, Xu J, Hauswirth W W, et al. 2014. Genetically targeted binary labeling of retinal neurons. J Neurosci, 34(23): 7845-7861.

Zingg B, Hintiryan H, Gou L, et al. 2014. Neural networks of the mouse neocortex. Cell, 156(5): 1096-1111.

图版

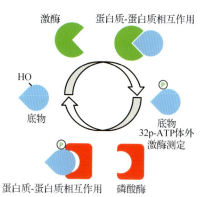

图版 I 图 1 G 蛋白可逆蛋白磷酸化反应示意图

蛋白激酶以ATP作为磷酸基团的供体,通过与底物结合进行磷酸基团传递,将其转移到底物的羟基上;而蛋白磷酸酶可以通过水解反应去除底物羟基上共价结合的磷酸基团。在这个过程中,由于激酶、磷酸酶都可以与底物结合,可以通过检测蛋白互作的方法,也可以通过放射性核素^{32}P标记的ATP(γ-^{32}P-ATP)在试管中模拟磷酸化反应证明激酶与底物的反应可否直接发生

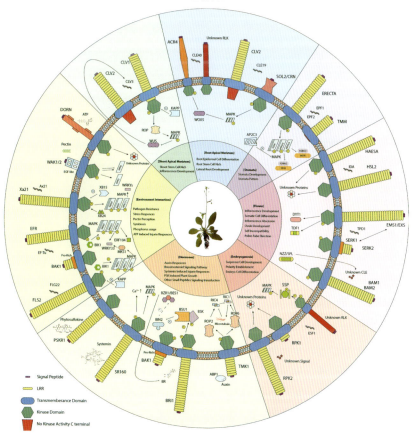

图版 I 图 2 植物类受体激酶(RLK)广泛参与植物生长发育和环境响应

类受体激酶在植物生长发育和环境响应的各个方面发挥重要作用:CLAVATA1、CLAVATA2和ACR4对维持植物茎顶端分生组织和根顶端分生组织十分重要;ERECTA调控叶片气孔细胞的排列;BAM1、BAM2和EMS/EXS与雄蕊体细胞绒毡层分化有关;TMK1、BRI1和SR160能够识别植物激素信号;FLS2、Xa21和EFR等在植物应对有害生物侵袭方面十分关键(详见文中所述)

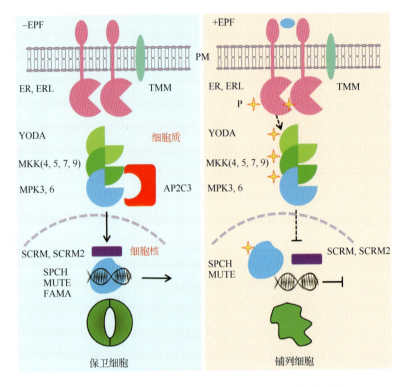

图版Ⅱ图1 ERECTA介导的植物气孔发育途径模式图

在没有分泌性肽EPF1、2结合时，或者由EPFL9竞争与其受体ER、ERL1、ERL2结合后，SCRM-SPCH、SCRM2-SPCH促进拟分化母细胞（MMC）向拟分化细胞（M）转化，SCRM-MUTE、SCRM2-MUTE进一步促进M向保卫母细胞（GMC）分化，再由YODA－MKK7、9－MPK3、6激活SCRM-FAMA、SCRM2-FAMA促进分化为保卫细胞（GC），形成气孔（左）。在周围细胞中，EPF1激活ER、ERL1、ERL2后，进一步激活YODA－MKK4、5－MPK3、6，抑制SCRM-SPCH、SCRM2-SPCH与SCRM-MUTE、SCRM2-MUTE的功能，抑制M与GMC分化（右）

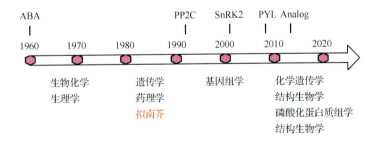

图版Ⅱ图2 PP2C-SnRK2所介导的核心ABA信号通路发现示意图

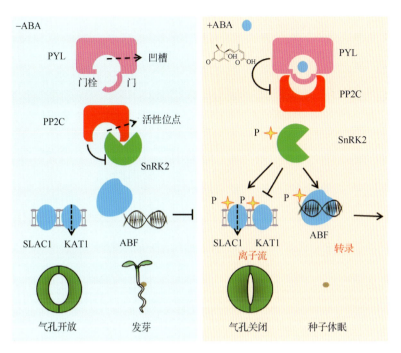

图版Ⅲ图 1　PP2C-SnRK2 介导的 ABA 信号转导

没有ABA或ABA水平较低的情况下，PP2C与SnRK2结合并抑制SnRK2自磷酸化，导致下游离子通道SLAC1、KAT1和转录因子ABF2等都不被磷酸化，SLAC1关闭，KAT1打开，ABF2不能结合到下游基因的启动子，下游基因不能表达，因此气孔开放、种子正常萌发和生长（左）。在逆境、胁迫或种子成熟过程中，ABA的量大幅上升，ABA结合到受体PYL后，诱导PYL与PP2C结合，解除PP2C对SnRK2的抑制作用，SnRK2自磷酸化而激活，并进一步磷酸化下游的通道蛋白和转录因子，SLAC1开放，KAT1关闭，ABF2等转录因子激活下游基因表达，导致气孔关闭、种子萌发受抑、种子后期成熟后脱水休眠等（右）

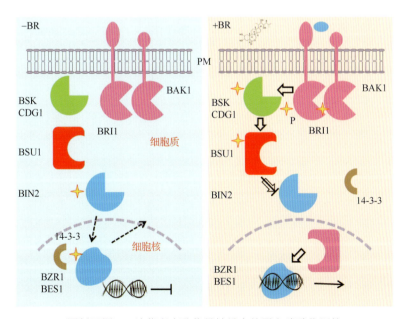

图版Ⅲ图 2　油菜素内酯信号转导中的蛋白磷酸化调控

没有BR时，BZR1/BES1被BIN2磷酸化后，丧失DNA结合活性，由14-3-3结合运到细胞核外降解，不能发挥功能（左）。当BR与受体BRI1结合后，通过与BAK1相互作用，作为共受体，相互磷酸化而激活，进一步激活胞内激酶BSK1、CDG1，这两类激酶磷酸化磷酸酶BSU1后，BSU1对BIN2去磷酸化，从而抑制BIN2的功能。BZR1可以被PP2A去磷酸化，激活下游基因表达（右）

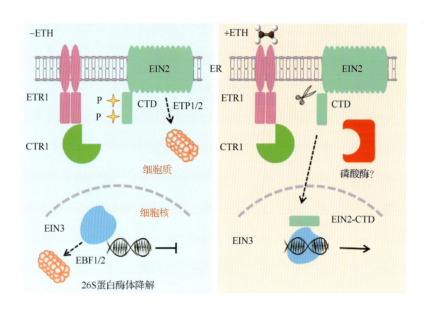

图版Ⅳ图1　磷酸化在乙烯信号转导中的作用

没有乙烯（ETH）或乙烯浓度较低的情况下，内质网上的受体ETR1等与CTR1结合，促使CTR1磷酸化EIN2；同时，EIN2可以被ETP1和ETP2两个F-box类泛素连接酶泛素化而降解（左）。当乙烯浓度较高时，乙烯与受体结合后可以抑制CTR1的激酶活性，同时EIN2不能被泛素化而大量积累，随之非磷酸化（可能被某种磷酸酶去磷酸化）的EIN2被剪切后，其C端进入细胞核中与EIN3结合，激活下游基因表达，启动乙烯反应（右）

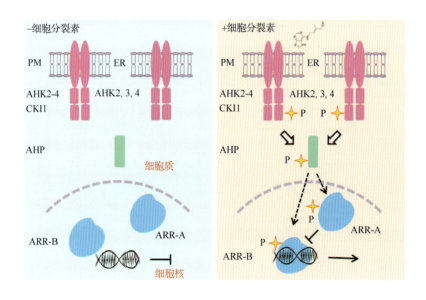

图版Ⅳ图2　二元组分系统构成的细胞分裂素信号转导

无细胞分裂素或其浓度较低时，信号通路被阻断（左）。在细胞分裂素浓度较高时，与受体AHK2、3、4等结合，通过磷酸传递，将磷酸基团转移到传递蛋白AHP上，进一步AHP进入细胞核，将磷酸基团转移到B类ARR（ARR-B）上，ARR-B被激活，启动下游基因表达。在此过程中，ARR-A也被诱导表达且激活，可以抑制ARR-B的活性，作为一种反馈调控机制（右）

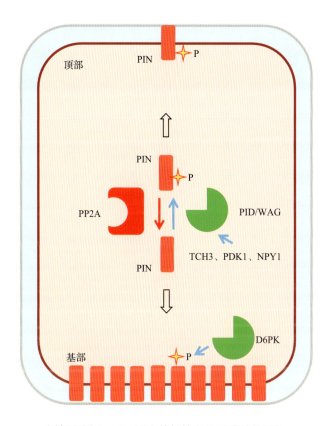

图版Ⅴ 图1 PIN蛋白的极性定位受磷酸化调控

PP2A去磷酸化PIN蛋白，PID等激酶磷酸化PIN蛋白，调控PIN蛋白的极性定位。PID过量会导致PIN蛋白的极性定位改变，影响生长素极性运输。PID的活性受到TCH3、PDK1和NPY1等的调控。此外，D6PK激酶定位于细胞基部，通过磷酸化调控PIN蛋白的活性

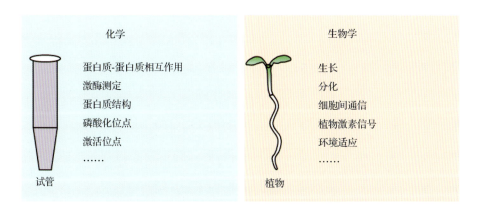

图版Ⅴ 图2 蛋白磷酸化机制与功能研究的一般模式

要证明某个蛋白激酶或磷酸酶通过对底物磷酸化调控而参与到某个生命过程中，一方面需要通过"试管"中的一系列体外试验，证明酶与底物之间的相互作用、磷酸化修饰关系、三维结构、磷酸化位点、活性中心位点等，说明某个酶与底物具有直接的关系。另一方面，要证明体外实验室的结果是植物体内，乃至生物体内发生的生命活动的原因，需要通过对酶与底物的表达水平、相互作用、修饰位点、相互作用区段进行细致研究，检测其对相关生命活动的影响。体内、体外实验室结果的相互验证和补充才能充分说明该酶与底物的修饰关系确为特定生命活动的调控机制

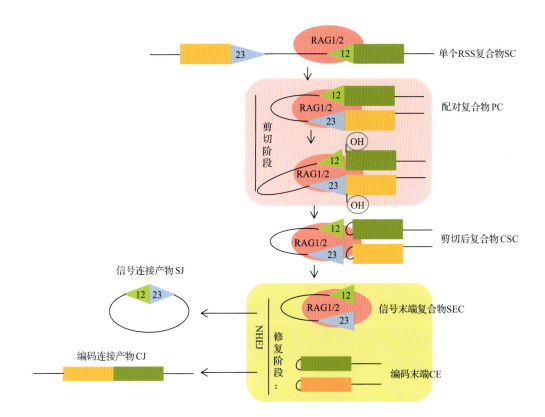

图版Ⅵ　V(D)J基因重排过程

蓝色三角形表示23-RSS,绿色三角形代表12-RSS,红色椭圆形表示RAG复合物。V(D)J基因重排分为两个阶段。第一个过程被称为剪切阶段：RAG1与RAG2以复合物的形式识别并结合12-RSS,形成单RSS复合物SC,SC捕获了23-RSS后形成配对复合物PC,RAG1作为内切酶在一对RSS的七聚体与编码序列的边界剪切单链DNA,产生缺口,游离的3′-OH攻击互补链造成DNA双链断裂,形成信号末端SE与编码末端CE,RAG复合物与DNA断裂组成的复合物被称为剪切后复合物CSC。第二个阶段被称为修复阶段：编码末端首先脱离CSC,然后通过NHEJ途径被重新连接。RAG与信号末端组成的信号末端复合物稍后也通过NHEJ途径被修复

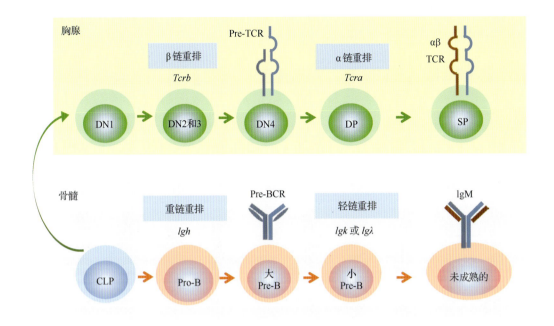

图版Ⅶ图 1　V(D)J 基因重排的细胞特异性及分化阶段特异性示意图

T淋巴细胞在胸腺中分化成熟，CLP从骨髓中迁移至胸腺开始分化，在DN期经过TCRβ链基因重排，正确表达了β链的细胞通过pre-TCR信号作用继续分化至DP阶段并开始TCRα链基因重排。表达了αβ TCR的DP细胞要经历阳性选择和阴性选择分化成熟为单阳性细胞。B淋巴细胞在骨髓中经历pro-B阶段的重链基因Igh重排，重排成功的重链与替代轻链组成pre-BCR，pre-BCR信号促进pre-B细胞的轻链Igκ与Igλ的重排。轻链重排成功后与重链一起组成IgM表达在未成熟的B淋巴细胞表面

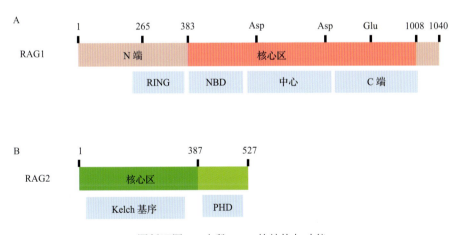

图版Ⅶ图 2　小鼠 RAG 的结构与功能

A. RAG1蛋白全长1040个氨基酸，1-383aa被称为N端非核心区，包含了一个具有泛素连接酶活性的指环结构（ring finger motif：265-383aa）。384-1008被称为核心区（core-RAG1），core-RAG1的N端负责结合RSS九聚体的一段被称为NBD（384-458aa）。Core-RAG1包含了由D600、D708及E962构成的酶活中心，负责剪切RSS；core-RAG1的中央区（central：528-760aa）主要参与结合RAG2以及RSS的七聚体；core-RAG1 的C端负责结合锌离子并介导自身二聚化（761-979aa）。B. 小鼠RAG2蛋白全长527个氨基酸。其核心区为1-387aa，有包含6个Kelch重复基序，主要通过结合RAG1促进RAG1的构象变化从而剪切RSS的七聚体；360-408aa被称为铰链区，负责维持DNA末端走向NHEJ修复途径，防止alternative-NHEJ发生；414-487aa为PHD区域，主要结合组蛋白H3K4Me3，帮助RAG2定位并激活重排反应

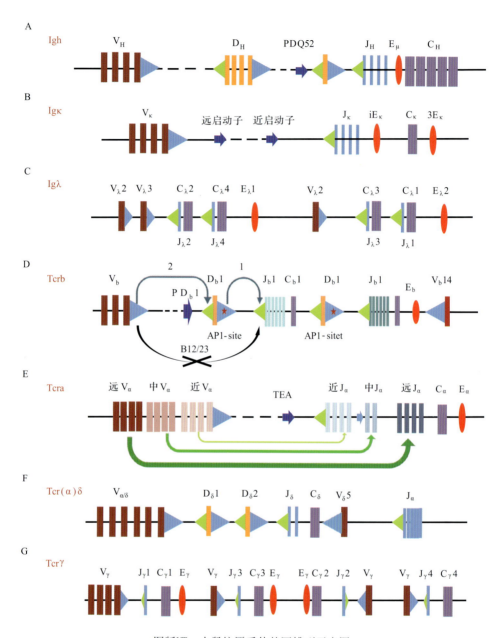

图版Ⅷ 小鼠抗原受体基因排列示意图

（A）小鼠 *Igh* 包含了152个 V_H 片段，其中能够参与重排的大约为97个；17~20个 D_H 片段，具有功能的约14个；4个 J_H 片段都能参与重排。（B）小鼠 *Igk* 包含了174个 $V_κ$ 片段，其中约有94~96个 $V_κ$ 能参与重排；5个 $J_κ$ 片段中4个参与重排。（C）小鼠 *Igλ* 被划分为两个类群，$V_λ$ 全都参与重排，5个 $J_λ$ 中有三个具有重排功能。（D）小鼠的 *Tcrb* 包括了35个 $V_β$ 片段，大约21~22个参与重排；2个 $D_β$ 片段均参与重排；14个 $J_β$ 片段中约11~12个具有重排功能。（E）小鼠的 *Tcra* 包括了98个 $V_α$ 片段，大约73~84个参与重排；60个 $J_α$ 片段中约38个具有重排功能。按照 $V_α$ 与 $J_α$ 彼此距离的远近可以将而者划分为远端，中间，近端三部分。（F）小鼠的 *Tcrd* 嵌在 *Tcra* 基因簇内，一共包含16个 $V_δ$，其中的10个 $V_δ$ 是与 *Tcra* 共用的，剩余的6个 $V_δ$ 中5个具有重排功能。两个 $D_δ$ 片段与两个 $J_δ$ 片段均参与重排。（G）小鼠的 *Tcrg* 被划分为四个类群，第三组为反向定位。所有的 $V_γ$ 与 $J_γ$ 片段都具有重排功能

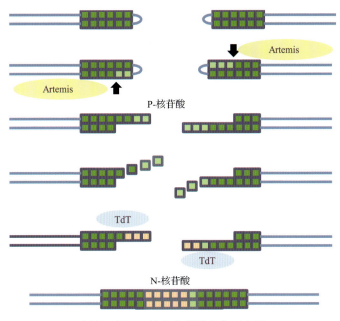

图版Ⅸ图1 连接多样性产生示意图

Artemis通过剪切编码末端的发卡结构将P-核苷酸引入DNA断裂末端；在DNA断裂处由TdT随机引入的核苷酸被称为N-核苷酸。DNA外切酶会随机剪切黏性DNA末端造成碱基缺失。这三种方式造成了重排编码连接产物的连接多样性

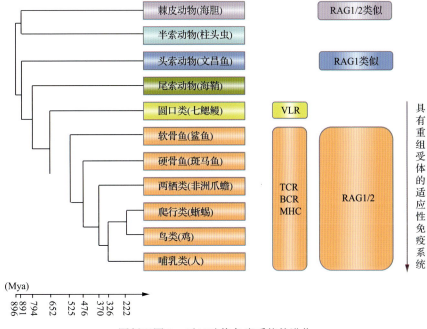

图版Ⅸ图2 后口动物免疫系统的进化

该图标示了谱系进化各个阶段出现的关键免疫分子。颜色标注：在脊椎动物有颌下门中被发现的分子被标为橙色；在脊椎动物无颌总纲中被发现的分子被标注为黄色；在头索动物亚门中被发现的分子被标为蓝色；在棘皮动物门中被发现的分子被标为紫色。RAG类似基因在海胆和文昌鱼中都被鉴定到。物种分叉时间（Blair et al., 2005）以Mya（million years ago, 百万年前）为单位

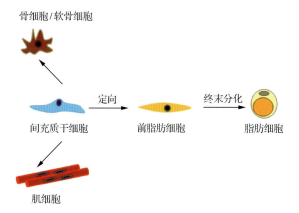

图版X图1　脂肪细胞的分类及功能

图版X图2　白色脂肪细胞的发育阶段

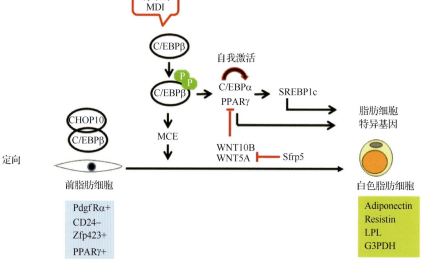

图版X图3　白色脂肪细胞发育分化的调节

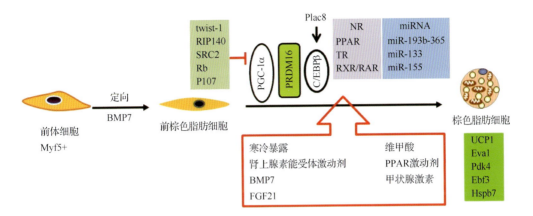

图版 XI 图 1　棕色脂肪细胞发育分化的调节

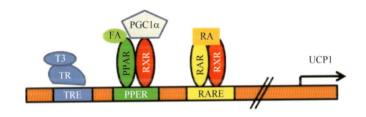

图版 XI 图 2　核受体对棕色脂肪细胞发育的调节

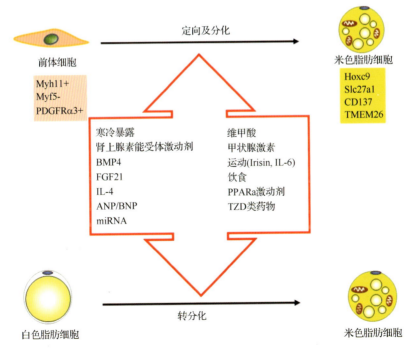

图版 XI 图 3　米色脂肪细胞发育分化的调节

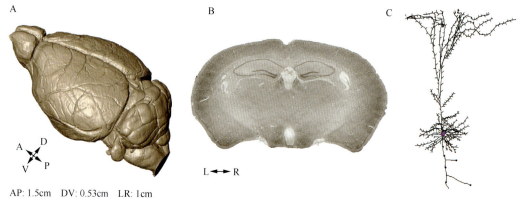

图版XII图1 小鼠全脑、脑冠状面、单个神经元的结构示意图

a. 显微光学切片断层成像系统获取的C57小鼠全脑的三维面绘制图,小鼠全脑在头尾(A-P)、背腹(D-V)、左右(L-R)方向尺度均为厘米量级;b. 尼氏染色的小鼠脑冠状断面图,其中可见的海马及皮层厚度等尺度在毫米量级;c. 单个神经元的精细结构示意图,其胞体及树突、轴突的直径都在微米量级甚至更小

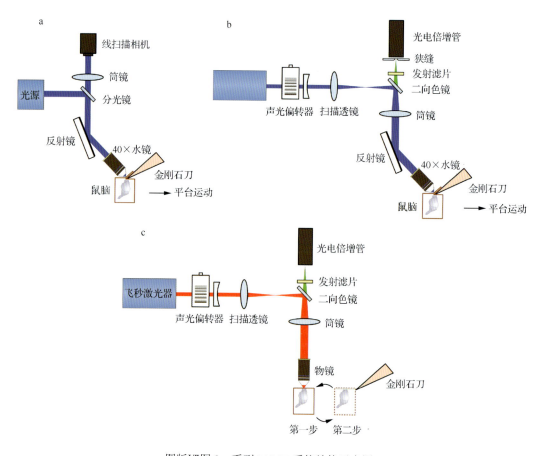

图版XII图2 系列MOST系统结构示意图

a. 为宽场照明MOST系统,适用于对Golg染色、Nissl染色的样品进行成像;采用成像与样品切片同步进行的方式;b. 为基于共聚焦原理的荧光MOST系统;c. 为基于双光子成像的荧光MOST系统,成像和切削被分为两个独立步骤,循环进行

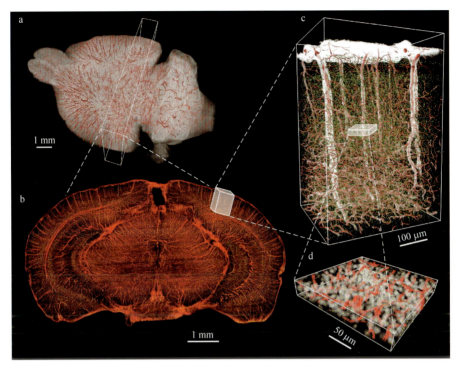

图版XIII图 1　小鼠脑中神经 - 血管结构图

a. 三维渲染小鼠脑中的血管分布情况，灰色为脑轮廓，红色为血管；b. 小鼠脑中冠状断面的血管分布，可以观察到很多毛细血管；c. 桶状皮层区域的血管和神经细胞分布，其中绿色小点为神经细胞的中心点；d. 皮层第Ⅳ层中的细胞和血管的空间分布，灰白色为细胞，红色为血管结构

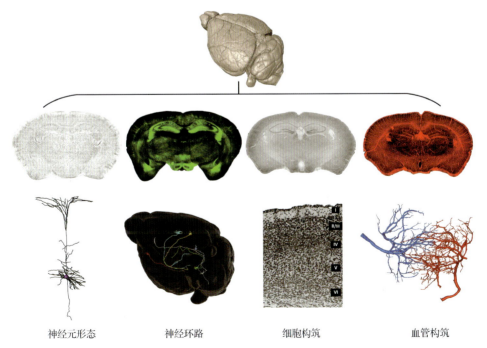

图版XIII图 2　用 MOST 系列技术获得的小鼠全脑及神经元形态、神经环路、细胞构筑和血管结构的可视化结果